One Dimensional Cellular Automata

Harold V. McIntosh

One Dimensional Cellular Automata

Luniver Press
2009

Published by Luniver Press
Frome BA11 3EY United Kingdom

British Library Cataloguing-in-Publication Data
A catalogue record for this book is available from the British Library

One Dimensional Cellular Automata

Cover image courtesy of Dr. Genaro Juarez Martinez.

ISBN-10: 1-905986-20-3
ISBN-13: 978-1-905986-20-0

Contents

Chapter 1

History

Up to now, the history of cellular automata seems to revolve around three outstanding events, or periods of heightened interest. The first of these was John von Neumann's self-reproducing automaton, the second, Martin Gardner's popularization of John Conway's game of *Life*, the third, Stephen Wolfram's classification of automata into four general categories. Even with all that, there has been a long prehistory wherein automata were conceived as mechanisms capable of performing sequences of movements or actions, maybe even capable of taking initiatives. Beyond that, there has been no lack of parallel developments whose immediate connection with cellular automata may not have been so evident, but which were eventually recognized as belonging to the same stream of ideas; many applications of cellular automata fall into this category. And of course there are many parallel disciplines, such as neural nets, genetic algorithms, or lattice gas theory.

1.1 Early origins

Automata theory itself has an ancient history, if one thinks of automata as mechanisms capable of performing intricate movements; but if the actual apparatus is discarded in favor of the activity itself, such a theory more properly begins with the neurophysiological abstractions of McCulloch and Pitts. Somehow, the complexity of an apparatus never seems to have been so much of a qualification for an automaton as its capability for independent or unassisted operation. To judge from past experience, each technological episode which has produced substantial new computing power has brought with it some new development in automata theory. The industrial revolution allowed Babbage to envision a mechanism which transcended dolls having the style of a Swiss music box operated by cams, levers, and gear wheels; even so, inadequacies of technology and project management kept his design of a functioniong machine from becoming a reality. Consider that calculating machines also have a long history, but only when Charles Babbage and Ada Lovelace thought of programming one of them did the true versatility of their prospects begin to appear.

Their refinement into the theory of regular expressions by Kleene constitutes one of several viewpoints, which have gone on to include semigroups (or monoids) of mappings of a set into itself, or even the theory of grammars.

1.2 von Neumann's initiative

Cellular automata have been studied as a part of the abstract theory of computation since the time that John von Neumann became interested in the possibility of constructing self-reproducing automatic factories. Since actual factories and physical machinery involve a myriad of practical but non-essential details, he eventually followed a suggestion of Stanislaw Ulam that an abstract mathematical model would be more amenable to a demonstration of the possibilities of universal construction and self reproduction. He worked out a scheme for such an automaton, in terms of a cellular space occupying a two dimensional grid, in which each cell would be found in one of twenty nine states.

The details of von Neumann's construction remained unpublished at the time of his death in 1957, but were subsequently edited and published by A. W. Burks [177]. Even as he was working on his model, von Neumann realized that it was too literal an interpretation of the computing machines of the era, but he himself never attempted to carry out a complete revision of his original design. About a decade later, in the years 1964-65, E. F. Codd [48] worked out a variant which required only eight states per cell, still using the original five cell neighborhood.

Ulam's [220] work on functional iteration and his experiments on nonlinear mappings were reported in conference proceedings, and in the course of time cellular automata became a topic in the theory of abstract machines, along with the work of Edward F. Moore, Claude Shannon , and others. The principal results of the time were the demonstrations of the existence of universal constructors, and Moore's "Garden of Eden" theorem which showed the necessary existence of configurations in automata of a rather representative type which could only be initial states. Such a pattern could never again be repeated during the course of the automaton's evolution.

Of course, there were even earlier beginnings to automata theory, for example in the studies of Warren S. McCulloch and Walter Pitts in 1943 on neural nets, followed in 1951 by an interesting mathematical abstraction to regular events by S. C. Kleene, and even in general ideas about the new subject of Cybernetics introduced by Norbert Wiener [241] in his famous book of 1948.

An interesting parallel development arose from an even older source, Henri Poincaré's emphasis of the qualitative aspects of classical mechanics in terms of stability, ergodic properties, and the recurrence of orbits; such topics nowadays constitute measure theory, topology, or symbolic dynamics, all of which are an outgrowth of various of Poincaré's ideas.

The latter, symbolic dynamics, elaborated by George Birkhoff [25] and others, was described by Walter Gottschalk and Gustav A. Hedlund in an American Mathematical Society *Colloquium Publication* [97] in 1959, and continues to be a subject of intense mathematical interest. Hedlund's very abstract summary [114] of 1969 contains a wealth of results applicable to cellular automata, although its orientation is entirely different and the relationship between the two concepts has not always been well appreciated.

1.3 Automata theory

The subject of cellular automata has passed through several stages of evolution during the fifty or so years during which it may be thought to have existed. It is interesting to see what has been accomplished during this time, and not only to speculate as to what remains to be done but also to wonder to what degree the subject has applications, and how successful it will be in meeting the expectations which may be held for it.

If an automaton is supposed to be a mechanism capable of performing intricate tasks or exhibiting complex behavior, the concept is surely as old as antiquity; but it was not until the mastery of electricity at the end of the last century that really delicate and subtle constructs could be imagined. Far beyond that, it required an advanced degree of electronics, and even of microelectronics in the form of reliable vacuum tubes, transistors, and nowadays integrated circuits, before truly elaborate automata could become a reality.

For this reason, the innovative ideas of Warren McCulloch and Walter Pitts [161] and the designs of John von Neumann [177] are more properly considered to be the foundations of automata theory than, say, Charles Babbage's [9] proposal for his difference engine or the more advanced analytical engine which made use of ideas taken from the Jacquard loom. As it is, there is a wide variety of sources from which to choose, whatever one's point of view. Little of this will be taken into further account here, but for those interested there are excellent references which can be consulted [192, 8, 94, 83].

McCulloch and Pitts' investigations concerned the possibility of constructing a model of mental processes. While derived from their knowledge of the physiology of nervous systems, their model was intended mostly to reflect characteristics such as the connectivity of the neural net or the interactions in Boolean terms of signals propagating through it. Neither the strict form of the signals, nor the material composition of the neurons were considered to be primary issues.

In similar fashion, von Neumann thought of different models, some of them quite physical, through which his plans for automatic construction could be realized. Nevertheless, as is quite the custom among mathematicians, the abstract approach proved to be the formulation which seemed best to capture the central ideas.

From such a beginning voluminous results have followed, both in the form of mathematical theorems and in the design of specific circuits, all of which have been duly recorded in the scientific literature. Cellular automata achieved substantial popular notoriety when John Conway took up the subject [49] and Martin Gardner reported one of his most interesting discoveries in *Scientific American* [90]. The evangelical efforts of Stephen Wolfram [249] have given a whole new dimension to the topic in recent years.

Cellular automata are but a specialized instance of the general theme of automata theory; the difference lies in the fact that automata are driven by input singnals and produce output signals. Cellular automata enjoy all the symmetries, mostly translational, inherent in their crystallographic layout; but they use the states of selected neighbors for input signals and are not generally considered to produce output.

The semigroup aspect has been exploited by Kenneth Krohn and John L. Rhodes [142]; that was followed up with monoid theory by Samuel Eilenberg [72, 73]. Semigroup theory is much more intricate than the theory of groups, whose classification has been one of the more important mathematical accomplishments of recent times; but a general knowledge of the principles involved is still very useful for automata. These principles include the definition of one sided and two sided ideals, and their use to give a semigroup a standard structural form.

The grammatical approach is primarily a creation of the MIT linguist Noam Chomsky [42, 43, 44, 45], referring primarily to the sequence of transformations which a system can undergo. The emphasis is more on characterizing a set of symbolic equations describing the transformations, than on the transformations themselves, which is the province of semigoup theory. The details to be resolved consist principally in establishing the existence and possibly the uniqueness of the solutions.

The decades of the fifties, sixties, and into the seventies, saw a tremendous amount of research on automata, languages, and similar topics; cellular automata as such were not given much special attention, although sometimes they were utilized as examples or illustrations.

Some of the themes treated, although not necessarily unique to cellular automata, but often emphasizing one dimension, were the "busy beaver" problem in which long lasting activity from simple initial configurations was sought, the "firing squad" problem in which evolution to a specified configuration was the goal, or the ability to achieve "universal computation" via the emulation of a Turing machine.

1.3.1 The concept of an automaton

Whether or not the ideas of McCulloch and Pitts have any bearing on neurophysiology, there is no doubt that their proposals stirred up a considerable amount of interest, some of which can fairly be seen as having resulted in the present theory of automata. Indeed, von Neumann made quite direct and extensive reference to their ideas in the process of designing the EDVAC as well as in his work on self reproducing automata.

Another direct outgrowth of their work was origin of the theory of regular expressions, which was formulated by Stephen Kleene [136] as a rigorous logical and algebraic foundation for their ideas. In turn, regular expressions have been found to lie at the foundation of the theory of formal languages; indeed they constitute just about the simplest of all languages, because the others always include them. This is a subject which can be pursued in great detail, but which is not strictly a part of the theory of *cellular* automata. Nevertheless the circle closes when it is found that regular expressions are invaluable for describing paths through graphs, one of the preferred representations by which the evolution of cellular automata may be described.

All of these early ideas have undergone extensive development since they were first proposed. To start with, von Neumann was deeply involved with a critical phase in the development of computers, which coincided with his interest in cellular automata. As a mathematician he had an interest in symbolic logic and the theory of computation which permeated his approach to computer design. Setting out to equip a computer with the organs required to realize a certain philosophy of computation is a rather different proposition from uncovering all the capabilities of a complex circuit which one has designed.

The logical issues which preoccupied von Neumann would not have concerned all designers, nor did he necessarily cast computer design into a mold which would not otherwise have shaped it. Nevertheless decisions ranging from the choice of the binary number system to the systematic use of stored programs evolved during that formative era. Previously computers had not been complex enough to make the issues important, although Ada Lovelace touched on some of the points in her description of Babbage's Analytical Engine; subsequently computers worked well enough and were sufficiently well understood that after 1956 or thereabouts they tended to be copied rather than redesigned as they went into commercial production.

Another direction which the early ideas took was the growth of the general theory of automata as well as the study of formal languages. In fact the McCulloch-Pitts "neurons" were only one example of digital circuitry, as distinguished from analog circuitry; naturally the use of such circuits required a theory to explain them. Boolean algebra served for combinatorial circuits; the incorporation of memory elements or time delay elements extended the theory to take into account sequential circuits, the kind that could be used in computing machines.

General automata differ from cellular automata in two important respects. Ordinary automata are concerned with input and output, that is, the transformation of signals through the intervention of the automaton. Moreover, design problems are typically concerned with constructing the precise automaton which will produce a given transformation, ascertaining the possible equivalence of two automata in their signal handling abilities, finding the simplest automaton for a given purpose, and so on. The second characteristic is that they are typically small, in any event finite, being intended for use in actual construction.

In contrast cellular automata are required to form lattices, infinite in principle if not in practice, and do not work with signals. Rather, the "signal" which activates each cell is its knowledge of its own state and the states of a certain number of its neighbors at any given moment. Were the lattice not infinite, there would be no discussion of Turing machines, universal computation, or the like. On the other hand, just because of its large scale regularity, a cellular automaton lends itself readily to implementation in terms of microelectronics.

In any event, it could hardly be expected that it would be possible to discuss cellular automata without first taking a certain amount of the general theory into account.

1.3.2 The evolution of finite systems into cycles

It is not surprising that the reasons for studying automata change from time to time, and that the amount of detail required keeps increasing. Basically a cellular automaton is a lattice formed from cells, each of which has k states, conveniently numbered from 0 to $k-1$. As a consequence of forming part of the lattice, each cell may be associated with several nearby cells, which form its neighborhood. The geometrical shape of the neighborhood may vary, but typically only the closest neighbors up to some radius r are included. Neighborhoods are supposed to have the same form for all cells, and are therefore translates of one another. Consequently there is a single function φ to specify the state of a cell at the next generation in terms of the states of its neighbors (generally including itself) in the present generation.

Although φ relates cells to neighborhoods, its systematic application throughout the lattice defines a mapping Φ from the configuration—assignment of states to lattice sites—of the automaton in one generation to that of the next. Of course, the central problem of the theory of cellular automata is obtaining a satisfactory description of Φ.

If an automaton, such as von Neumann's universal constructor, had been carefully constructed to serve a special purpose, the greatest interest in its operation would naturally lie in whether it performed according to specification or not. On the other hand, if an arbitrary automaton were presented for examination, the most natural reaction would be to observe the course of a typical evolution, that is, the result of repeated iteration of Φ. Different concepts of "typical" would be likely to produce somewhat different results. The initial configuration might be a simple pattern, such as a group of cells in one state with the remainder in another. Alternatively, some short sequence of cell states might be repeated indefinitely, giving a periodic initial configuration. If still greater variety were desired, the states of the individual cells might be assigned with the help of a random number generator.

The best way to reach a compromise between finite and infinite automata would be to combine the second alternative with the third. But then one would soon find that mappings of finite sets into themselves have one fundamental property which turns up over and over again in practically any context whatsoever. Their iterations must eventually become periodic. Evolution being the pertinent mapping for cellular automata, the inescapable conclusion is that the sequence of configurations through which the automaton evolves must eventually become cyclic, oftentimes even static. Thus one is advised to begin the analysis of an automaton by ascertaining all of its periods, both temporal and spatial.

Knowing whether an automaton is large enough to be considered infinite, distinguishing between periodic and aperiodic extensions, and resolving other philosophical dilemmas have a practical importance. Recognizing an entity such as the square root of two so that the diagonal of the unit square can have a length only requires that that such a number can be represented to an arbitrary accuracy. Likewise, although a Turing machine must never be denied additional tape, a generous allotment might suffice for all the calculations required by some project. Thus it is convenient to work with finite structures in such a way that their scale does not have to be stated explicitly, making them implicitly infinite. It then remains to identify any additional structure which may have inadvertently been introduced.

Describing the asymptotic behavior of infinite, but not periodic, configurations within an automaton is a relatively recent development, much of whose motivation arises from trying to identify the computational powers of automata. At the same time, it has been realized that the transient portion of their evolution, particularly when it persists for an exceptionally long time, is not entirely uninteresting, and that its statistical properties warrant investigation. Also, there is interest in locating some or all of the cycles without the necessity of performing all the trial evolutions, the possibility of which is part of the theory of computational complexity.

Whatever the degree of sophistication of the presentation, it is not likely that one is going to find a discussion of a cellular automaton which omits all reference to its periodicity, from its static configurations onward.

1.3.3 Self-reproducing automata

Crystal growth does not qualify as a reproductive process, whilst the real process involving DNA molecules and a whole chemical environment remains far from being fully understood; nevertheless one is skeptical that it will be found to contain some complex computing mechanism. Certainly comprehension of biological reproduction has advanced considerably since those days when chromosomes were known but genes were merely conjectured on the basis of empirical observations, which was still the case when cellular automata came into being.

Questions of reproduction, be it sexual, asexual, or self reproduction, tend to be rather philosophical in nature, along with questions of consciousness, the borderline between the living and the nonliving, and similar issues. A robot is easy enough to imagine, not to mention the possibility of its having a certain dexterity and the ability to follow instructions of a sort. Never mind that functional examples may be in short supply, but it was just this gap between fanciful speculation and actual practice which led von Neumann to settle instead for self patterning wallpaper. The suggestion came from Stanislaw Ulam, who had been experimenting extensively with functional transformations and means of representing them.

By working with a medium which could support a variant of Alan Turing's hypothetical computer, the constructor which von Neumann sought could avoid the monotony of reproduction through crystal growth by calculating the nature of its intended construct, opening quite a few possibilities of variability, adaptation, and evolution. Presumably the use of a full Turing machine would leave the constructor with the utmost versatility possible in choosing its products, or the strategies for carrying out its constructions. Consequently it would appear that whatever such a model lacked in realism might be compensated by the possibility of establishing theorems similar to those which Turing proved for general computations, namely the existence of a universal machine and of undecidable calculations.

The resulting object was cumbersome, containing twenty nine state cells arranged on a two dimensional square lattice, and a finished universal constructor that would occupy an area thousands of squares on a side if it were ever to be "built." Given that the construction originally contrived to prove a point is not necessarily the simplest or most elegant route to the same conclusion, it is not surprising that there are other alternatives, such as one with eight states found by Edgar Codd [48]. Nor is it surprising that von Neumann himself found that as he worked out the details of more and more of his proposed machine, he encountered ideas that would simplify the construction of the parts of the machine which he had already designed. Earlier this trap had thoroughly ensnared Babbage, who repeatedly laid aside plans for one machine in favor of another, still more elegant one, while gradually running out of financing.

The requirements which von Neumann laid on his machine may have been much more severe than necessary. If only the emulation of a Turing machine is required, Alvy Ray Smith III [212] has shown a fairly straightforward way to do so; if only reproduction is required, Christopher Langton [144, 145] has shown some simple variants of Codd's automaton which will grow, neither using a computer to guide the expansion, nor even their own description. Langton does argue, however, that a description of sorts is inherent in the form of the simplest reproducing unit of the organism.

The search for automata with elegant computational properties still goes on, and automata continue to be judged either by the power of the computation that they are able to carry on, or the power of the computer required to understand their behavior.

1.3.4 The Garden of Eden

The evolution of an automaton is nicely represented by a transition diagram or graph; every configuration is represented by a node, the nodes are linked according to the rule of evolution. Chains formed from the links describe long term evolution. The deterministic character of evolution implies that there is exactly one outgoing link for each node. In the forward direction, chains must close into loops; they could also continue indefinitely when the number of configurations is infinite.

In the backward direction the additional possibility exists that a configuration has no ancestor. For a finite diagram that is a necessary alternative if any of the chains are confluent. The reason is simply that after counting exactly one link per node in the outgoing direction, a node with two incoming links must deprive some other node of a compensating incoming link.

Simple counting does not work for infinite diagrams, but the fact that the neighborhoods defining the evolution have to overlap to some degree allowed Edward F. Moore [164] to show that a similar conclusion nevertheless follows, namely that whenever some configurations have multiple ancestors, there must be others which have none; the term Garden of Eden seemed appropriate to describe the situation.

The result is an interesting limitation on the kind of constructions which an automaton can perform. But a universal constructor is not expected to construct *everything*, and according to this theorem, clearly cannot. Generally it is considered adequate that it can construct objects at least as complicated as a Turing machine, or slightly more so, so as to be able to make copies of itself.

There are straightforward procedures for ascertaining the Garden of Eden configurations of a one dimensional automaton, as well as the actual ancestors for the remaining configurations. Jen [130] has shown one such construction, based on a symbolic de Bruijn diagram. It is typical that the corresponding calculations are undecidable for arbitrary two dimensional automata or beyond, generally because of conflicts arising from the Post correspondence principle. Nevertheless two Garden of Eden configurations for *Life* have been published [91, page 248],[20, page 829], and a third cited [91, page 248], all of them were evidently encountered after a long series of exhaustive eliminations.

Garden of Eden configurations represent one end of the evolutionary tree, its leaves, which lie at the opposite extreme from the cycles, which form its roots. Discovering them requires more than simply following out trial evolutions, but their inclusion nevertheless forms an interesting part of the description of any given automaton.

1.4 The Gardner era

Public awareness of cellular automata can mostly be attributed to John Conway's interest in finding a simpler configuration than von Neumann's and exploring its capabilities. Some of his results were presented in 1970 as an ecological game called *Life*, at a time when such concerns were popular, in Martin Gardner's monthly *Mathematical Games* column in Scientific American. For a period of about three years Robert T. Wainwright maintained a quarterly newsletter [239] disseminating discoveries made by Martin Gardner's readers, some of which were followed up in later columns in Scientific American. Many of the more interesting results were obtained at MIT's Artificial Intelligence Laboratory with the help of the graphics facilities of their PDP-6 computer [1].

When microcomputers began to attract popular attention, Conway's game of *Life* became one of the early inspirations for an application; Cromemco's "Dazzler," a color video controller and one of the earliest peripherals, was frequently used to display the evolution of *Life* configurations. Early issues of Byte magazine contained some material on *Life*, but in one memorable issue many results which had appeared in Wainwright's newsletter [239] a decade earlier, but still not reached mass circulation, were presented by some of their discoverers. Other magazines, such as Omni, also revived the topic [111, 167], and in recent years Scientific American has returned to the subject, most recently in A. K. Dewdney's *Computer Recreations*, the current successor to Martin Gardner's column.

All of Martin Gardner's columns of the early 1970's have been reissued in a recent reprint collecton [91], together with some of his reminiscences. He identifies three places as having been particularly active centers of interest, at all of which people were mainly concerned with collecting examples of one type of behavior or other, many of them reported in Wainwright's newsletter. Many fanciful names were invented to describe the variety of configurations which were found, including two (glider gun and puffer train) due to Conway himself.

Conway had devised the rules of evolution of *Life* carefully, to avoid the extremes in which live cells proliferated and grew without bound, or in which live cells dwindled and eventually died. He was still not sure about the ultimate fate of his delicately balanced creation; there could always be some fairly uncommon combinations nevertheless capable of unlimited growth. Several small patterns were known which delayed thousands of generations before their final behaviour became evident.

A five-cell figure called a "glider" capable of diagonal movement had been discovered in the early stages of experimentation, as well as larger figures – "space

[1] for an account, see Levy's book [147]

ships" – capable of horizontal or vertical movement. An isolated "glider gun"that produced gliders periodically would be one structure with unlimited growth, a puffer train formed by a space ship which left permanent debris behind as it moved would be another. Whether such structures existed was not clear at first, but it was evident enough that regular structures would be required before theorems could be proved. Otherwise there were simply too many courses of evolution open to study them all.

At MIT glider guns were quickly found; also glider collisions were studied in great detail. Interestingly enough, puffer trains were also found, even some that included gliders in their residues. Through this combination a mass was found which, not really dense but still reasonably compact, violated Conway's conjecture against an infinitely growing configuration in the worst possible way. That is, it occupied a diamond shaped region of constant density whose borders expanded at a uniform velocity – a quarter of the "velocity of light" – because of the prevalence of gliders in its makeup.

Glider guns created a steady stream of objects - gliders - which could be used as signals in something resembling an electrical network, which enabled Conway to design a configuration which fulfilled von Neumann's original goals of creating self-replicating automata. In the process it was made clear that many questions about automata theory were undecidable because their answers would also have to describe the Turing machines which could be associated with the process. However, that conclusion can be reached much more directly by embedding linear cellular automata which emulate Turing machines in two-dimensional environments. However such machines use many more than the two states per cell which Conway's construction achieves.

Work at MIT also revealed some interesting schemes for constructing very large configurations of period two, even some which completely vanished in the process of creating the new generation and which in turn disappeared as they recreated the original generation. Another group of students in Canada worked out large numbers of oscillators of different periods, on the basis of the strategic placement of structures which came to be called "eaters."

It was evident from the first that some structures were much more stable than others, and that they could be used as building blocks to regulate the growth of more erratic or less stable combinations. Indeed, glider guns seem to have been a byproduct of having first constructed oscillators by confining colonies with promising but undisciplined growth, and then checking out collisions of the moving parts of such oscillators.

Of course, given moving objects such as gliders or space ships, it was natural to look into their collisions with stationary objects or with each other. Rather than simply admiring the spectacle, one can take a scientific approach, systematically varying the relative positions of the reactants and recording the results. In one way or another, some extremely varied and highly interesting configurations were discovered.

Wainwright undertook a classification of all the noteworthy *Life* configurations which his correspondents reported, and one could fairly say that the "Gardner era" of cellular automata was characterized by an intensive search for "interesting" configurations in Conway's *Life*. Some modest variants were considered by various persons; but *Life* itself was sufficiently challenging, and the variations exibited by the variants insufficiently spectacular, to result in any great volume of reports.

The extent of the diverse results which were obtained was indeed impressive, but it was inevitable that the field would become saturated as the readily imaginable concepts of "interesting" were explored, and the combinations accessible to the computer technology of the day were exhausted. The next steps depended not only upon a new computer technology, but upon advancing concepts in information theory and even in formal language theory.

1.4.1 Conway's *Life*

For some time the theory of cellular automata consisted of the inevitability of evolution into cycles, the existence of pristine initial configurations, and a respectful reverence for the complexity of von Neumann's detailed plans and the philosophical implications of his efforts. In the practical direction, a variety of efforts with image processing and photographic interpretation were understood to be related as much to cellular automata as to Fourier or Walsh transforms [190], and some thought was given to designing circuit arrays as automata.

One way or another. the lore of automata theory has gradually spread around. Studying cellular automata has become a popular computer game in addition to being a learned academic subject in several stages. Perhaps the initial step resulted from John Conway having decided to review the area just at the time that computers acquired versatile visual display equipment.

By taking the approach of first defining the automaton, then examining its capabilities, Conway [90, 91] encountered a two-dimensional binary cellular automaton of incredible elegance and beauty. Christened *Life*, it was eventually shown to have the same powers of universal computation and self-reproduction as von Neumann's automaton [20]. Even so, much work and several years were required to acquire familiarity with *Life*; many of the results were reported as they were being discovered in a quarterly newsletter published for nearly three years by Robert T. Wainwright [239]. Later some of the information was published more formally in a book written by William Poundstone [188]. The substance of all the notices in Martin Gardner's columns was collected in one of his reprint volumes [91].

From the outset Wainwright undertook to catalog the *Life* configurations that were being discovered, classifying them roughly along the lines of the general theory. Conway knew about some travelling configurations called gliders, but was apparently unprepared for the discovery of first "glider guns" and then "puffer trains" which between them permitted the assembly of incredibly complex artifacts. Nevertheless he was able to use them to achieve von Neumann's results, although in terms of an equally arduous construction spreading over a fantastic area if it were ever to be realized in practice. The number of contributors to Wainwright's newsletter was considerable, but the contributions of William Gosper and several associates at MIT's Artificial Intelligence Laboratory were fairly outstanding. Some of their adventures have been recounted in Stephen Levy's book, *The Hackers* [147].

Conway's choice of a particular rule and lattice resulted from careful experimentation, giving *Life* a relatively distinguished setting. Poundstone [188] reports experiments with an alternative rule, Packard and Wolfram [180] surveyed numerous two dimensional automata; in both cases reporting that their automata fail to meet Conway's criteria. It would appear to be an interesting challenge, either to account for the singular nature of *Life* or to encounter additional specimens.

Besides offering a new and interesting automaton to the existing repertoire, Conway's contribution was to begin with specific automata with the intention of ascertaining their potential behavior, rather than beginning with the application and searching for the automaton which would fulfil the requirement.

1.5 The Wolfram era

Professional scientific interest in cellular automata received a considerable impetus from the investigations of Stephen Wolfram [249], who undertook a computer based search through the properties of one dimensional automata, guided by some concepts from the realm of nonlinear dynamics and statistical mechanics. For anyone, in fact, the microcomputers, programming languages, and video displays which are currently available are sufficient for many experimental studies of cellular automata, not a few of whose results have considerable artistic merit.

A recent article exploring the aesthetic side of automata theory was entitled *Abstract Mathematical Art*, by Kenneth E. Perry [185], published in the December, 1986, issue of Byte. The article included a Basic program for use on IBM/PC compatible computers, with indications that a Pascal version was available from the magazine, and a statement that the author himself had used machine language to quickly seek out the hundred examples which he selected for the readers' experimentation. Several of them were shown in striking color photographs illustrating the article.

The idea of a one dimensional cellular automaton is quite simple, and its evolution in time is ideal for a two dimensional presentation, as on a video screen. To start with, a cell is a region, even a point, with differing forms, called states. For convenience, these states are usually numbered with small integers beginning with zero, rather than described. For the purposes of automata theory the nature of the states does not matter, only their relation to one another, and the way they change with time according to their environment. Since they are abstract, they can just as well be represented by colored dots on a video screen, which is what makes them so dramatic when interpreted as an abstract artistic design.

Although both von Neumann and Conway were aware that alternative rules of evolution existed, incredibly large numbers of them in fact, they concentrated on one single rule which served their purposes, exploring its consequences in detail. Wolfram, by contrast, was one of the first to compare the evolutionary histories of large numbers of different rules, with the intent of classifying them according to their long term behavior. Indeed he seems to have been inspired by work in dynamical sustems theory, particularly Stephen Smale's discovery of "strange attractors," and the possible parallels they might have in automata theory.

In any event, he noticed that the evolutionary histories of linear cellular automata were quite varied, for which he proposed four classes. The first contained automata evolved automata evolving, usually fairly rapidly, into a constant field.

Quiescence, of course, was a basic assumption of von Neumann, Conway, and others who needed a static background against which to perform their constructions. Wolfram took this class to be the analogue of those dynamical systems which evolve toward a point of stable equilibrium.

The second class was supposed to correspond to the limit cycles of nonlinear mechanics, that is, limiting behavior which is cyclic. With automata, many limiting fields are not quiescent, but nevertheless show little, no, or completely predictable, activity. *Life*'s still lifes are a typical example, as are the blinkers, gliders, and other isolated periodic structures which characterize its long term evolution.

The third class is quite the opposite, consisting of chaotic fields whose behavior is never predictable; these would correspond to ergodic behavior in dynamical theory or even strange attractors.

The fourth, and remaining, class is the one which may be the most interesting; Wolfram wished to associate it with universal computation, but it is essentially the same class which held Conway's interest; the reasoning was quite similar. In spite of the ordering implied by Wolfram's numbering, the fourth class is probably best regarded as an interface between the second and third classes as suggested by Christopher Langton [146]. In any event, it is a class characterized by "islands of chaotic behavior in an ocean of quiescence." One does not find such behavior in binary automata of small radius; it is always fairly uncommon, yet far from rare.

1.5.1 Wolfram's classes

As computer power steadily increases, there have been ever increasing opportunities to perform experimental mathematics, both by professionals and by amateurs. An excellent example of this tendency has been all the recent work on fractals, nonlinear differential equations, dynamical systems, and similar topics. Many extremely classical results, such as the results of Pierre Fatou and Gaston Julia concerning functional iteration, some very esoteric topological results such as Stephen Smale's strange attractors, or even such advanced theorems relating to differential equations such as Kolmogoroff, Arnol'd and Moser's results in celestial mechanics, acquire an entire new perspective when their consequences can be followed through a numerical calculation.

A rather interesting confrontation of this nature took place when Stephen Wolfram began to experiment with the evolution of one dimensional cellular automata, and felt that he saw a certain amount of analogy between the phenomonological characteristics of the evolution of the automata and some classifications of limit sets which had been found relevant in nonlinear dynamics. As a result of numerous observations, he proposed a system of four classes, which reflected the same number of different kinds of evolution which he had been observing. Roughly speaking, cellular automata seem to settle down to a constant field (Class I), isolated periodic structures (Class II), uniformly chaotic fields (Class III), or isolated structures showing complicated internal behavior (Class IV).

In addition to a series of computer experiments in one and two dimensions, with varying numbers of states per cell and sizes of neighborhoods, Wolfram, occasionally working with some coauthors, has analyzed the mathematical aspects of the rules of evolution of their automata. Thus [251] discusses the relation of automata to formal language theory, including the observation that the evolute of a configuration described by a regular expression is still a regular expression, but may fail to remain so in the limit. Lyman P. Hurd has taken up the explicit question of the limiting behavior in his thesis [123] and related publications [122, 124].

Wolfram also discusses some measures of complexity, such as the size of the automata associated with the regular expressions.

Martin, Odlyzko, and Wolfram [155] have worked out the evolution of automata for which φ is defined by a linear combination of the states of the cells of the neighborhood, assuming them to be elements of the corresponding finite field; an example would be evolution according to the rule of parity, in which each cell transforms into the sum of the three neighbors, *modulo* 2. Although such automata may not be entirely typical, the fact that quite complete and exact results can be obtained makes for useful comparisons.

Wolfram has edited a collection of most of his own papers and a good assortment of others, including several useful appendices containing a wide variety of data [252].

Notwithstanding the fact that assigning an automaton to one of Wolfram's classes is an undecidable proposition [56], his investigations represent the first time that a systematic and exhaustive study had been made of large classes of one dimensional linear cellular automata, rather than selecting a single automaton for concentrated attention. At the same time the classification has a strong visual appeal and has been widely adopted by other investigators, along with a few notational details such as his numbering scheme.

1.6 One dimensional cellular automata

To make a one dimensional cellular automaton, a series of cells is strung out along a line, the simplest assumption being that all the cells have the same number of similar states, and that the rules of evolution will be the same for all of them. The idea of forming the cells into a line implies a linear order, but of course other arrangements are possible, both in terms of the dimension and the connectivity of the cells. This is the second element of definition for a cellular automaton – the relationship between the cells, or the kinds of neighborhoods which they form. Again, the simplest neighborhood would consist of a cell and its nearest two neighbors; generally speaking we would take a neighbborhood to consist of r neighbors on each side, giving $2r+1$ for the total number of cells in a neighborhood.

There are some small quibbles to be made. If a chain is finite, it has ends, which surely do not have the same neighborhoods as the interior cells. Either they can be treated differently, or the chain can be imagined to close into a ring, which would preserve the uniformity of all the neighborhoods. Also, it is considered preferable to work with symmetric neighborhoods, each one centered on its own cell, rather than worrying about irregular neighborhoods. An interesting exception would be to work exclusively with a single neighbor, always on the same side, but that is another story.

Thus there arises Wolfram's notation (k, r) for a linear cellular automaton which has k states within each cell, and such that it, together with r cells on either side, is considered to form a neighborhood. In reality, a neighborhood does not have to be centered on its cell; for example when its length is an even number of cells, succeeding generations might be staggered by half a cell, and the radius assumed to be half-integral.

There is one final ingredient in the definition of a cellular automaton, which is the rule of transition by which the cell changes state from one generation to the next, conventionally assumed to be the same rule for each neighborhood. It is the judicious selection of a rule as much as anything that makes a particular automaton interesting or not.

Conway's game of *Life* was the result of a particular choice of rule for a two dimensional binary automaton – two states per cell – whose neighborhoods contained the cell in the center and the eight cells touching it, four of them laterally and four of them diagonally. The announced critera by which that particular rule was chosen were that a field of cells should neither dwindle away to nothing – all zeroes – or eventually fill up completely – all ones. Reportedly, he examined many different rules before choosing the particular one which gave us his famous game; even so, so much variety was encountered with that one particular rule that years passed before many others were studied.

Wolfram's further work, mostly done at the Institute for Advanced Studies in Princeton, systematically examined all the possible rules for one dimensional automata. His recent book [252], a collection of reprints comprising much of his work on automata, contains an extensive appendix tabulating diverse properties of all 256 $(2, 1)$ automata.

In two dimensions there are far too many possibilities – vastly more so even than in one dimension – for there to be a chance to try everything, even with a very fast computer. Notwithstanding, Dewdney's column in a recent issue of Scientific American describes a three dimensional variant of *Life* devised by Carter Bays [14, 15, 16]. With twenty seven cells per neighborhood, its analysis has to be far beyond the reach of either present methods or present and foreseeable computers. However, it is not necessary to choose such an ambitious model to gain useful insight and obtain interesting results. So, even the one dimensional automata of type $(2, 1)$ constitute a reasonable starting point for systematic studies.

1.7 The vocabulary of automata theory

One of the heritages of the interest in *Life* is a rich vocabulary of such picturesque terms as still lifes, gliders and so on, describing typical artifacts encountered during evolution; even though of two-dimensional origin, they are freely used for other automata.

Similarly, Wolfram has created some basic descriptive terms by which automata can be classified according to their rules of evolution, not to mention their classification into four phenomonological evolutionary types.

For $(2, 1)$ automata, a neighborhood contains three cells, each of which exists in one of two states; altogether eight possible neighborhoods. Since there is nothing to require evolution of a given neighborhood to lead to one value of the new cell or another, there are 256 possible ways that each set of eight different neighborhoods can evolve into the next generation, starting with the possibility that everything evolves into zero and ending with the possibility that everything evolves into one.

The choice of the words "zero" and "one" assumes a decision to number the states of an automaton, which is certainly convenient for computer processing, or even a plain mathematical discussion. Yet the choice is quite arbitrary; Conway must have thought that the more vivid dichotomy between "live" and "dead" or "inactive" would make his game more appealing. Naturally the states of an automaton displayed upon a color television screen would be the colors of the pixels themselves.

As suggested by the choice of a numerical representation, the easiest way to enumerate $(2, 1)$ neighborhoods is to make up a binary number whose eight digits tell how the neighborhoods 000, 001, 010, and so on evolve. Every evolutionary rule gets its own serial number in the process; it is convenient to always do this in a standard way and refer to the rule by the decimal equivalent of the resulting number. We might call the following choice its *Wolfram rule number*: Set out the possible sequences of cells in reverse lexicographic order:

111	110	101	100	011	010	001	000
0	0	0	1	0	1	1	0

The second row shows how the three-cell neighborhoods evolve for Wolfram's Rule 22, the decimal equivalent of the eight-digit binary number 00010110. This ordering of neighborhoods is natural if one thinks in terms of "high order first" but is just the opposite of the one used in Perry's article [185].

The working of the rule of evolution can be seen in the following sample of ten generations of evolution of Rule 22 from a randomly chosen initial configuration. A cyclic ring is assumed, so the first cell in each row is the right neighbor of the last cell, just as the last cell is the left neighbor of the first cell.

Every neighborhood is repeated numerous times throughout the diagram; evolution is referred to central cells, whose new values can be found in the next line, just below the old cell.

1	·■·■··■■·■■·■■··
2	■■·■■■········■·
3	······■······■■·
4	·····■■■····■··■
5	■···■···■··■■■■■
6	·■·■■■·■■■■·····
7	■■·········■····
8	··■·······■■■··■
9	■■■■·····■···■■■
10	····■···■■■·■···

A table of transitions explicitly describes the function mapping each neighborhood of a cell into the cell's next value, but for general discussions or proving theorems it is better to use common mathematical notation. The symbol $\varphi(a, b, c)$, or simply φ, is often reserved to denote this *transition function*. Of course, such a definition serves only for nearest neighbors; the function would have $2r+1$ arguments if it belonged to a (k,r) automaton.

Either Perry's or Wolfram's representation makes clear how many rules there are, because, given a (k,r) automaton, any one of the k states can result from the evolution of each of the k^{2r+1} different neighborhoods. Because of the exponent involved, the number of possibilities increases drastically if either the number of states for a single cell increases, or the number of neighbors increases. Thus a $(3,1)$ automaton – three states with nearest neighbors – has $3\times3\times3$ or 27 neighborhoods, and the total number of possible rules would be 3^{27}, somewhere on the order of 10^{12}, so they are not soon going to be studied one by one. Alternatively, the combination $(2,2)$ would have 32 different neighborhoods, and thus 2^{32} different rules, which is "only" in the range of 10^{10}.

To obtain a reasonable sampling of even the smaller linear automata, Wolfram used the idea which was already implicit in Conway's statement of his rule, that the evolutionary criterion should depend on the number of cells in the neighborhood, but not on their particular arrangement. Talking in such terms reveals some hidden assumptions about our vocabulary. In Conway's binary game, zero represented a dead cell, one a live cell, and his rules were stated in terms of the number of live cells in the neighborhood. It is a simple extension of this idea to assign numbers (weights, if you wish) to the states, and make the transition depend only on their sum. A rule gotten this way is called a totalistic rule; not all rules are totalistic, but they lead to a more manageable sampling of all the possible rules.

Curiously, rule 22 cited above is also a totalistic rule. Three binary digits can add up to zero, one, two, or three. If these possibls sums are arranged in descending order as before and paired with the corresponding value of the evolved cell, a *totalistic rule number* can be derived (which in this case would be decimal 4).

sum	3	2	1	0
new state	0	1	0	0

Even the use of the number of distinct sums can lead to very large numbers; also there is a question of whether the evolving cell should have its own weight included or excluded from the sum. For binary automata (but for higher dimensions) Bays has introduced the notation $w/x/y/z$ to mean: if the cell is zero and there are between w and x neighbors, a cell will be born; while if is one (alive) and has between y and z neighbors, it survives. In this notation, Conway's original game was a 2/3/3/3 *Life*.

Doubtless other notations will evolve as significant new combinations requiring them come to light.

1.8 Probability theory

The enthusiasm which greeted Martin Gardner's exposition of *Life* was initially focussed on tracing evolutions of two-dimensional configurations that struck one's fancy, or in trying to design configurations which would have some interesting characteristic. The original challenge was to either produce a configuration that would grow without limit, or to demonstrate that none existed. Conway proposed two mechanisms, a glider gun or a puffer train and examples of each were soon found – contrary to his expectations in spite of having conjectured them, apparently.

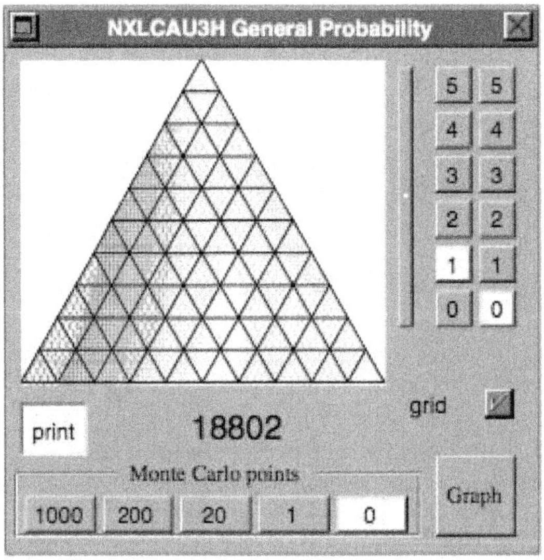

Figure 1.1: Contours for fixed point probabilities.

However, as David J. Buckingham [37] remarked is his Byte article, interest eventually turned to information-theoretic aspects of automata theory – something which would probably surprise anyone not familiar with this branch of engineering (or mathematics.) Even from the beginning, there was attention paid to the stastical properties of *Life*, Martin Gardner having reported some results of Robert Wainwright on the behaviour of "primordial soups" and their eventual evolution at fairly constant density [91, p. 237].

Indeed, any prolonged observation of a *Life* field leads to the conclusion that there are three phases of evolution: The first phase is a relatively short transient phase – at most ten or tens of generations – in which excessively high or low initial densities adjust themselves; the second phase may last for thousands of generations in which nothing seems to be definite; followed by the third and final phase in which isolated groups of cells go through predictable cycles of evolution.

Simple ideas of probability yield approximately correct but undeniably discrepant results. The evident explanation that assumptions about independence and exclusivity – after all, if the neighborhoods of two adjacent cells overlap by 50% or more, their evolutions might not be independent – identifies the likely source of error, but says little about what to do about it. M. Dresden and D. Wong [67] made some probabilistic calculations concerning the evolution of *Life* in 1975; L. S. Schulman and P. E. Seiden [202] slightly later in 1978.

It has recently become apparent that old ideas about probability nests, quite highly developed in other branches of mathematics, are ideally suited to contending with this problem. It is a theory which can be especially well developed for one dimensional systems; W. John Wilbur, David J. Lippman and Shihab A. Shamma [242], then Howard A. Gutowitz, Jonathan D. Victor and Bruce W. Knight [107] have all published interesting results using these techniques.

Not only has there been considerable interest in resolving the statistical properties of cellular automata, there is a very definite current running in the opposite direction, by which known automata are taken as the starting point for exploring the statistcs of other systems whose rules of evolution are much complicated. In this way one returns to Poincaré's views of approximating the properties of a dynamical system. The underlying hope is that valid statistical conclusions will arise from dynamical systems that are much simpler than those governed by differential equations.

Regarding a discretized differential equation as a cellular automaton is too much of a simplification unless the cells have a rather large number of states; still there are some significant aspects of nonlinear systems which can be modelled by relatively small automata. In one well studied case, three characteristic regions are recognized for the variable governed by a differential equation — quiescent, active, and transitional.

Such a system is usually quiescent; but an activity can be initiated which must procede to completion before calm can be restored; especially in two or more dimensions some interesting self-sustaining activation chains can be found, which have been taken as models of chemical or neurological behavior [100, 154].

1.9 Graph theory and de Bruijn diagrams

Looking in another direction, the cells of one dimensional automata, by definition, form linear chains. But the neighborhoods themselves form another kind of chain, wherein they necessarily overlap. The result is an arrangement which has a very close connection with shift register theory [95], which is a systematic study of the properties of overlapping strings and how the overlaps happen to come about. In particular, the form of graphical representation known as the de Bruijn diagram [191] enters into many discussions, and can be used to organize a major portion of the theory.

The application to automata theory arises from labelling links in a de Bruijn diagram in two ways – as a neighborhood N or as the evolutionary image of that neighborhood, $\varphi(N)$. Including or excluding links according to various properties of either the neighborhood or its image results in a considerable variety of subdiagrams, or graphs, to be used for further study.

Graphs provide a finite frame of reference for describing the multitudinous paths corresponding to the actual configurations of an automaton; an economy of presentation which fully justifies linking graph theory with even the theory of general automata. Graphs in turn have such diverse representations as their connectivity matrix or a system of symbolic equations expressing their connectivity, not forgetting simple paper sketches of graphs which are not overly complex.

In turn the collection of paths through a diagram is readily described by such entities as polynomials in the connectivity matrix or even by regular expressions. The former permit the definition of a zeta function; the latter follow from the symbolic connectivity equations, and can be seen as an opportunity to use formal language theory to obtain further results. Both approaches have been discussed in the literature.

Still further aspects of graph theory are available to describe automata and their evolution – such as dual diagrams, cartesian product diagrams, or mappings between diagrams. For example, product diagrams serve to compare two or more paths through the underlying graph, a comparison which can profitably be used to decide questions of uniqueness or of ambiguity.

So in one way or another it pays to be familiar with the standard results of graph theory. As it is, Masakazu Nasu [169] refers the properties of injective and surjective evolutionary functions to de Bruijn and related diagrams; Wolfram [251] himself used them to express evolutionary properties, and Erica Jen [130] has used them to calculate ancestors.

Other examples can be found in the literature. Moreover, it is always possible for a retrospective knowledge of the appropriate diagrams to facilitate the understanding of those publications whose authors did not avail themselves of graph theory. Or at the very least least, to hope that the resulting insight will lighten the burden of obtaining further results.

1.10 Reversible automata

Supposing that von Neumann's goal of exhibiting an automaton capable of universal construction or of universal computation has been realized, additional questions come to mind. For example, an automaton which contains an embedded Turing machine has to suffer some of the same limitations as the Turing machine, the most notorious of which is undecidability.

It would appear that a cellular automaton ought to be more versatile than a Turing machine; for example it could have numerous read heads and work on many computations simultaneousy. Nevertheless, since any automaton can presumably be modelled within some Turing machine, the computational powers of the two artifacts must be coextensive. One seeming paradox arose when it was found that some automata had configurations which could not be the product of any evolution, the "Garden of Eden" states of Edward F. Moore [164, 166].

The reconciliation of such states with a universal constructor lies in the realization that the constructor is not required to fill space with arbitrary designs, but rather to create specific objects (including copies of itself) according to the instructions which it has received. Universality refers to whether arbitrary descriptions can be followed, not whether arbitrarty constructs can be produced.

Nevertheless, there has been considerable interest in ascertaining whether or not there are restrictions on the long term behavior of an automaton, both in the remote past and in the remote future. Such restrictions could manifest themselves in unattainable configurations such as the Garden of Eden, as well as in certain limiting configurations which would only develop fully as limits.

But there is also a middle ground, consisting of rules or even of configurations within a given rule, which never end up in some inaccessible region, either past or future. For this to be true, it is especially important that there be no barrier such as the Garden of Eden, devoid of a previous existence in any form. Equally, although the future always exists in some form or other, there might be reasons to avoid an approach to extremely complicated limits.

In other words, there is an interest in time reversal symmetry, or even a simple equivalence between past and future, whereby past configurations would be just as recoverable from a given configuration as the future configurations that are deducible from the same data. Quite trivial instances in which the states shift sideways, remain unchanged, or complement themselves between generations readily come to mind; naturally real interest centers on whether there are others, besides.

S. Amoroso and Y. N. Patt [6] searched for possible reversible rules and discovered eight nontrivial $(2, 3/2)$ rules in 1972; Tommaso Toffoli [218] found a way to generate whole classes by an increase in dimension in 1977, and Edward Fredkin discovered another scheme which has been reported in Toffoli and Margolis's book [219]. Fredkin employed evolutionary functions extending over two generations, evidently extendible within the same framework to three or more generations. Yet another alternative employs some of the states in a cartesian product to create a sort of memory.

It is interesting to observe that the whole idea of reversible automata falls within the province of dynamical systems (once the connection is realized) just as it was already reported in great detail by Hedlund [114] in 1969. A recent survey article by Roy Adler and Leopold Flatto [3] contains a good exposition of the relationship of symbolic dynamics to flows through a graph.

The reason that symbolic dynamics has such a special relationship to reversible automata lies in the fact that a dynamical system can be given a topology, with respect to which it is much easier to discuss infinite systems and the existence of limits. This in turn brings up such concepts as continuity, the mutiple valuedness of mappings, and the existence of inverse functions.

An essential element of the discussion hinges upon the fact that the evolution function for automata is not invertible in and of itself, but the discrepancies between counterimages can be pushed to remote regions of the automaton. With the help of the topology they can then be made to vanish in the limit, providing a context in which the reversibility of evolution can be discussed.

1.11 What has and what hasn't been done

Research on the subject of cellular automata is surveyed, with the intention of distinguishing between what *has* and what *has not* been accomplished during the course of its history, and to try to foresee something of its future.

1.12 What has been done

When certain subjects are mentioned, they are immediately associated with a certain body of results; others are more tentative or may even be in the process of revision. There is probably a certain consensus as to what the theory of cellular automata comprises, overlapping somewhat with automata theory in general, and touching upon the theory of probabilistic automata. The following synopsis is a mixture between a catalog of topics and an account of the historical development.

1.12.1 Probability measures

For various reasons, the statistical properties of the evolution of cellular automata have attracted attention. That there are statistical properties to be investigated is immediately apparent whenever there are graphical means available to display the evolution, such as can be done readily with most computers nowadays. Typically, if an initial configuration is randomly chosen, it quickly develops a characteristic texture which persists until the inevitable cycles make their appearance and dominate the evolution; thus there are usually two inherent time scales and three textures associated with the evolution of any given automaton.

The first phase can be orderly or random, in either event reflecting the details of the initial pattern or random number generator used to produce it. Its duration gives some idea of the incompatibility between the evolutionary rule and the initial arrangement; for random patterns the time is typically quite short, but for quiescent automata with a recognizable "velocity of light," a reasonable propagation time must be allowed for quiescent regions to feel the influence of their neighbors.

One would like to think that there might be a "self consistent probability" associated with any particular evolutionary rule, but calculations based on the simple rules applicable to independent probabilities produce indifferent results, traditionally disclaimed by invoking unaccounted correlations. Evidently an appeal must be made to more elaborate statistical concepts if better results are required; the crude theory usually produces results which match observations at the level of 10% to 20% error, which is often considered adequate.

Otherwise it is necessary to investigate measures which commute with the evolution or are otherwise realted to it. Related concepts such as entropy are also relevant, and can also be studied. One of the first studies to be made was that of Schulman and Seiden [202], who tried to explain their measurements of Conway's *Life*. Later Dresden and Wong [67] made corrections for correlations, but the most comprehensive studies have been made by Wilbur, Lippman and Shamma [242], and then by Gutowitz, Victor and Knight [107].

There are actually three or four levels at which self consistent probabilities can be approached. All of them involve comparing the probability that a cell lies in a certain state with the probability that the cells in the neighborhood are in just the states in which they are found. Self consistency requires that the two probabilities agree, but the methods differ in the way that the probabilities are estimated.

The simplest comparison is to ask how many of the possible neighborhoods can evolve into the state in question; the answer may range from 'none' through 'an equal share' to 'all of them'. The uniform fields resulting from the extremes produce no surprise, a state with scarcely any ancestors would not be expected to be very common, while a democratic assortment of rules would lead one to expect a representative mix of states. In general terms these expectations are seem to be confirmed by experience and are thus well founded.

It is a general statistical principle that the variance in averages decreases inversely as the square root of the sample size; as applied to the evolution of automata it is possible to obtain very stable frequencies for the occurence of the different states by collecting them from an automaton with a very large number of states – say a thousand or more. When this is done, discrepancies of tens of percent are found with respect to estimates by rule of thumb, such as counting transition rules.

The next level of self consistency consists in estimating the probability of a neighborhood by combining the probabilities of its cells according to the traditional rules, while assuming the validity of the traditional assumptions of independence. Called mean field theory, it provides polynomial equations to be solved for the self consistent probabilities. Indeed, the simpler estimate is just the probability that would be assigned the standard distribution – the one in which all states have the same probability – and thus could be the starting point for an iterative solution of the mean field equations.

Empirically mean field theory provides better frequencies, but not all automata perform equally well and descrepancies still range on the order of 10% or so, a useful but hardly a precision result. So the next level concentrates on taking correlations and other factors into account. The block structure theory of Gutowitz *et.al.* is based on Bayesian extension; empirical studies have shown that by taking into account probabilities of blocks of six, eight, ten or more cells, frequencies can be matched to two or more figures. The self consistency equations involve rational fractions rather than polynomials, and can also be solved by iteration.

The final level would be to obtain the invariant measure of the automaton, but having to describe it numerically leaves one on the level of local structure theory with very long blocks. Whatever the level of detail at which they are attempted, it would seem that the probabilistic approach is a significant alternative, and a complement to, the point of view of formal language theory.

1.12.2 De Bruijn diagrams

Whether working with probabilities, measures, or simple evolution, the overlapping which occurs between different neighborhoods comprises the greatest technical obstacle to computations; unless this problem is resolved adequately, further progress is almost impossible. Fortunately, for one dimensional automata, a diagrammatic technique which lies at the heart of shift register theory [95] saves the situation; the diagrams are called de Bruijn diagrams, but they are just simple graphs showing the possible ways in which different neighborhoods can overlap [191]. Erica Jen has shown how many properties of automata can be extracted from such diagrams, especially the static and periodic configurations on a cylinder of fixed circumference [131].

In principle, such a diagram could be extended to automata of higher dimensions, but a problem arises from selecting partial neighborhoods that will join to form full neighborhoods in all directions. The straightforward approach of building up strips of successively higher dimension runs afoul of Post's correspondence principle when arbitrary intermediate strips have to be matched to form the strips of the next higher dimension. If only periodic solutions are required, the problem is still soluble, but again the conflict between large systems and unbounded systems arises, tending to leave the generic properties of aperiodic systems undecidable.

At least in one dimension, there is nothing difficult about a de Bruijn diagram; as applied to cellular automata it is simply a graph in which partial neighborhoods are the nodes, with links connecting those which may overlap to form a full neighborhood. Given this correspondence between links and full neighborhoods, each link is also associated with the evolved cell belonging to the neighborhood. Consequently characteristics of the evolution can be used to select subgraphs of the de Bruijn diagram; for example, there is a subgraph composed of the neighborhoods whose central cell never changes. Global properties of the automaton can be read off in terms of the chains to be found in such a subdiagram; for the present example, the chains determine all the static configurations.

The advantage of using a de Bruijn diagram is that many problems concerning automata are thereby transformed into known problems regarding the tracing of paths through a graph. For instance, no loop can be longer than the total number of nodes in the graph without repeating some segment; but then there must exist still other loops in which the repeated segment is traversed an arbitrary number of times. For example, a binary automaton depending upon nearest neighbors has eight distinct neighborhoods, representable as eight links connecting four nodes, it follows that no static configuration can be more than four cells long without repeating some two-cell partial neighborhood. Thus the static configurations are rather severely constrained.

Sometimes the de Bruijn diagram reveals information about localized aspects of a configuration. For example if an acceptable path terminates at a node in which *all* the outgoing links are acceptable, it need continue no further. Likewise if all the incoming links are acceptable, the path may begin just as though it had been part of a loop. Thus semi infinite structures may be located, or even finite ones if both ends have such universal terminations. This leads to the phenomonon of membranes and macrocells which Wolfram noticed during the course of his investigations. That is, an automaton may have patches which are isolated from one another by static regions, whose evolutions procede quite independently.

The converse process is also possible, to define the rule of evolution of an automaton by postulating that the de Bruijn diagram have prescribed properties; for example that the unit cell (0101) must be a static configuration. To the extent that the requirements are not contradictory, and all the possibilities are covered, an automaton may practically be designed to order.

Enumerating the paths through a graph is a classical task, to which many papers have been devoted, but which has a particularly elegant solution in terms of regular expressions. Conway's book on regular algebra [49] expounds the technique; a later article of Backhouse and Carré [10] gives a very thorough presentation.

Probabilistic versions of both the de Bruijn diagrams and the evolutionary diagrams exist, being useful for studying correlations between blocks of cells, or in the numerical calculation of self consistent block probabilities. All the standard theorems regarding positive matrices and stochastic matrices apply. All told, the introduction of graphical techniques to automata studies is very profitable.

1.13 What has not been done

It is hard to know whether there are likely to be any surprising new developments in automata theory, in order to be able to predict them.

A century later, the first completely electronic computer inspired von Neumann to speculate about the possibilities of automatic factories, leaving a huge gap between his mathematical symbolism and the ability to design specific cellular automata, much less their translation into performing mechanisms. Greatly improved computers, interactively operable, permitted following up his original designs and simplifying them. Still further advances, in the form of symbolic programming languages and visual displays, were waiting when Conway's discoveries attracted attention. Finally, Wolfram's forays into one dimensional automata would seem to have coincided with the advent of a generation of widely accessible computers, endowed with ample memory and rapid operation.

When circuits become available, for which large numbers of cells with large numbers of states can rapidly evolve in parallel according to programmable rules, there is no doubt that new insights will be found, very likely through empirical observation. In the meantime it is always possible to say that more details could be filled in with respect to any of the facets of the theory which have already been studied. So the problem is to take some feature which seems to show promise, to see which of its aspects make it interesting, anticipate what results are expected, and try to find the best way to go about establishing them.

1.13.1 Convergence of block probabilities

Kolmogoroff's theorem asserts that a family of consistent measures converges to a limit measure. Gutowitz *et.al.* [107] use this result to assure a meaningful construction of a system of equations for self consistent neighborhood probabilities, wherein extrapolation via Bayes' theorem yields probabilities for the counterimages of individual neighborhoods. They incidentally remark that the Bayesian extension yields maximum entropy; in other words, the least additional information about the extension.

However, each different length of neighborhood, or order, possesses its own set of equations defining self-consistency, making the measure deduced vary from one order to the next. It would be desirable to have an estimate of how much these measures can differ from one another, to know whether the sequence of measures deduced from successively longer neighborhoods can be demonstrated to converge to a unique limit, and at what point a desired degree of accuracy has been reached.

The approach of Gutowitz *et.al.* was to select rules for study, keep using longer blocks until the results stabilized, and apply statical tests for goodness of fit between the result and empirical observations. In its context, this is an entirely valid methodology. But each rule needs its own analysis; some rules seem to be much more amenable to the procedure than others.

As an idea of the hazards which are to be encountered, consider the Garden of Eden states. They can be described in terms of excluded sequences, which necessarily have probability zero. Nevertheless a zero probability cannot be obtained via Bayesean extrapolation from the probabilities of shorter blocks not from the

Garden of Eden, so that explicit block probabilities at least as long as the shortest excluded sequences are required. However, there is not necessarily a "longest shortest excluded block," so that there may be no finite approximation at all which is logically correct. In practice, the excluded sequences may already be assigned such small probabilities by local field theories of low order that their nonvanishing would never be detectable.

In point of fact, each periodic configuration of the automaton gives rise to a discrete measure concentrated on the orbit of that particular configuation, so that without further restrictions, the determination of a measure is not at all unique. Likewise there are rules of evolution based on the linear algebra of finite fields for which periodicities of one sort or another are so scrupulously observed that the measures also retain a thoroughly discrete structure.

No wonder that it would be nice to have a more comprehensive understanding of invariant measures and their relationships to one another. Especially with respect to establishing error bounds.

1.13.2 Automata which are necessarily infinite

The very most exotic applications of the theory of computation require computers which are infinite in extent, at least in principle. Otherwise one is dealing with a finite state machine which can presumably always be evaluated by an exhaustive enumeration. In practice, a large enough extent is sufficiently close to infinite, and supposedly any quirks of infinite machines can be approximated up to a point by making the finite machine large enough. Nevertheless, there are certain theoretical considerations which are practically equivalent to choosing a definition of infinity, for which finite approximations will definitely not suffice. In any event, think of how much more tractable analysis becomes when real numbers can be used freely rather than rational approximations.

To a certain extent, the problem which arises is not with the limiting behavior of an automaton as with describing that behavior. Some limits are complicated and have a complicated description, but others have an exceedingly simple description with respect to other terms of reference. The recognition of balanced parentheses is the classical example; their description by regular expressions is infinite, substantially a listing of cases, while their description by a "context free language" is very short. No matter that the automaton required in either case is the same, it is the difference in the length of the descriptions required by the two schemes which counts.

Then again, it is likely that the majority of limit sets will be horrendously complicated, no matter what representation is chosen for them. Wolfram repeatedly emphasizes this point.

1.13.3 Computational requirements

Numerous variants on the theme of cellular automata have been proposed, such as stochastic evolutionary rules, varying the rule from one cell to another or from time to time, or even permitting such variations in the definitions of the neighborhoods.

Unfortunately the quantity of computation required to simulate an automaton increases exponentially, and usually with a rather large exponent, with the size of the parameters involved. Thus any increase in the number of parameters will only aggravate a situation which is already rather difficult. If a significant practical application of one of these variants were to be discovered, no doubt the means would be found to compute its properties. In the meantime the regular version of celluar automata has enough unsolved problems to keep one occupied.

There are two visible tendencies. One is the approach of Conway, or of those who already have an automaton and want to find out what it does. The other is the original approach of von Neumann, to go ahead and construct the automaton which one requires, regardless of the cost in cells and states. Given that a Turing machine can be embedded in cellular automata in a fairly standard way, there will always be *some* automaton which will realize a given calculation; typically such a straightforward realization will be neither aesthetic nor efficient.

In either event enormous computer power is required. In the former, the de Bruijn diagrams are enormous; for two dimensional binary automata, it is hardly even possible to compute the period 2 states for *Life*, for instance. For one dimensional binary nearest neighbor automata, the calculation of cycles of length 34 or periods of length 15 strain present microcomputer capacity. Yet computer experiments reveal numerous configurations which are even bigger or repeat with longer periods. So, although much surer, systematic computation does not yet extend to revealing many fairly common experimental results.

Self consistent probabilities likewise require exponentially growing facilities, as the number of nonlinear equations to be solved increases with order. In the second case these concerns may be set aside, but it is still necessary to manage an extremely large table of transitions and to ensure its own internal consistency, especially in the face of backtracking from changed decisions as the rule is gradually elaborated.

1.13.4 Practical applications

So far there has been a shortage of physical, biological, social, mathematical, or whatever sort of systems, of independent interest, which have been shown to follow a rule of evolution which would qualify them as cellular automata. There is a result from the theory of symbolic dynamics that cellular automata define the continuous mappings with respect to a particular topology, but it is not known whether this has led to the detailed examination of any particular automata.

Some examples have been proposed; Preston and Duff [190] describe image processing applications, Wolfram's reprint collection [252] includes applications to nonlinear equations describing chemical reactions. It has often been proposed that the discretization of partial differential equations will render them into cellular automata. However, one has to wonder whether this approach gives any new insight or produces useful methods of solution. That is, the utility of the typical theorems of celular automata theory such as the evolution into cycles, or the existence of Garden of Eden configurations, has to be contrasted with the existence theorems, stability criteria, and so on of traditional differential equation theory.

There is no doubt that if an actual cellular automaton were built mimicing Laplace's equation, for example, that it would be enormously useful. Still, each cell would have to have enough states to represent a real number to a significant accuracy and would have to have a capacity for at least addition and shifting. The number of such cells would have to be large enough to interpolate a plane area to a reasonable degree of accuracy. So, the cost of the design of an appropriate integrated circuit, and subsequent volume of sales, would have to be balanced, on the one hand, against the cost and utility of designing some other comparably complex circuit, and on the other against the ease of performing the same simulation in a contemporary single or parallel computer.

1.14 Conclusion

With any body of knowledge, there is a set of definitions and axioms, followed by a series of results—the theorems. One would like to know that the basic assumptions lead to a well defined structure which then exhibits certain remarkable characteristics which set it apart from other structures.

In the case of cellular automata, the definitions lead first to the association of "cells" with "neighborhoods" via a function φ defining cellular evolution. Then the association is repeated uniformly throughout some lattice, leading to the function Φ mapping configurations from one to another, and thus the evolution of the lattice. The theory of cellular automata is essentially the theory of the interrelation of these two functions, and especially the behavior of Φ under iteration.

In the forward direction, the theorems relate to the limit sets of the evolution, including the rate at which limits are approached and the nature of the limiting configurations, often from the point of view of formal language theory. "Crystallographic" limits, cyclic or shifting in space and periodic in time, are of a particular practical interest; de Bruijn diagrams lead directly to their determination in the great majority of cases. Theoretically and philosophically important are the aperiodic limits, particularly those whose linguistic description is especially complex. In this latter category may be included the universal constructors of von Neumann, the emulators of Turing machines, and the like.

In the backward direction, the overlapping of neighborhoods allowed Moore to prove the Garden of Eden theorem, which does not apply to arbitrary mappings.

The existence of a probability measure μ compatible with Φ permits the calculation of equilibrium frequencies and entropies for the lattice and so is an important adjunct to Φ itself; consequently the determination of either μ or reasonable approximations thereto play an important role in theory of cellular automata.

Finally, the existence of an extensive collection of well studied examples is a delightful part of the theory, particularly because many of them have proved to have a considerable entertainment value.

Chapter 2

What to look for

Before studying the properties of linear cellular automata in detail it may be worth a quick review of a few of their most visible features, as well as a summary of some of their more important features. This includes the Wolfram classes, some traditional elements of mathematical structure, specialized rules of evolution, and relationships between different automata. Everything will be discussed again at greater length, very likely in its own chapter.

2.1 General characteristics

To watch the evolution of an arbitrary linear automaton from a random initial configuration is to see a great deal of confusion. Gradually – in some cases quite quickly – it becomes apparent that each rule of evolution has its own personality, and that as rules and types of automata are varied, similarities are as apparent as differences. Presumably this led Conway to seek out rules for which random configurations eventually settled down to simple activity rather than disappearing entirely, remaining motionless, or filling up the entire space. However, there seem to be automata for practically every taste.

Wolfram laid down a serviceable classification into four categories

- Class I – evolution to a uniform state
- Class II – evolution to isolated cyclic states
- Class III – evolution to comprehensive cyclic states
- Class IV – evolution to complex isolated states

which were derived from some classifications in nonlinear mechanics. His attention was particularly attracted to the Class IV states. It seems that these are to be found for rules whose de Bruijn diagrams contain certain loops. These can be readily detected, for short periods at least.

A good starting point, having selected a specific rule, is to work out a table of periods (time repetition) and cycles (space repetition) in which a given row shows the number of cycles of given period in rings of length given by the columns. It may be a bit disappointing that so few rows can be obtained within the limits of computer memory and running time that presently exist. The exponential growth of resources required ensures that rows or columns will only be added one at a time, and gradually at that, as computer power increases. Still, the first few rows and columns can actually be done; the information obtained can be quite informative.

2.2 Symmetry

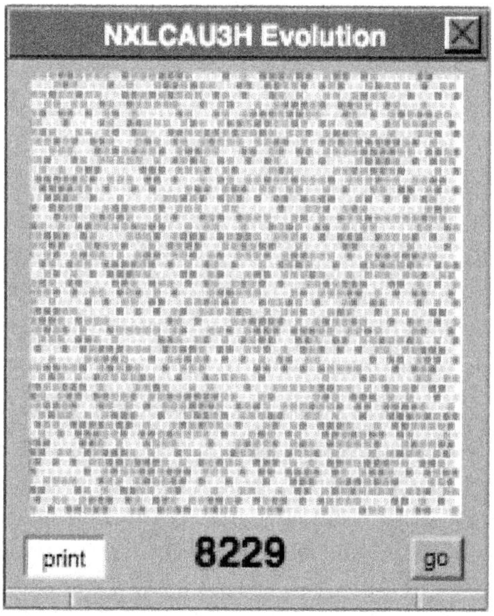

Figure 2.1: A fully symmetric (3,1/2) automaton.

Since there are large numbners of automata of any given type, there are good reasons for assorting them into classes which admit of a simpler description. Symmetry is a useful principle, and can be applied to cellular automata in at least three ways. Of course, translational symmetry is always an underlying characteristic of cellular automata, given that all the neighborhoods are uniform and that the same rule of evolution is applied to each of them.

Beyond the implicit symmetry, the first thing to consider is the effect of of the neighborhood's internal symmetry. In one dimension, the only symmetry is reflection, but for higher dimensions both reflections and permutations of the axes produce closely related rules when applied to the lattice of cells. For some rules, evolution will be indistinguishable from what originally prevailed before the lattice was adjusted, while for others differences in orientation will be readily discernible. In intermediate cases, there may be symmetry subgroups which leave the evolution unchanged, while their cosets exhibit modified evolution.

The second category of symmetries arises from the essential arbitrariness in labelling the cell states. Historically, a *quiescent* state was sometimes distinguished by supposing that a configuration consisting exclusively of those cells would remain fixed, providing a background against which the remainder of the evolution could be viewed. If the states of an automaton are to be identified with a finite field – convenient for some analyses – the zero of the field would be the quiescent state.

In generality, an automaton may have several states meeting that criterion, requiring one of them to be singled out, or the discourse modified to include the possibility of multiple backgrounds. All in all, including quiescence in a discussion may be more of a hindrance than an advantage. In addition, there is the consideration that when an automaton is visualized through colored pixels, their hues are entirely arbitray, even though one ore more of them may be representing quiescent states.

Suppose that $\varphi(x, y, z)$ is the evolutionary function of a three-neighbor automaton, and that π is any permutation of the state set. The condition of full state symmetry would then be

$$\pi(\varphi(x, y, z)) = \varphi(\pi(x), \pi(y), \pi(z)).$$

In other words, permuted values arise from permuted arguments; the placement of π in the equation must be exactly as shown.

Both state permutations and neighborhood isometries have simple, readily observable effects on the evolution of any automaton. There are still other symmetries whose effects tend to be much more obscure and less intuitive. For example, since a permutation (such as complementation in a binary automaton) is invertible and a function of a single variable, one could be composed with the evolution function; in general, one could write

$$\psi(x, y, z) = \pi(\varphi(x, y, z)).$$

Although this is a completely new function, it could still be taken to be related to φ via a symmetry. Even if that were the supposition, the actual evolution would be likely to appear to be quite different.

2.3 Cycles

There are two ways to obtain the cycles for a given automaton. The first is to enumerate all the rings of the desired length, and follow up the evolution of each. In doing so various shortcuts can be taken, such as generating the configurations in Gray code order so that only a single cell changes state from one to the next. Still lifes can be detected very quickly this way. Numerical comparison of successive generations means that whenever the new generation is smaller, it will already have been examined and need not be pursued further.

The second way is more systematic and is worth the bookkeeping effort involved. A graph whose links are determined by evolution is prepared, following which a path enumerating procedure is followed to locate all the loops, whose lengths will give the periods of all the cycles of that length. Cycles of length up to ten can be obtained easily, twenty with effort, but passing thirty requires dedication; for binary automata it is slightly easier, increasingly more difficult for others.

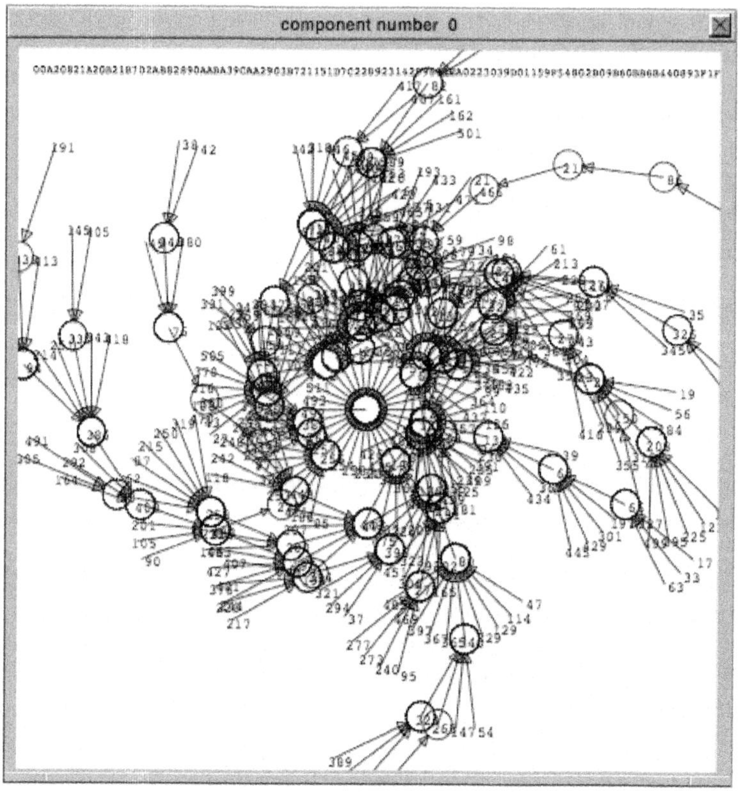

Figure 2.2: Evolution tree for a (2,4) automaton.

Certain theoretical conclusions do not depend on practical limitations. For example, we might select an initial ring and examine successive generations of its evolution. The new generations can all be distinct, or at some point we might find that an earlier generation has repeated itself. For a finite ring the first possibility cannot occur, because there are only k^n possible different combinations of states in a ring of length n – 32 in a binary ring of length 5, for example. If no repetitions were allowed, the supply of distinct rings would eventually be used up, so repetition is the only alternative. In practice, the number of generations elapsing before one of them repeats may be fairly large, but is usually much smaller than the exponential bound of the worst possible case.

Once repetition has occurred, we need to recall that evolution is uniquely defined – there is just one successor to a given ring in each new generation – so the tail end of the evolution has to repeat over and over again in a fixed cycle, starting with the generation which first repeated.

Statistics of interest concerning the cycles include: the number of cycles and their lengths, the height of the transient trees leading into the cycles, and the convergence factor at each node of these trees.

2.4 Periods

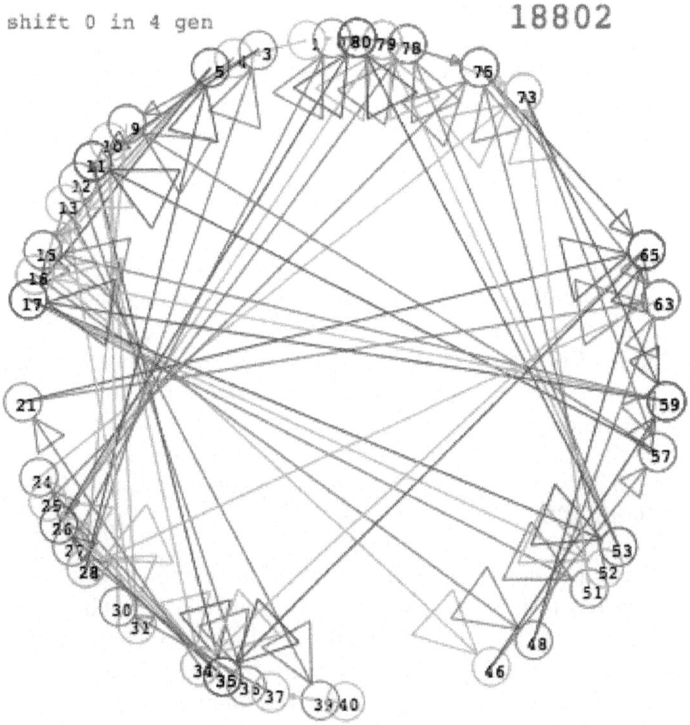

Figure 2.3: de Bruijn diagram for a (3,1/2) automaton.

The rows of the period-cycle table can be found from de Bruijn diagrams in the same way that the cycles can be found from the evolution diagram; since $2r+1$ cells are needed to deduce a generation of evolution, only about half as many periods as cycles can be worked out for a given amount of effort. This anomaly is really an artifact of the way r parameterizes the neighborhood, and would disappear if half-integral increments were taken for r.

Similar theoretical conclusions are possible, since the periods are gotten from a subset of the de Bruijn diagram. A $2r$-stage de Bruijn diagram for k symbols has k^{2r} nodes; k times as many links. Once this number of links has been used up in constructing a path through the diagram, one of them would have to be repeated. Thus there is also an exponential upper bound in the rows of the period-cycle table. For example, if an automaton has a cycle of period 2, it must already show up in some short ring; if it has not appeared in rings below a certain limit, it will never appear in longer rings.

Similar statistics can be compiled for the de Bruijn diagrams: the number of periods and their lengths, the number of transients leading into loops, and the corresponding convergence and divergence factors. Failed loops are still interesting; they correspond to strings which are periodic, but which dwindle away with each generation, eventually leaving a residue which is no longer periodic.

Both the cycle diagrams and the period diagrams may have intersecting loops; this simply means that choices are present at certain junctures in working up a chain of cells with a certain property, leading to a greater variety of sequences than would otherwise occur. This choice is particularly vivid in the de Bruijn diagram when one of the loops consists of a single node of identical cells embedded in another loop, since these can be strings of Wolfram's Class IV.

Not only can the de Bruijn diagram can be used to reveal the periodicities of a given cellular automaton, it can also be used in automaton synthesis. That is, desired loops can be marked out first in the diagram, and then a rule chosen which respects the links. For period 1 properties the mapping has to be direct, since each link in the de Bruijn diagram corresponds to a distinct neighborhood in the automaton. Including or excluding it from the period diagram either determines or limits the value of the transition for that neighborhood. For longer periods the evolution corresponds to a composite rule, which may or may not be factorizable in the required manner.

2.5 Ancestors

Another use of the period diagram is to obtain ancestors of a given chain. The simplest application is to find the ancestors of uniform chains, which follows readily from the fact that each link represents the evolution of the central cell in one neighborhood. All the links evolving into zero determine the chains which must evolve into zero; conversely demanding that given loops evolve into zero determines the rules for which such an evolution is possible. For a binary automaton, that is enough information to determine the rule uniquely.

An interesting exercise is to show that for totalistic automata, the ancestor diagram for constant chains consists of pure loops, without any transients at all. One consequence of this result is that any finite chain which maps into zero can be embedded in a still longer chain which also maps into pure zeroes.

Only slightly more complicated is the determination of the static chains, or still lifes. Rather than choosing neighborhoods which evolve into a constant value, neighborhoods for which the central cell evolves into itself must be chosen. Quite a few other variations on the theme are also possible.

Historically, de Bruijn diagrams were created to solve the problem of finding all the distinct sequences of certain symbols [191]. This idea can be applied to a period diagram, by asking whether all possible sequences of cells can appear as possible evolutions. Since the period diagram is a restriction of the de Bruijn diagram, it may be suspected that they may not; this confirms the existence of "Garden of Eden" states for cellular automata. These are chains of cells which can only be seen as intital configurations for an automaton, because they have no ancestors and cannot arise during the course of evolution from any other states.

Searching for a path in a diagram can be a tedious process, but if no more is required than knowing of its existence, Moore's subset diagram [163] provides a way to systematize the search. It is a new diagram, whose nodes are subsets of states; each subset is linked to the union of the states which are linked to its members. Let the evolutionary function, which links nodes in the de Bruijn diagram, be φ. Further let A be a subset whose typical member is the node a. Then

$$\Phi(A) = \bigcup_{a \in A} \{\varphi(a)\}$$

defines the linkage between subsets.

One usually starts from the full set, supposing that it does not matter where a path begins, continuing as long as possible. Having arrived at the null subset, it is certain that there was no path of the same characteristics in the original diagram. An even more elaborate subset diagram retaining all details can also be constructed, from which the original paths with all their multiplicities can be extracted.

2.6 Subautomata

Binary automata may be judged to be less interesting because they "don't do anything" or fall into Wolfram's Classes I and II; but there are other ways in which automata with larger numbers of internal states can fall into a pattern of restrictive behavior. For many rules, watching the screen display for a while will reveal that one of the colors has disappeared. This would be especially noticable for a rule in which a certain value never appeared in the rule, because it would have to be be absent in all lines after the first.

A mathematician would describe this situation by saying that he was dealing with a subautomaton – one for which a subset of states could be found which was closed under evolution. What that means is that states in the given subset would evolve only into each other and into no others. Just as mathematical definitions tend to include many apparent quibbles as extreme cases of some general proposition, it might be remarked that a good many automata actually exhibit one extreme example of subset behavior. For these automata dead states evolve only into dead states, using Conway's biological metaphor.

Here, the extreme subset is the one consisting of just the quiescent state, so that the subset automaton, strictly speaking, would be a monary automaton; a category which we might have thought that we would never need to use. In reality we have happened upon the concept of an automaton with a quiescent state (a more elegant adjective than "dead"), and seen its characterization by saying that the quiescent state belonged to a subautomaton.

At the opposite extreme, equally a convenient quibble, the whole set of states can be considered as a subset of states forming its own automaton. The value of setting up such a vocabulary becomes apparent when we have more complicated automata in which a whole hierarchy of subautomata can be perceived and we want to make systematic comparisons between the members of the hierarchy.

2.7 Xenophobic, or clannish, automata

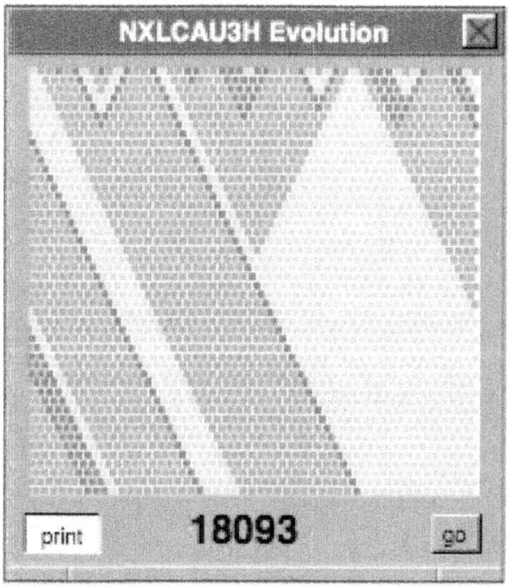

Figure 2.4: A clannish (3,1/2) automaton.

An extreme example of subautomata is comprised of those for which every subset of states is closed with respect to the evolution. Their rules could be called *xenophobic*, or from a more sympathetic point of view, *clannish*. The concept makes little sense for binary automata, since the only thing that it requires, aside from the trivial cases of the empty set and the full set, is that zeroes evolve into zeroes, or that ones evolve to ones.

Ternary automata exhibit greater diversity, because of the diversity of two-element subsets. The longer the neighborhood, the more such subsets will be found; however with only two neighbors, the following table contains all possibilities.

x	y	$\varphi(x,y)$	x	y	$\varphi(x,y)$	x	y	$\varphi(x,y)$
0	0	0	1	0	0,1	2	0	0,2
0	1	0,1	1	1	1	2	1	1,2
0	2	0,2	1	2	1,2	2	2	2

In other words, every state is quiescent, but there are six combinations offering a binary choice, for a total of 64 different clannish (3,1/2) rules. In general, a clannish rule can be built up recursively from the rules which apply to successively subsets of states. Given three neighbors, the following possibilities arise.

x	y	z	$\varphi(x,y,z)$	x	y	z	$\varphi(x,y,z)$	x	y	z	$\varphi(x,y,z)$
0	0	0	0	1	0	0	0,1	2	0	0	0,2
0	0	1	0,1	1	0	1	0,1	2	0	1	0,1,2
0	0	2	0,2	1	0	2	0,1,2	2	0	2	0,2
0	1	0	0,1	1	1	0	0,1	2	1	0	0,1,2
0	1	1	0,1	1	1	1	1	2	1	1	1,2
0	1	2	0,1,2	1	1	2	1,2	2	1	2	1,2
0	2	0	0,2	1	2	0	0,1,2	2	2	0	0,2
0	2	1	0,1,2	1	2	1	1,2	2	2	1	1,2
0	2	2	0,2	1	2	2	1,2	2	2	2	2

This time, three quiescent combinations are forced, there are 18 binary choices, and six opportunities to make three choices, for a grand total of $2^{18}3^6$ different rules. The recursive hierarchy begins with the obligatory (1,1) rules, continues with chosing suitable (2,1) rules for the three subsets of consisting of pairs, and finally encompasses deciding the fate of all six permutations of three distinct elements.

Once the length of the neighborhood exceeds the number of states, the recursion becomes saturated, without further restrictions but leaving many more choices at all levels but the first. This is clearly seen in the fate of binary automata, wherein quiescence of both states is the only restriction. "Clannish" an apt description of one fourth of all automata with two states; generally speaking, clannishness is only a moderately restrictive attribute for automata.

2.8 Xenophilic, or cosmopolitan, automata

A class of subautomata complementary to the clannish automata originates with those for which the new state is excluded from belonging to its neighborhood. Such automata could be called *xenophilic*, or perhaps *cosmopolitan*. The concept does not work for binary automata because there is nowhere for a distinct pair to turn. Two-neighbor ternary automata are candidates, even though the choices are limited.

x	y	$\varphi(x,y)$	x	y	$\varphi(x,y)$	x	y	$\varphi(x,y)$
0	0	1,2	1	0	2	2	0	1
0	1	2	1	1	0,2	2	1	0
0	2	1	1	2	0	2	2	0,1

The definition precludes any state from being quiescent, leaving pairs of distinct states to force the choice of the remaining state. Altogether there are three binary choices, or 8 automata of this class. Turning to automata with three neighbors, there results

x	y	z	$\varphi(x,y,z)$	x	y	z	$\varphi(x,y,z)$	x	y	z	$\varphi(x,y,z)$
0	0	0	1, 2	1	0	0	2	2	0	0	1
0	0	1	2	1	0	1	2	2	0	1	none
0	0	2	1	1	0	2	none	2	0	2	1
0	1	0	2	1	1	0	2	2	1	0	none
0	1	1	2	1	1	1	0, 2	2	1	1	0
0	1	2	none	1	1	2	0	2	1	2	0
0	2	0	1	1	2	0	none	2	2	0	1
0	2	1	none	1	2	1	0	2	2	1	0
0	2	2	1	1	2	2	0	2	2	2	0, 1

This time, the constant configurations yield binary choices, though all distinct pairs are foreordained, leaving the distinct triples to confront an impossible situation. Strictly speaking, there are no genuinely cosmopolitan automata whenever there are more neighbors than states. However, the rule could be modified, by conserving a central state, choosing the least popular state, or making a random choice, all of these alternatives leading to interesting rules.

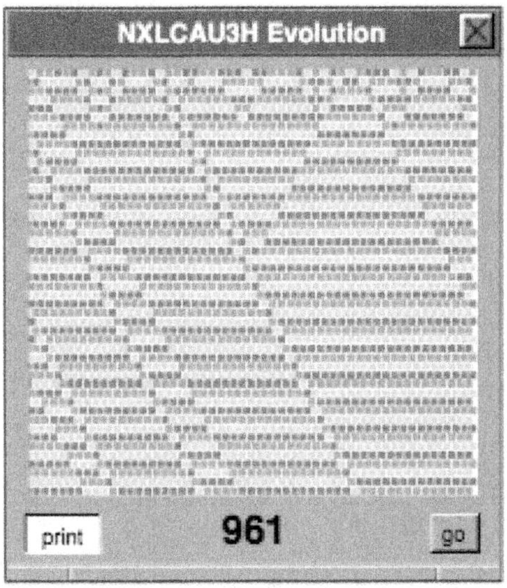

Figure 2.5: A cosmopolitan (3,1/2) automaton.

There is a greater variety of cosmopolitan (4,1) automata, but the strict definition runs out at (4,3/2), as it does in turn with even greater numbers of states.

x	y	z	$\varphi(x,y,z)$	x	y	z	$\varphi(x,y,z)$	x	y	z	$\varphi(x,y,z)$	x	y	z	$\varphi(x,y,z)$
0	0	0	1, 2, 3	1	0	0	2, 3	2	0	0	1, 3	3	0	0	1, 2
0	0	1	2, 3	1	0	1	2, 3	2	0	1	3	3	0	1	2
0	0	2	1, 3	1	0	2	3	2	0	2	1, 3	3	0	2	1
0	0	3	1, 2	1	0	3	2	2	0	3	1	3	0	3	1, 2
0	1	0	2, 3	1	1	0	2, 3	2	1	0	3	3	1	0	2
0	1	1	2, 3	1	1	1	0, 2, 3	2	1	1	0, 3	3	1	1	0, 2
0	1	2	3	1	1	2	0, 3	2	1	2	0, 3	3	1	2	0
0	1	3	2	1	1	2	0, 3	2	1	3	0, 3	3	1	3	0, 2
0	2	0	1, 3	1	2	0	3	2	2	0	1, 3	3	2	0	1
0	2	1	3	1	2	1	0, 3	2	2	1	0, 3	3	2	1	0
0	2	2	1, 3	1	2	2	0, 3	2	2	2	0, 1, 3	3	2	2	0, 1
0	2	3	1	1	2	2	0, 3	2	2	3	0, 1, 3	3	2	3	0, 1
0	3	0	1, 2	1	3	0	2	2	3	0	1	3	3	0	1, 2
0	3	1	2	1	3	1	0, 2	2	3	1	0	3	3	1	0, 2
0	3	2	1	1	3	2	0	2	3	2	0, 1	3	3	2	0, 1
0	3	3	1, 2	1	3	3	0, 2	2	3	3	0, 1	3	3	3	0, 1, 2

2.9 Factor automata

Setting up a relationship between subsets to which the arguments of an evolutionary function may belong and those from which their values are taken leads naturallly to the idea of equivalence relations. According to this concept, two or more states of the automaton might be regarded as being interchangeable. Even though they were not strictly identical, there might not be much reason to find any essential difference among them.

Watching the evolution of certain automata on the screen, there sometimes seems to be a wash of color laid over an underlying pattern. The pattern seems to endure, while the color overlay has a life of its own. This is an example of a factor automaton, in which the overlaid colors are equivalent. They form one equivalence class, black (the general background color) another; the automaton could just as well be binary and could be viewed on a monochromatic screen.

Finally, there are mappings from one automaton to another. One of the simplest examples would be to complement all the cells of a binary automaton. The complemented automaton would probably not evolve according to the same rules, but it might. For automata whose states are television colors, interchanging the colors would be such a mapping. Aesthetically the difference is striking; mathematically it is the same automaton.

Let f be a function between the state sets of automaton A and automaton A', with equal neighborhood size and evolutionary functions φ and φ'. Functions which satisfy the condition

$$\varphi'(f(x), f(y), f(z)) = f(\varphi(x, y, z))$$

are of a special kind, evidently more compatible with the two automata than arbitrary functions. Counterimages of individual states by such a function are the prototypes of equivalence classes; equivalence means having the same image.

2.10 Product automata

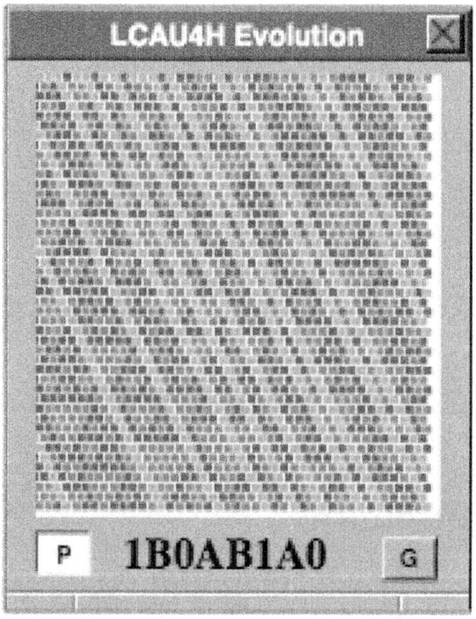

Figure 2.6: The cartesian product of (2,1/2) Rules 4 and 6.

The identification of subautomata and factor automata is an analytic process, in which structure is to be sought in a previously existing system. Synthesis, the converse process, seeks to build up more complex objects by combining simpler parts in prescribed ways. The most common synthetic procedure is to try to endow a cartesian product with structure analogous to that of its constituents. As would be expected, the process works well with automata.

Suppose that A is one $(k, 1)$ automaton, with transition function α, and that B is a second, with transition function β, and that (x, y) is a pair of states. We could define the new transition function φ for paired states by setting

$$\varphi((a, b), (c, d), (e, f)) = (\alpha(a, c, e), \beta(b, d, f)).$$

In other words, corresponding members of the pairs are neighbors of one another, and transitions in the two halves take place independently, even according to different rules if such is the case.

With pairs for states, a common neighborhood width, and the product transition function, we have created a new automaton, $A \times B$ which is called the product (or cartesian product) of the two automata A and B. Although the definition given applies to $r = 1$, a similar definition would apply for automata with wider neighborhoods. It is apparent that such definitions multiply the number of states in the automaton, while keeping the same neighborhoods. Thus, a $(2, 1) \times (2, 1)$ automaton would really be a $(4, 1)$ automaton, if the states were relabelled properly.

Two applications of product automata come to mind at once. One is to see how the same data evolves according to two different automata; we would set $x = y$ to see simultaneous evolution by two different rules. The second is to set $A = B$, to see how different intial data evolve for one single automaton.

It is a standard result of structure theory that a product automaton has factor automata corresponding to each of the two automata used to construct it, and that their equivalence relations are complementary. Likewise, we would want to use the notation $\varphi = \alpha \times \beta$.

2.11 Automata with memory

There is another way to use the cartesian product to create a composite automaton. Begin with the $(2, 1)$ automaton whose transition rule is α. Define, for pairs such as (x, y) ($\otimes$ is exclusive or),

$$\varphi((a,b),(c,d),(e,f)) = (d, c \otimes \alpha(b,d,f))$$
$$\varphi^{-1}((a,b),(c,d),(e,f)) = (\alpha(a,c,e) \otimes d, c).$$

The left part of the pair simply remembers the right part from the old generation while the right part combines the new and old values of the cell. Thus both items of information are always present and can be extracted if needed. The inverse rule, whose structure is similar, follows the same procedure.

States a and e are of no consequence in defining the transition, serving simply as a one-generation memory which will be forgotten by the second generation. Nevertheless their presence is essential, likewise the fact that φ must ignore them.

As with the cartesian product, we have a $(4, 1)$ automaton with the square of the number of states and the same width. Still another way to induce an automaton in the cartesian product would be to omit the exclusive or:

$$\psi((a,b),(c,d),(e,f)) = (d, \alpha(b,d,f)),$$

from which the previous generation with respect to α can always be recovered, but not with respect to ψ. By contrast all of φ's previous generations can be recovered, but it must not be thought that the same applies to α, even though it participates in the definition of φ.

For example, suppose that $\alpha(b, d, f) = 0$, which is not reversible at all. Then $\varphi((a, b), (c, d), (e, f)) = (d, c)$, indeed reversible, but of no help whatsoever for obtaining α^{-1}.

2.12 Idempotent rules

For some automata it can be seen that states repeat themselves; indeed this behaviour is a very prominent feature of Wolfram's Class II automata. Formally, a $(k, 1)$ automaton with transition rule φ is idempotent if

$$\varphi(\varphi(a,b,c),\varphi(b,c,d),\varphi(c,d,e)) = \varphi(b,c,d).$$

Even though the algebra is messy, it can be verified that for such a transition rule, a cell remains unchanged from the second generation onward.

It is a temptation to state the idempotency condition in the form $\varphi^2 = \varphi$; in a sense it is true that iterated φ acts like φ; but φ has three arguments and only one value, and so it is better to give the precise definition shown above.

It is too strong a requirement to insist that $\varphi(a,b,c) = b$ always, because only the identity rule fulfils this requirement. There are also many evident variations on the theme – evolution could stagnate after the third generation rather than the second, for example. Likewise, the states could shift sideways rather than stagnating:

$$\varphi(\varphi(a,b,c),\varphi(b,c,d),\varphi(c,d,e)) = \varphi(a,b,c),$$

for example. Other variations would have the states undergoing permutations leaving the entire configuration stagnant only in those generations in which the full cycle of the permutation had run its course.

The evolution of idempotent automata, or those equivalent by one of the variations, is not particularly interesting from a dramatic point of view. Nevertheless, they form a class susceptible to a closed analytical treatment, and they are of rather common occurrence when all the possible types of rules are taken into account. There is an interesting borderline class of automata whose rules are not idempotent, but monotonic. For example, one of the $(2,1)$ automata has the rule (& is boolean AND)

$$\varphi(a,b,c) = a\&b\&c$$

which is the logical AND of the three binary neighbors. The rule is Wolfram's # 128, and has the property that zeroes persist. Unless one began with an infinite or cyclic chain of ones, all ones must eventually disappear, and the rule looks idempotent to the zeroes left behind. It is convenient to call these rules *assymptotically idempotent* and classify them together with idempotent rules.

2.13 Membranes and macrocells

More subtle than Wolfram's Class IV automata are another class which seem to exhibit natural barriers, not just the quiescent regions which characterize Class IV. Areas of independent evolution are observed, which seem to go about their business independently of what is happening in other regions [35].

Nor are the barriers always invariable; they may go through their own relatively simple cycle of evolution while still separating regions of more complex activity. Let us call such regions *macrocells* to distinguish them from the individual cells of which the automaton is composed. When their width is short their behaviour must be periodic of relatively short duration. There is the slight difference that their evolution will be subject to different boundary conditions than if the macrocells were cyclic or part of an infinite cyclic pattern. Wider macrocells have a correspondingly longer – exponentially longer – waiting time for repetition.

An explanation can be found by consulting the de Bruijn diagram; it is found that there are certain nodes, for which all incoming or outgoing links survive in the period diagram, even if they do not continue on to form closed loops. It is only necessary that there is an unbroken chain from a node with complete incoming links to another (possibly the same) with complete outgoing links. These chains form the prospective cell membrane; the environment of the terminal cells ensures that the membrane will evolve properly, no matter the content of the macrocells which they enclose.

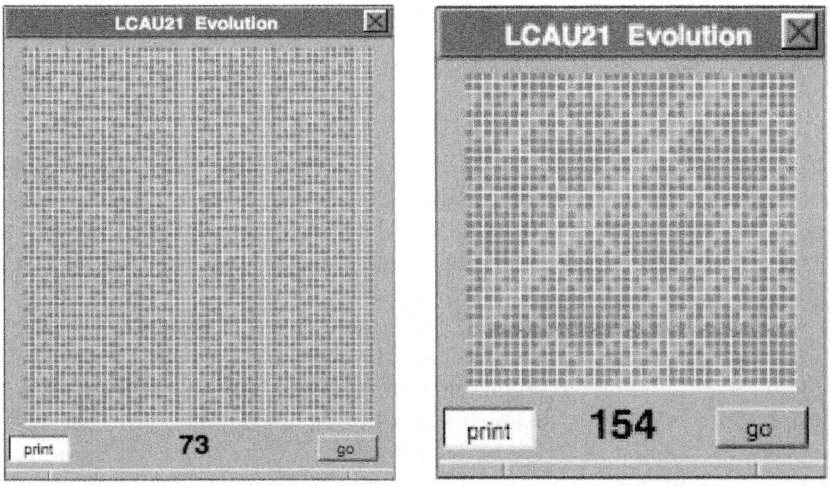

Figure 2.7: Two examples of membranes and macrocells. For (2,1) Rule 73, the membrane is static, whilst for Rule 154 it always shifts to the left.

Slight perturbations are possible; a membrane may be stable in most environments and attacked by configurations in others. Once it starts to dissolve, the membrane will diminish by two cells each generation, and no long term declarations can be made about the evolution. Membranes may form spontaneously when conditions are right; it is also possible to have periodic membranes or shifting membranes. With moving boundaries, the de Bruijn diagram need only be adjusted so that all the pertinent links correspond to a consistent translation. However, the positioning of the guard strings must be changed; instead of flanking the boundary membrane they must run ahead of it ensuring an adequate anticipation of the glider which they are protecting.

A final variation, which is also often encountered, is to find that not all the links beyond the terminal nodes in the de Bruijn diagram are guard links, but that the macrocell will only place cells near the membrane whose links are guarding. As an extreme case, a wall of one color could confine cells of different colors, but still dissolve when confronted with cells of its own color.

Since the critical requirement for a membrane is the presence of guard links at its extremes, its internal structure can be simple or complex as the occasion demands. Thus, there could be loops in the chain defining the membrane, allowing macrocells to be separated by membranes of varying structure.

2.14 Totalistic rules

A *totalistic rule* is one for which the transition depends upon the sum of the weights of the neighborhood. Such a rule, for a $(k, 1)$ automaton, would take the form

$$\varphi(a, b, c) = \tau(a + b + c) \bmod k$$

where τ is a function of the integers modulo k. In fact, the simple modular sum is a good example of a totalistic rule, yielding a rule for which the computations of linear algebra suffice to determine its evolution.

While the choice of a totalistic rule simplifies evolutionary calculations, it is clear that a given sum can be formed in many different ways. Thus every neighborhood possessing a fixed sum must necessarily evolve into the same state. This should not cause undue concern; since there are always more neighborhoods than states, any rule will necessarily have clusters of neighborhoods mapping into the same state. It is just that the multiplicity follows the degeneracy of sum formation for totalistic rules, giving them a statistical property which is useful.

There are more ways to form sums in the middle range than for the extremes; one might think in terms of the binomial distribution. Thus the values assigned the middle range will be relatively influential in determining the overall behavior of the automaton, while the extremes can be used for fine adjustments. One extreme determines what happens to long sequences of zeroes, the other to long sequences of $k - 1$'s, and in both cases to sequences in which these extremes dominate.

In practice, a sum of zero can only arise in one way, from a string of zeroes. Equally, the high sum can only be formed from a string of cells of the highest weight. Whatever value is assigned to these extremes will influence the color of the background, just as will the values assigned to the evolution of any other constant strings. Thus the sequencing of long strings of solid color can be read off directly from the totalistic rule. In prompting for the rule, the programs mark off the sums which would correspond to constant values.

A bit more effort is required to work out the sequencing of repeated pairs, but it is not too hard to do mentally. It is an interesting observation that for totalistic rules, the sequencing of iterated triples follows the same sequencing as for constant sequences, but in terms of the sum corresponding to the triple.

In conclusion, the values assigned the extreme sums influence the cycling of background colors and can be deliberately chosen for this effect, but the drama of this change should not overshadow the fact that the remainder of the evolution is little changed and that the change will hardly be noticable unless the rule tends toward large areas of constant color.

The next-to-extreme sums tend to determine what happens to the fringes of regions dominated by the middle values, and it is often preferable to step through these values rather than the extemes themselves to get small changes in the patterns of evolution.

Although totalistic rules form a special class of rule, they are general enough to be representative in the sense that any other rule can be transformed into a totalistic rule, although the new rule may have a considerably larger number of states. Albert and Culick have demonstrated [4] that all that is really required is to express the (k, r) neighborhood as a r-digit number relative to the base $k+1$, but in such a way that the overlap between neighborhoods is taken into account. Thus, suppose that the configuration

$$\ldots \quad a \quad b \quad c \quad d \quad e \quad f \quad \ldots$$

is rewritten

$$\ldots \quad 000a \quad 00b0 \quad 0c00 \quad d000 \quad 000e \quad 00f0 \quad \ldots.$$

Now, suppose that the sum U of three four digit numbers $X + Y + Z = U$ has the base $k+1$ representation $U = wxyz$. The reason for changing to $k+1$ is that the new representation needs a zero which cannot be confused with one of the states of the original automaton. If the original states were numerical, they should be shifted to accomodate the new base; if they were not numerical, they should now be assigned numerical values. Then, define the auxiliary transition function τ to be

$$\tau(U) = \begin{cases} 00\varphi(z, y, x)0 & w = 0 \\ 000\varphi(w, z, y) & x = 0 \\ \varphi(x, w, z)000 & y = 0 \\ 0\varphi(y, x, w)00 & z = 0 \end{cases}$$

and finally, the new transition function Φ by

$$\Phi(X, Y, Z) = \tau(X + Y + Z).$$

With this definition, we have embedded a $(k, 1)$ automaton with an arbitrary rule in a $((k+1)^4, 1)$ automaton with a totalistic rule, although it is necessary to code the cells before making the simulation and to decode them afterwards. If strict adherence is made to only necessary values for cell states, one non-zero digit in the four-digit base $k+1$ expansion, the number of states could be considerably reduced, improving the state economy. Otherwise, transitions for the unused combinations get arbitrary definitions.

2.15 Two-cell neighborhoods

Amongst the variety of transformations which can change the appearance of cellular automata are those which modify the neighborhoods. For example, it is possible to show that linear cellular automata do not need to have large neighborhoods; in fact, as Albert and Culick [4] showed,it is sufficient to have just two cells in a neighborhood. The essential idea is to replace the entire neighborhood by a single cell; of course the new cell must have enough states to describe every possible content of the original neigboorhood.

Even if the cells are grouped in clusters, there will still be some interaction at the boundary, so that a new rule of evolution has to be devised. Also, since the interaction has to be one-sided if there are only pairs of cells in the new neighborhood, the original two-sided interactions can be recovered by extracting information from alternate generations. Since three cells would form a second generation neighborhood, it can be contrived that we think of the central cell as evolving, rather than one of the edge cells.

Thus, for lack of a better notation, let us define a $(k(k+1), 1/2)$ automaton A with the rule Φ in terms of a given $(k, 1)$ automaton a whose rule of evolution is φ. The states of A are to be the $k \times k)$ ordered pairs (i, j) together with the k states i. We shall only define $\Phi(I, J)$ partially through the table

	(j', k)	j
(i, j)	$\begin{cases} \varphi(i,j,k) & j = j' \\ \text{undefined} & \text{otherwise} \end{cases}$	$\cdots$
i	$\cdots$	(i, j)

.

There is an enormous number of possible definitions of A, since the mixed transitions involving pairs and singlets are left undetermined. In a given case, some guidance may be available to complete the table, but for the present construction the completion is irrelevant.

To recover the operation of a from the initial configuration $\ldots ijk \ldots$ we need to see that A generates first $\ldots (i,j)(j,k) \ldots$ and then $\ldots \varphi(i,j,k) \ldots$. Thus every second line contains the evolution of a, but we have to be observant of how it is positioned inasmuch as A always leaves its result in the left cell.

In order to guarantee the substitution of a two-cell neighborhood for a three cell neighborhood it has been necessary to augment the states by state pairs. For a binary automaton this means that we have six states rather than two. We defined the rule of evolution Φ for just enough arguments to ensure the recovery of the automaton a, which means that there are really many automata capable of making the simulation. There may be some reason for selecting the remaining transitions in a certain way, perhaps to simplify the automaton somehow. It also shows that one automaton may be capable of simulating another if only one can detect the proper embedment.

2.16 Blocking transformations

There are no doubt countless ways of embedding an automaton in another whose neighborhoods are constructed differently, or which has a different state set; but one mapping which might be among the first to be tried out would be to group several cells together and treat them as a single unit. It is not as useful a procedure as it might seem at first because there are necessarily redundancies in the evolutionary rules. This is because cells which lie beyond the range of interaction after the grouping cannot influence the evolution, so that blocked neighborhoods in which they differ must have the same rule of evolution. Of course there is a small possibility that this might turn out to be an advantage going in the opposite direction, whereby the number of states might be reduced by dividing the cells into pairs.

For simplicity, suppose that a $(k, 1)$ automaton undergoes an evolution described by the following sequences of cells

$$\begin{array}{cccccc} \ldots\ x & a & b & c & d & y\ \ldots \\ . & . & \phi(a,b,c) & \phi(b,c,d) & . & . \end{array}$$

Then the transition rule for the blocked $(k^2, 1)$ automaton is defined by

$$\Phi((xa),(bc),(dy)) = (\varphi(a,b,c),\varphi(b,c,d))$$

Conversely, any $(k^2, 1)$ automaton for which a coding of its states could be exhibited satisfying this rule could be regarded as a $(k, 1)$ automaton by splitting its states.

2.17 Tailor made automata

Two contrasting ways to approach automata theory are to begin with an automaton to see how it evolves, or to think of an interesting evolution and try to select the rules accordingly; the latter were von Neumann's and Conway's approaches. From either aspect it is convenient to have a random number generator available; in the first instance for following the evolution of typical configurations, in the second it can even provide the rule to study.

A more deliberate approach begins specific plans for the evolution, and attempts to deduce a suitable rule. Suppose that the objective is a slow glider, that advances two cells in four generations. There is at *least* a 5×5 rectangle to be considered, to accomodate the generations, the displacement, and the neighborhood length. The evolution of at least 25 neighborhoods must be defined, but fewer may suffice on symmetry grounds; at least five if every row is the same.

Neither $(2, 1/2)$, $(2, 1)$, nor $(3, 1/2)$ automata look promising with 4, 8, and 9 distinct neighborhoods, respectively; but any other state-radius combination begining with the 16-neighborhood $(2, 3/2)$ ought to have rules with such a glider. These figures may not be precise, but they lend plausibility to the conjecture that there is an automaton of some minimal complexity for almost any task that can be described by a sample evolution.

One typical task which is quite instuctive is to emulate a Turing machine; Albert and Culick [4] suggest one approach. Just as there are simpler Turing machines than the universal machine, there are numerous simple automata to be designed. Binary counters are a good example; their existence shows that there are automata with very long transients, also with very long cycles. A challenge might be to come as close as possible to the theoretical upper limit, a sort of variant on the "busy beaver" problem for Turing machines.

Another traditional exercise is the "firing squad problem," wherein there is a reserved state, "fire" which all the cells of an automaton of arbitrary length are to enter simultaneously, none having occupied it previously [165, 240]. It is a whimsical one-dimensional variant of one of the components of von Neumann's constructor; which was supposed to activate each finished object, as a finishing touch when it was finally completed and turned loose on its own.

Chapter 3

Cycles in space

Fundamental to finite automata is eventually periodic evolution, prolonged to the full number of states in the exceedingly rare event that their sequence is cyclic. The number of "states" of a cellular automaton is the number of configurations (not states per cell), providing an exponentially large bound relative to the automaton's length. In practice, many short cycles usually predominate over a few long ones, almost always reached through transients.

Longer automata admit longer cycles and longer transients too; the infinite limit may lack cycles. Cyclic boundary conditions locate behavior repeating over a finite range, leaving truly aperiodic configurations for a separate study.

3.1 Cycles in *Life*

When Martin Gardner first announced Conway's game of *Life*, there was much experimentation with simple designs, some of the simplest consisting of just a single row, column, or diagonal of live cells. Columns tended to grow shorter and fatter until they formed a diamondlike configuration which in one remarkable case—a column of fifteen cells—retreated back into a column and became periodic. It was one of the first complex oscillators to be discovered.

However, instead of forming diamonds, very long columns exhibited a rather curious behaviour not unlike a binary counter. That is, except for the ends, a single column survived while two new live columns flanking it were born. The next generation saw the birth of still another pair of flanking columns, but the central three could no longer survive, leaving the new pair of columns isolated by a gap of three cells.

Then in the fourth generation two triple columns separated by a vacant column produced a single pair of columns separated by a distance of seven cells. In subsequent generations these columns repeated the behaviour already established until a collision occurred between the flanks expanding into the central region, This time a single pair of columns separated by a gap of fifteen cells was left in the ninth generation. Continuing this sequence on through further generations inevitably suggests a binary interpretation of the evolution.

Infinitely long constant columns reduce a two dimensional automaton to one dimension; all the cells in any given column behave alike, leaving all the significant information to be gleaned from a row acting as a cross section. Such configurations were called "ripples" in Wainwright's newsletter. Adaptation of Conway's rule shows that Wolfram's Rule 22 for a $(2,1)$ automaton gives the appropriate description of the result, which might reasonably be referred to as *Life in one dimension*.

A great advantage of working with one dimensional cellular automata is that their time evolution can be shown on a two dimensional chart, whereas the evolution of a two dimensional structure would require a third dimension. While not impossible to show, there is too much information involved to keep the presentation from becoming extremely cluttered. Even so, an infinite, or even a very long, line is hard to manage. A systematic study could easily start with short lines, folded around to form a ring, avoiding end cells whose rules of evolution would differ from the interior cells. For short enough rings, the complete evolution of all possible configurations can be calculated.

3.2 Evolution for Rule 22 - one dimensional *Life*

Let us study some examples.

The shortest ring has one cell, which is consequently always its own neighbor. A zero must evolve into zero, while a one evolves into zero, which must repeat itself therafter.

This result applies specifically to Rule 22, but it is equally applicable to any other rule for which uniform neighborhoods have the same transitions. General binary rules admit other possibilities, four altogether, since a field of zeroes can either be quiescent or switch over to ones. A field of ones could do the same, so either: 1) both fields are quiescent, 2) they alternate parity in a cycle of length two, 3) zero is quiescent and one vanishes in a single generation, or 4) one is quiescent but absorbs zero after the first generation.

Naturally the more states there are, the more sequences of evolution that could be followed by uniform fields; ascertaining which one should be the first order of business in analyzing any given automaton.

There are four rings of two cells: 00 evolves into 00, 01 into 01, 10 into 10, but 11 evolves into 00. Again there are many formats for the general case to follow, the more so the more states in the automaton; it is a matter of how many graphs can be constructed using k^2 nodes with an outgoing link for each.

There are eight rings of three cells, the shortest ring for a $(2,1)$ automaton in which left and right neighbors can be distinct. Shorter rings always lack certain neighborhoods, in this case, those which are nonsymmetrical. Therefore classifying short ring evolution is useful in establishing common features that could eventually distinguish classes of automata.

The transitions for three cells are:

000	→	000	100	→	111
001	→	111	101	→	000
010	→	111	110	→	000
011	→	000	111	→	000

It is already evident that there are four classes with cyclic symmetry, namely {000}, {001,010,100}, {011,110,101}, and {111}, and that it would be sufficient to describe the mappings of the classes into one another.

For this purpose it is convenient to select one typical element from each class, but a class representative will not often evolve into another representative. Therefore the mapping between representatives should be supplemented with an indication of the discrepancy. A sequence of cells is naturally interpretable as a binary number, making the least number in the class a convenient representative.

{000}	→	{000}	{011}	→	{000}
{001}	→	{111}	{111}	→	{000}

Of course, there are sixteen rings containing four cells. They form six symmetry classes: {0000}, {0001, 0010, 0100, 1000}, {0011, 0110, 1100, 1001}, {0101, 1010}, {0111, 1110, 1101, 1011}, and {1111}. Individually, the transitions are

0000	→	0000	0100	→	1110	1000	→	1101	1100	→	0011
0001	→	1011	0101	→	0101	1001	→	0110	1101	→	0000
0010	→	0111	0110	→	1100	1010	→	1010	1110	→	0000
0011	→	1100	0111	→	0000	1011	→	0000	1111	→	0000,

so that the economy of listing the transitions by class becomes more and more evident:

{0000}	→	{0000}		{0101}	→	{0101}
{0001}	→	{0111}	(rotated right)	{0111}	→	{0000}
{0011}	→	{0011}	(rotated two right)	{1111}	→	{1111}

Writing down lists of transisitions, either between rings or between symmetry classes, does not show the structure of the transitions to very good advantage. It is better to group them into transitive chains, in which an evolutionary sequence is shown until it repeats. Only maximal chains should be shown, which means beginning with a ring which has no ancestor if the chain has one. Sometimes there are cycles which have no ancestors outside the cycle. Since evolutionary sequences tend to converge, it may be necessary to mark some chains as continuing in another chain that has already been presented.

If we give this treatment to the thirty two chains of five elements, the result is surprisingly compact; there are eight symmetry classes, seven of which form a sequence terminating with pure zeroes, and another which evolves to zero independently (enclosing a string of classes in parentheses signifies that the string forms a cycle).

11111	(00000)					
01011	00001	00111	00011	01001	01111	(00000)

There are sixty four chains of length six, which fall into thirteen symmetry classes (the confluence of one chain with another is indicated by enclosing the junction in square brackets).

010111	000001	000111	000101	011011	(000000)
001111	000011	001001	111111	(000000)	
001011	[000111]				
(010101)					
011111	(000000)				

Several formal representations of sequences of this kind exist; two of them are through graphs and their connectivity matrices.

3.3 Evolutionary diagrams and matrices

These evolutionary sequences could be represented by diagrams, which makes them considerably easier to visualize. According to the presentation, either the rings of cells or their symmetry classes are represented by nodes in a diagram. Nodes are linked according to whether the node at the head of the arrow has evolved from the node at the tail of the arrow or not. The result is what would technically be called trees rooted on cycles. Zero, one, or more arrows may enter a node according to the number of its ancestors. Only one arrow can leave each node, because evolution is unique, and so is the symmetry class of the descendant. However, cycles are possible because a ring could evolve either into itself or one of its ancestors.

A numerical, or at least matricial, representation of these sequences is also possible. The dimension of the square matrix is equal to the number of nodes of the diagram; its elements are to be zeroes or ones. Call the $N \times N$ matrix M, its matrix elements $M_{i,j}$; then its elements will be zeroes or ones according to whether the row nodes are linked to the column nodes. A variant of the Kronecker delta represents this alternative numerically.

$$\begin{aligned} M_{i,j} &= \begin{cases} 1 & i \rightarrow j \\ 0 & \text{otherwise} \end{cases} \\ &= \delta(i \rightarrow j) \end{aligned}$$

Such matrices are sparse, meaning that the non-zero elements are usually few and far between. Nevertheless, even if they are not constructed explicitly, they are very useful for expounding certain properties of diagrams. To begin with, the well-known formula for matrix multiplication, together with the property that zero annihilates any product in which it participates, shows that

$$\begin{aligned} M^2_{i,j} &= \sum_{k=1}^{N} M_{i,k} M_{k,j} \\ &= \sum_{k=1}^{N} \delta(i \rightarrow k)\delta(k \rightarrow j) \\ &= \sum_{k=1}^{N} \delta(i \rightarrow k \rightarrow j). \end{aligned}$$

Thus M^2 has zeroes except for those elements for which the row index is connected to the column index by two links. Generally, the elements of the pth power of M show where there are connecting paths of exactly p steps. Of course, the non-zero elements of some of these powers may be integers greater than one, but that simply means that there are multiple paths between those pairs of nodes.

Other properties of M and its powers:

- the row sums give the number of chains leaving a node,
- the column sums give the number of chains entering a node,
- the diagonal elements tell how many loops contain that node,
- the trace gives the total number of loops.

Normally neither M nor its powers are symmetric matrices, which complicates using Sylvester's theorem to represent them. Nevertheless, since all their elements are non-negative, the classical results of Frobenius and Perron [89] apply. The principal conclusions are that:

- a largest positive eigenvalue is bounded by both row sums and column sums;
- there is an nonzero eigenvector belonging to the largest eigenvalue, none of whose elements is negative;
- under certain circumstances this eigenvalue is unique, with a unique normalized eigenvector.

These results can be used for estimating the behavior of large powers of M, and thus the general characteristics of long chains.

3.4 Evolution of a seven-cell ring

We could then set up the matrix M for the eighteen symmetry classes of the seven-cell rings. To obtain managable indices, let us express each binary class representative in decimal. Since these numbers are not consecutive anyway, we might as well improve the appearance of M by ordering the indices as they appear in maximal chains. Figure 3.1 shows the evolutionary diagram together with its connectivity matrix corresponding to a ring of length seven.

There are seven classes without ancestors, corresponding to the zero columns. There are two static classes, represented by ones on the diagonal. Every class has just one successor, shown by the single one to be found in each row. Seven classes have unique ancestors; these are the ones with a single one in their column. Multiple ones within the same column mean multiple ancestors. All of this information can be obtained by inspection from any evolution matrix; much of it is discernible numerically by examining thr product of M with a vector all of whose components are ones.

The lengths of transients can also be read off from the matrix, as well as the lengths of cycles, but this information is harder to perceive. If, as has been done here, the row indices follow the evolutionary squence insofar as possible, a chain of ones will be found on the superdiagonal of the submatrix corresponding to the cycle or transient. Otherwise the chain may be fairly well hidden among the other matrix elements.

	15	5	27	3	9	63	0	21	47	1	7	23	43	19	31	55	127	11
15	.	1	.	.	.	.	.	.	.	.	.	.	.	.	.	.	.	.
5	.	.	1	.	.	.	.	.	.	.	.	.	.	.	.	.	.	.
27	.	.	.	1	.	.	.	.	.	.	.	.	.	.	.	.	.	.
3	.	.	.	.	1	.	.	.	.	.	.	.	.	.	.	.	.	.
9	.	.	.	.	.	1	.	.	.	.	.	.	.	.	.	.	.	.
63	.	.	.	.	.	.	1	.	.	.	.	.	.	.	.	.	.	.
0	.	.	.	.	.	.	1	.	.	.	.	.	.	.	.	.	.	.
21	.	.	.	.	.	.	.	.	1	.	.	.	.	.	.	.	.	.
47	.	.	.	.	.	.	.	.	.	1	.	.	.	.	.	.	.	.
1	.	.	.	.	.	.	.	.	.	.	1	.	.	.	.	.	.	.
7	.	.	.	.	1	.	.	.	.	.	.	.	.	.	.	.	.	.
23	.	.	.	.	1	.	.	.	.	.	.	.	.	.	.	.	.	.
43	.	1	.	.	.	.	.	.	.	.	.	.	.	.	.	.	.	.
19	.	.	.	.	.	.	.	.	.	.	.	.	.	.	1	.	.	.
31	.	.	.	1	.	.	.	.	.	.	.	.	.	.	.	.	.	.
55	.	.	.	.	.	.	1	.	.	.	.	.	.	.	.	.	.	.
127	.	.	.	.	.	.	1	.	.	.	.	.	.	.	.	.	.	.
11	.	.	.	.	.	.	.	.	.	.	.	.	.	.	.	.	.	1

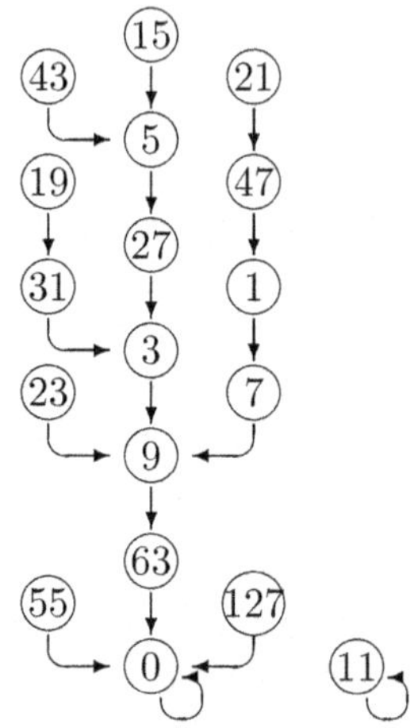

Figure 3.1: Evolution of a seven-cell ring under $(2, 1)$ Rule 22.

Powers of the connectivity matrix, being less sparse, are often easier to interpret because the remaining zeroes indicate disconnected groups of nodes more clearly. In any event, the elements M_{ij}^p specify the number of paths of length p running from node i to node j; the diagonal elements therefore disclose the number of loops passing through a given node. The trace in turn counts the total number of loops; each loop is counted once for each of its nodes, but not for multiple passes through the same node.

Some results which have been calculated for rings up to length 12 are

N	rings	symmetry classes	Gardens of Eden	longest chain	cycles
1	2	2	1	2	0
2	4	3	2	2	1
3	8	4	2	3	0
4	16	6	4	3	$2x1$
5	32	8	2	7	0
6	64	13	3	6	1
7	128	18	8	7	$1x7$
8	256	30	11	9	$1x1, 1x2$
9	512	46	19	13	$3x4, 6x6$
10	1024	78	34	9	$38x4$
11	2048	126	56	12	$1x1, 24x4$
12	4096	224	105	24	$42x5$

There are 2^N rings of length N, just the quantity of binary numbers of that length; but the number of classes is a more complicated group theoretical result[95, pp. 118-122]:

$$K = \frac{1}{^oG} \sum_{g \in G} I(g).$$

In this formula, the symmetry operations g of the group G of order ^oG have $I(g)$ fixed points. When the symmetry is rotational, the fixed points are sequences which repeat sooner than the full length of the ring; for reflections they are palindromes.

For rotational classes, Golomb transforms this formula into an expression involving Euler's function ϕ which counts divisors. The main point of interest is that the number of classes grows exponentially, although for longer rings it lags behind the number of configurations by a factor which fluctuates around $2N$, the size of the dihedral group of rotations and reflections.

The remaining statistics have to be obtained on a case-by-case basis for each rule and each cycle length, although some rules admit special techniques — for example if the rule of evolution is addition in a finite field. An extensive source of this kind of information is Wolfram's reprint collection [252], especially the tables of data in the appendix.

3.5 Cycles for a sixteen-member ring

The data eventually becomes so voluminous that it seems only to be worthwhile tabulating the cycles into which the rings of length N evolve, ignoring all the transients. Below, we show one particular case, $N = 16$, in detail. The ring is long enough to show a variety of interesting behaviour, but yet not so long that it cannot still be readily displayed.

The notation $x.y$ describing each cycle means that the period of the cycle is x, but that there are only y distinct phases within the cycle. This could either mean that the original pattern has reflected after y generations and will be completed by running through the mirror images, or that the pattern has been translated by a certain amount. Thus x must be a multiple of y, cycles of the form $2y.2$ almost always (but not exclusively) resulting from reflection. Sometimes an even more explicit notation, such as 16:14.7(ref) or $15 : 20.4(3\ell)$ is required, in which the length of the ring is shown along with an indication of the direction and distance of displacement. Since reflectivity could be inferred if a displacement were not shown, it is not always indicated.

First, we show the smallest numerical representatives for each of the eight distinct symmetry classes of cycles for a ring of length $N = 16$:

Next, the eight figures which follow show a full cycle of evolution for every one of these patterns. In preparing each illustration, every line is copied twice to improve perception of the region which would otherwise be broken up by the boundary.

3.5.1 Quiescent ring

A sequence of pure zeroes is a still life whenever zero is quiescent; the resulting redundancy is generally avoided by omitting the quiescent configuration from all

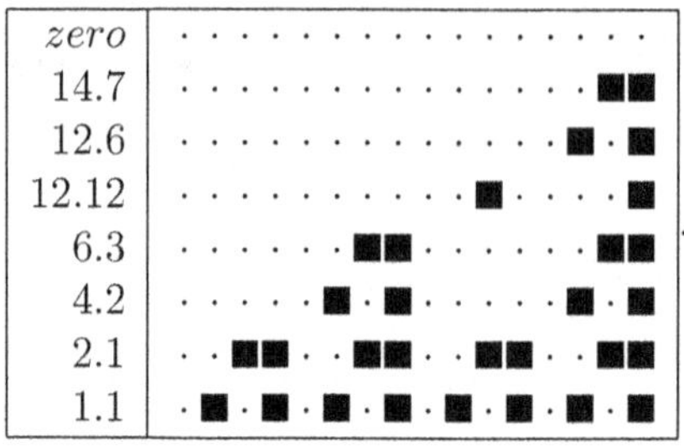

Figure 3.2: (2, 1) Rule 22 rings of length 16 have eight symmetry classes of cycles.

Figure 3.3: By definition, a quiescent configuration is a still life.

cycle censuses. Figure 3.3 contains the sixteen cell ring satisfying this description.

3.5.2 Period 14

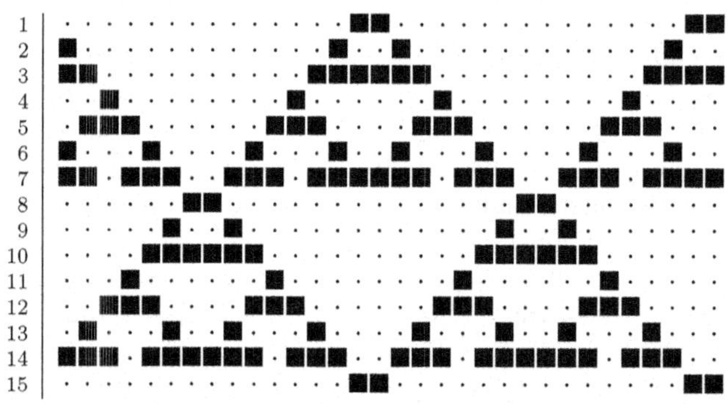

Figure 3.4: $(g.p) = 14.7$: Evolution from a joined (gap 0) pair.

Many evolutionary rules are distinguished by the presence of numerous inverted triangles in their space-time diagrams. The evolution of pure cycles shows them to especial advantage, often exhibiting a fractal structure [246]. Indeed, strings of any arbitrary format can be expected to show up occasionally; but quiescent strings are of particular interest because they can only disappear gradually, as live cells encroach on their boundaries. Triangles are the visual evidence of the eventual regrowth of the live region, evident in Figure 3.5.

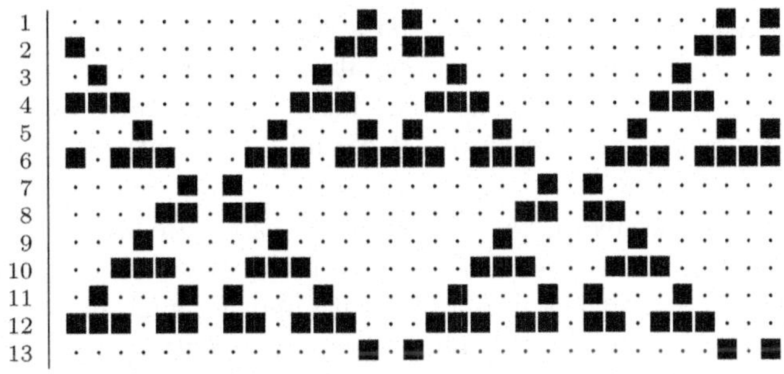

Figure 3.5: $(g.p) = 12.6$: Evolution from a gap 1 pair.

3.5.3 Period 12, 6 phases

Rule 22 is a rule which acts somewhat like a binary counter, in the sense that isolated cells expand, generating regions whose interiors periodically vanish, leaving the frontier cells to expand anew. These new sites of expansion eventually collide, possibly leaving the whole new interior vacant. Properly synchronized, the binary counter effect ensues. Rule 22 has an especial tendency towards binary counters, notably amongst configurations evolving from single pairs of live cells. Figures 3.5 and 3.6 show boundary collisions leaving gaps of 1 and 2 respectively

Compatibility with a finite cycle length implies constraints which exclude certain mutual separations; most work, the rest vanish after a while or convert into other patterns. The majority of the cycles of length 16 fit this pattern; scrutiny of the figures reveals a variety of interactions amongst advancing boundaries, also the systematic depletion of the interiors.

3.5.4 Period 12, 12 phases

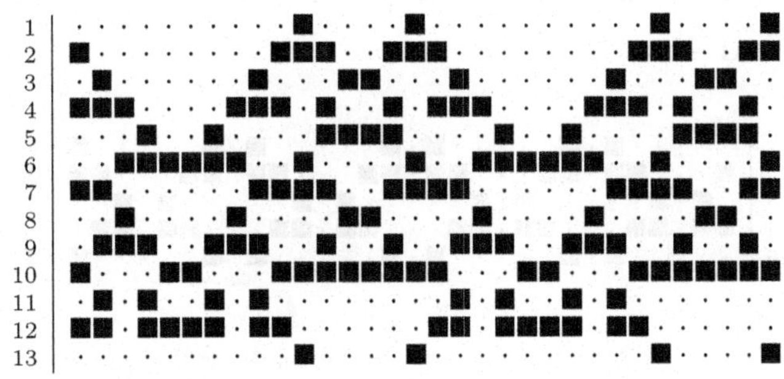

Figure 3.6: $(g.p) = 12.12$: Evolution from a gap 4 pair; no phase is shifted nor reflected relative to any other.

Some cycles run their course without returning to the same symmetry class; others may shift the original configuration several times before returning to their original form. Alternatively, they may pass through a reflected image before repeating; most cycles of the form $(2p.p)$ operate by reflection but a shift by half the length of the ring could produce the same result. Sometimes there is no distinction between translation and reflection. If N and p are relatively prime, quite long periods can result.

Amongst the last two examples, all the phases of Figure 3.6 are distinct, wheres Figure 3.5 manifests reflective symmetry. Given the way that boundaries interact, a pair with one gap may evolve into a pair with another gap, and then evolve back into the original pair, although this phenomenon is not present on a ring of length sixteen.

3.5.5 period 6

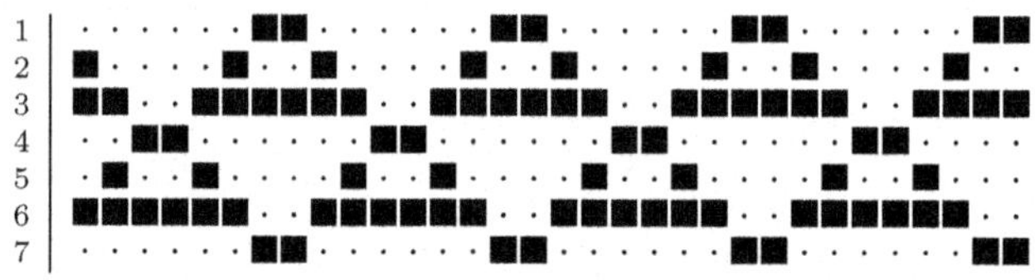

Figure 3.7: $(g.p) = 6.3$: A eight cell cycle repeated twice to get sixteen cells.

Any cycle of length N is automatically a cycle of length $2N$; or of any other integer multiple kN for that matter. Consequently censuses are prone to containing divisors unless it is agreed that they should be omitted from the tally, just as it is usually agreed to drop quiescent configurations (which are really just a special case). Figure 3.7 shows the cycle of period six which can also be found in rings of length eight.

3.5.6 Period 4

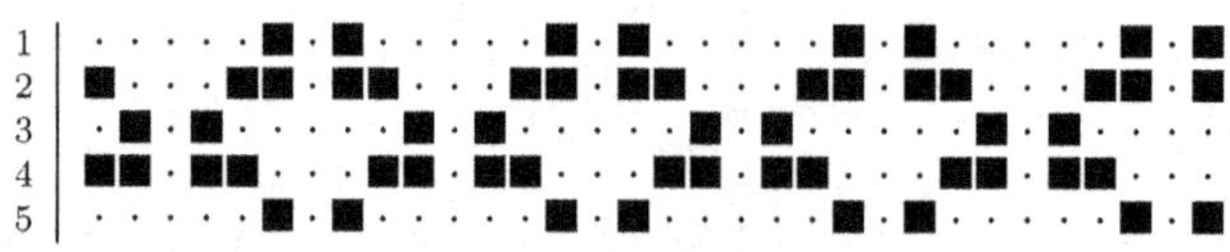

Figure 3.8: $(g.p) = 4.2$: Another cycle for an eight cell ring twice repeated.

Cycles of period four occur in a variety of forms during evolution according to Rule 22, although only one of them makes its appearance in the context of a sixteen cell ring (as a doubled eight cell ring), as shown in Figure 3.8. They all evolve from a pair of cells with small gaps, which can sometimes sum up to produce the same ring length in different ways.

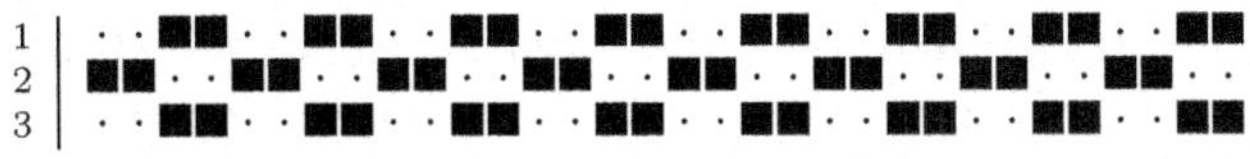

Figure 3.9: $(g.p) = 2.1$: A four cell ring repeated four times fills a sixteen cell ring.

3.5.7 Period 2

Since the cycle of period two has length four, it will recur in every ring whose length is a multiple of four. Cycles of very short period can usually be deduced by ad hoc methods, but their de Bruijn diagrams are sufficiently small that they can be rapidly identified from hand drawn graphs.

3.5.8 Period 1 (still life)

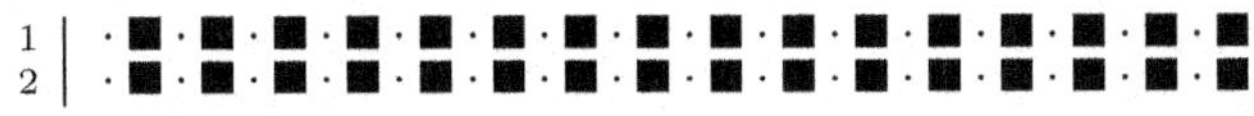

Figure 3.10: $(g.p) = 1.1$: A two-cell ring repeated eight times fills a sixteen cell ring.

Rule 22 has two still lifes, but the reason for discounting a uniformly quiescent field has already been given; Figure 3.10 shows the other one.

3.6 Cycles for Rule 22

In the process of surveying Rule 22, the following table of cycles N (row number) versus periods P (column number) was found. In counting the number of rings with each period, the zero ring was discounted because it occurs for every value of N; likewise a ring was not repeated if its period was a proper divisor of P, nor was redundancy due to shifting or reflection included (thus it is a table of cycle classes).

	1	2	3	4	5	6	7	8	9	10	11	12	13	14	15	16	17	18	19	20
1	.	1	.	1	.	1	.	1	.	1	.	1	.	1	.	1	.	1	.	1
2	.	.	.	1	.	.	.	1	.	.	.	1	.	.	.	1	.	.	.	1
3	.	.	.	.	.	.	.	.	.	.	.	.	.	.	.	.	.	.	.	.
4	.	.	.	.	.	.	.	1	1	2	1	1	.	.	.	1	1	3	3	5
5	.	.	.	.	.	.	.	.	.	.	1	1	1	.	.	.	.	.	.	.
6	.	.	.	.	.	.	.	1	.	1	.	.	.	.	.	1	.	.	.	2
7	.	.	.	.	.	.	1	.	.	.	.	.	.	1	.	.	.	.	.	.
8	.	.	.	.	.	.	.	.	.	.	.	.	.	.	.	.	.	.	.	1
9	.	.	.	.	.	.	.	.	.	.	.	.	.	.	.	.	.	.	.	.
10	.	.	.	.	.	.	.	.	.	.	.	.	.	.	.	.	.	.	.	.

	15	16	17	18	19	20	21	22	23	24	25	26	27	28	29	30	31	32	33	34
11	.	.	.	.	.	.	.	.	.	.	1	.	.	.	.	.	.	.	.	.
12	.	.	.	.	.	.	.	.	.	.	.	.	.	1	.	2	1	2	1	1
13	.	.	.	.	.	.	.	.	.	.	.	.	.	.	.	.	.	.	.	.
14	.	.	.	.	.	.	.	.	.	.	.	.	.	.	.	1	.	.	.	.
15	.	.	.	.	.	.	.	.	.	.	.	.	.	.	.	.	.	.	.	.
16	.	.	.	.	.	.	.	.	.	.	.	.	.	.	.	.	.	.	.	.
17	.	.	.	.	.	.	.	.	.	.	.	.	.	.	.	.	.	.	.	.
18	.	.	.	.	.	.	.	.	.	.	.	.	.	.	.	.	.	1	.	.
19	.	.	.	.	.	.	.	.	.	.	.	.	.	.	.	.	.	.	.	.
20	.	.	.	.	.	.	.	.	.	.	.	.	.	.	1	.	.	.	.	.
21	.	.	.	.	.	.	.	.	.	.	.	.	.	.	.	.	.	.	.	.
22	.	.	.	.	.	.	.	1	.	.	.	.	.	.	.	.	.	.	.	.
23	.	.	.	.	.	.	.	.	.	.	.	.	.	.	.	.	.	.	.	.
24	.	.	.	1	.	1	.	.	.	1	.	.	.	.	.	.	.	.	.	1
25	.	.	.	.	.	.	.	.	.	.	.	.	.	1	.	.	.	.	.	.
26	.	.	1	.	.	.	.	.	.	.	1	.	1	.	.	.	.	.	.	1
27	.	.	.	.	.	.	.	.	.	.	.	.	.	.	.	.	.	.	.	.
28	.	.	.	.	1	.	.	.	.	1	1	3	1	3	.	.	.	3	3	6
29	.	.	.	.	.	.	.	.	.	.	.	.	.	.	1	.	.	.	.	.
30	.	.	.	.	.	.	.	.	.	.	.	.	1	1	1	.	.	1	.	.
31	.	.	.	.	.	.	.	.	.	.	.	.	.	.	.	.	1	.	.	.
32	.	.	.	.	.	.	.	.	.	1	.	.	.	.	.	.	.	1	.	.
33	.	.	.	.	.	.	.	.	.	.	.	.	.	.	.	.	.	.	.	.
34	.	.	.	.	.	.	.	.	.	.	.	.	.	.	.	.	.	1	.	.

In addition, there were many cycles in the range $21 \leq N \leq 34$ with longer periods than convenient for a table, which are listed below.

N	P	N	P	N	P	N	P
21	77.11	25	55.55	27	1215.135	30	1070.107
23	138.6	25	150.6	29	72.36	31	248.8
24	50.25	25	150.6	29	667.23	31	248.8
24	54.9	26	78.6	30	40.4	32	41.41
24	72.6	26	78.6	30	70.35	32	84.42
24	72.6	26	90.45	30	86.43	33	138.69
24	80.40	26	546.21	30	120.60	34	52.52
24	384.32	27	162.6	30	240.8	34	4590.135
25	50.2	27	459.17	30	240.8		

3.7 The evolution matrix

The evolution of an automaton can be described in matrix form, as well as by the evolution function φ. The matrix required has to be rectangular, since there are boundary cells at the ends of the string whose evolution cannot be computed from the information available in the string itself. The simplest matrix in the series describes the cells of the second generation in terms of the cells comprising their neighborhoods; two values $\{0, 1\}$ can evolve from eight neighborhoods $\{000, 001, 010, 010, 011, 100, 101, 110, 111\}$, so the required matrix has the dimension 2×8; for Rule 22 it would take the form:

$$G(1:2) = \left[\begin{array}{c|cccccccc} & 000 & 001 & 010 & 011 & 100 & 101 & 110 & 111 \\ \hline 0 & 1 & . & . & 1 & . & 1 & 1 & 1 \\ 1 & . & 1 & 1 & . & 1 & . & . & . \end{array}\right],$$

where the evident formula for the matrix elements is

$$G(1:2)_{i,j,k,\ell} = \delta(i, \varphi(j,k,\ell))$$

Since the matrix is not square, successive generations of the evolution cannot be obtained by raising it to powers. However, by iterating φ it is possible to obtain the matrix for further generations, and thus to relate any pair of generations. Unfortunately such a procedure corresponds to none of the commonly recognized operations on matrices (such as the tensor product).

For example, we might work out, again for Rule 22, the matrix for one generation of evolution of the five-cell neighborhoods which will produce the three-cell neighborhoods which evolve into single cells. Such a matrix would be

$$G(2:3) = \left[\begin{array}{c|c} & 00000000000000001111111111111111 \\ & 00000000111111110000000011111111 \\ & 00001111000011110000111100001111 \\ & 00110011001100110011001100110011 \\ & 01010101010101010101010101010101 \\ \hline 000 & 10000000000001110000001100010111 \\ 001 & 01000000000000000000000000101000 \\ 010 & 00010000000000000000010010000000 \\ 011 & 00100000000000000000100001000000 \\ 100 & 00000011000100001000000000000000 \\ 101 & 00000000001000000100000000000000 \\ 110 & 00000100100000000001000000000000 \\ 111 & 00001000010000000010000000000000 \end{array}\right],$$

and finally, going one step further,

$$G(1:3) = \left[\begin{array}{c|c} & 00000000000000001111111111111111 \\ & 00000000111111110000000011111111 \\ & 00001111000011110000111100001111 \\ & 00110011001100110011001100110011 \\ & 01010101010101010101010101010101 \\ \hline 0 & 10101100111001110111101101010111 \\ 1 & 01010011000110001000010010101000 \end{array}\right].$$

According to this scheme, one has $G(1:3) = G(1:2)G(2:3)$, and a clear procedure for advancing through further generations. Since the dimension of these matrices increases rapidly – doubling each generation – and since they have very few non-zero elements, they are of more interest for theoretical discussions than as a practical means of computation. In keeping with the fact that they represent a function, there is exactly one non-zero element in each column, but since they are not square, some rows must necessarily have *several* non-zero elements. The exact distribution will vary from rule to rule and to a great extent will characterize the rule involved.

3.8 The reduced evolution matrix

The matrices $G(i.i{+}1)$ are rectangular rather than square because of the incomplete neighborhoods at the ends of a chain of finite length, which preclude calculating the evolution of all the cells of the chain. This incompleteness was fully understood in defining such matrices, but it would still be useful if the evolution of a block of cells could be related to the block itself. One approach would be to close the block into a ring. Variations on the same theme would involve extending the block in various ways before closing it into a ring, somewhat in the way that a Möbius strip is formed by giving a strip of paper a twist before joining its ends to form a cylinder.

Another approach to compensating the lack of knowledge of what lies beyond the limits of a finite chain would be to characterize the possible extensions of the chain numerically and either storing this information in a square matrix, or possibly annotating the links of a connectivity diagram.

For example, let the rows of the matrix be indexed by the 2^N sequences of N bits, and the columns by the 2^N central sequences of $N+2$ bits which can evolve into their indices via one of the (2,1) rules. The number of counterimages of each row index can range from zero to four, and this number could be recorded as the matrix element. Let us call such a matrix a *reduced evolution matrix,* denoted by E^N. Then

$$E^N_{i,j} = card\{a, b \ni \varphi^{-1}(i) = ajb\}$$

For example, choose $N = 3$, and use Rule 22; then,

$$E^3 = \left[\begin{array}{c|cccccccc} & 000 & 001 & 010 & 011 & 100 & 101 & 110 & 111 \\ \hline 000 & 1 & 1 & . & 3 & . & 1 & 2 & 4 \\ 001 & 1 & 1 & . & . & . & 1 & 2 & . \\ 010 & . & 1 & 1 & . & 1 & . & . & . \\ 011 & . & 1 & 1 & . & 1 & . & . & . \\ 100 & 1 & . & . & 1 & . & 1 & . & . \\ 101 & 1 & . & . & . & . & 1 & . & . \\ 110 & . & . & 1 & . & 1 & . & . & . \\ 111 & . & . & 1 & . & 1 & . & . & . \end{array}\right].$$

The column sums are always 4, but the row sums could range between 0 and 32, according to the rule involved. A zero sum would signify a Garden of Eden state, because the row index would then have no chains from which it could be formed. The high extreme would signify a pathological rule favoring one single sequence, for example Rule 0 mapping all chains into the zero chain. Diagonal elements do not portend chains of period one, but a still life could not exist if all diagonal elements were zero.

Unfortunately the square of E does not correspond to the number of two-generation ancestors producing a given chain even though it arises from two steps of evolution. The reason is precisely the correlations introduced by the order in which cells occur. Squaring E implies *some* ancestor for each step of the evolution, but does not guarantee that they coincide where they overlap. An easy counterexample is obtained from chains of length 1. E_1 for rule 22 is

$$\begin{bmatrix} 2 & 3 \\ 2 & 1 \end{bmatrix},$$

so its square is

$$\begin{bmatrix} 10 & 9 \\ 6 & 7 \end{bmatrix}.$$

The table for two generation evolution according to Rule 22 is

00000	00001	00010	00011	00100	00101	00110	00111
000	001	011	010	111	110	100	100
0	1	0	1	0	0	1	1
01000	01001	01010	01011	01100	01101	01110	01111
110	111	101	100	001	000	000	000
0	0	0	1	1	0	0	0
10000	10001	10010	10011	10100	10101	10110	10111
100	101	111	110	011	010	000	000
1	0	0	0	0	1	0	0
11000	11001	11010	11011	11100	11101	11110	11111
010	011	001	000	001	000	000	000
1	0	1	0	1	0	0	0

which produces a two-generation reduced evolution matrix

$$\begin{bmatrix} 10 & 11 \\ 6 & 5 \end{bmatrix},$$

which is clearly different from the square of the one-generation matrix.

There is an evident relation between the evolution matrices $G(i,j)$ and the reduced evolution matrices E_i; If we define a matrix R, for which

$$R(3) = \left[\begin{array}{c|cc} & 0 & 1 \\ \hline 000 & 1 & 0 \\ 001 & 1 & 0 \\ 010 & 0 & 1 \\ 011 & 0 & 1 \\ 100 & 1 & 0 \\ 101 & 1 & 0 \\ 110 & 0 & 1 \\ 111 & 0 & 1 \end{array}\right]$$

is typical, we have

$$E_j = G(j:j+1)R(j+1)$$

and the evident definition

$$[R(j+1)]_{p,q} = \delta(p, bin(q) = apb).$$

In such terms it is clear that $G(1:2)G(2:3)R(3)R(2) \neq [G(1:2)R(2)]^2$ as a general proposition.

Chapter 4

Periods in time

The constraint of a fixed spatial periodicity on a cellular automaton produces an assortment of time periods, into one of which the evolution must eventually fall. The converse problem would be to select a time interval, with the intention of enumerating the lengths of all the possible rings whose evolution would regenerate after the given delay.

4.1 Characteristics of cycles

The course of evolution of any finite automaton can be traced through a diagram constructed for the purpose, or such an equivalent representation as the connectivity matrix of the diagram. Features of interest surely include:

- the cycles in which all evolutions terminate,
- nodes with no ancestors – the Garden of Eden configurations,
- the length of the longest transient,
- the average and standard deviation of transient lengths,
- the amount and type of branching in the diagram.

The first item is the only one of importance for the operation of an automaton after an extremely long time, but it is still useful to understand the short term behaviour, particularly as it affects the choice among different long term alternatives. Furthermore, what is a long term for a ring of a hundred cells is still very short term for a ring of a thousand cells.

The technique of tracing out the full evolution of a ring of length N yields all the configurations in an infinite ring which repeat themselves after translation through a distance N, whatever their period of repetition in time; in other words all the patterns with a given spatial periodicity. In the process of observing the results one learns to recognize certain patterns and to be able to predict their periodicities. It would be nice to have a procedure which would yield all the configurations with a given period, irrespective of the length of the ring on which they might occur, beginning with a method for finding all the still lifes - patterns which never change with time.

4.2 Overlapping of neighborhoods

Neighborhoods in which the central cell does not change can be read off from the evolutionary rule. The complication in finding sequences which do not change lies in the overlap between neighborhoods of two successive cells. Fortunately there already exists a diagrammatic technique for dealing with overlapping sequences of symbols, which is commonly encountered in the theory of shift registers [95]. The underlying combinatorial theory goes back at least to the last century [191].

To see how this works, let us calculate the still lifes for Rule 22. The table below shows quite a bit more information than this; each transition is classified as to whether the new cell is the same as the old, as its left or right neighbor, whether it has evolved into zero or one, or the complement of its old value. In general the new cell could be tested against any Boolean function of the members of its neighborhood.

neighborhood	000	001	010	011	100	101	110	111
new cell	0	1	1	0	1	0	0	0
same	x	.	x	.	.	x	.	.
left	x	.	.	x	x	.	.	.
right	x	x	.	.	.	.	x	.
zero	x	.	.	x	.	x	x	.
one	.	x	x	.	x	.	.	.
complement	.	x	.	x	x	.	x	x

Three neighborhoods qualify for producing still lifes, namely 000, 010, and 101. 000 can overlap only with itself in either direction, so that it can only participate in a chain of pure zeroes. Such a chain does in fact remain zero, so it qualifies as a still life. The other two neighborhoods can overlap each other in alternation, leading to the sequence ...010101... which is also seen to be a still life. These two alternatives exhaust the possibilities, and answer the question "What are the still lifes of Rule 22?"

There are three neighborhoods whose central cell evolves into one, namely 001, 010, and 100. The first of these can only fit between the third and the second, in that order. The second can only fit between the first and third, and the third must follow the second and precede the first. Thus only the sequence ...100100100100... can produce pure ones. On inspection this is reasonable—contiguous ones necessarily force a zero, while a gap of zeroes longer than two can never be filled with ones in a single generation.

Similar arguments can be given regarding the remaining lines in the table. However, something more concrete than such verbal arguments is required, and is afforded by the subdiagrams of a map outlining all the possible ways in which the neighborhoods defining the evolution can overlap.

4.3 The de Bruijn matrix

The basic diagrammatic tool is called a de Bruijn diagram [95, 191], whose k^s nodes represent the possible sequences of s symbols chosen from amongst a collection of k; making the convenient choice of integers modulo k for symbols, the nodes become simply s-digit numbers to the base k.

Two nodes are to be linked if the first ends with the same $s-1$ digit sequence with which the second begins; in other words, if they represent overlapping sequences. Since there is no restriction on the symbols which can be placed at the free ends, each node will have k incoming links and k outgoing links. Having chosen numbers as symbols, the linkage rule is expressible in a simple arithmetic form; namely, that node i is linked to nodes $si, si+1, \cdots, si+k-1$. Since there are $S = k^s$ nodes in all, the foregoing sums are to be taken modulo S. Indeed the most straightforward representation of a de Bruijn diagram is through the vertices of an S-gon inscribed in a circle, chords marked according to the links present. It can also be represented by a connectivity matrix whose block diagonal structure models this circle.

Since those nodes are linked which differ by shifting their label, dropping the digit on the left and inserting a new digit on the right, de Bruijn diagrams are frequently used in shift register theory. Here, the shift register is simply a moving window which can be used to scan a long chain for the neighborhoods needed by the evolutionary rule of a cellular automaton. Any chain can be reconstructed from the windows by following a path through the diagram, recording the digit associated with each link as one moves along.

Now we need to examine de Bruijn matrices in detail, but to give concrete examples we will stick to low order matrices over the binary alphabet. Let $B_{k,n}$ denote the connectivity matrix for a de Bruijn diagram for k symbols and n stages; write B_n for $B_{2,n}$.

4.3.1 B_1

Let us start with B_1, which is a 2×2 matrix (showing a border identifying the 1-blocks which have been linked).

	0	1
0	1	1
1	1	1

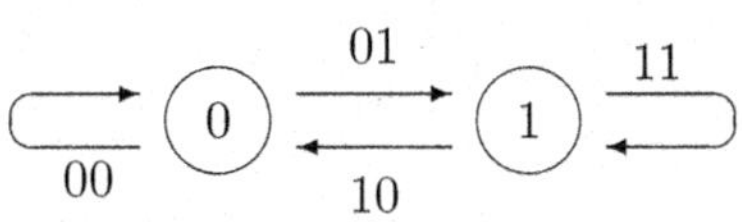

Figure 4.1: Single stage binary de Bruijn matrix and diagram.

The simple structure is due to the fact no matter whether 0 or 1 is dropped from a 1-block, or whether 0 or 1 is added to the block, the maneuver is feasible; thus all positions of the matrix are filled with a one. By inspection, we see that

$$B_1^2 = 2B_1,$$

which is the characteristic equation for B_1. It has eivenvalues $\lambda = 0, 2$. Since it is symmetric it has an orthogonal eigenvector matrix, which is:

$$\frac{1}{2}\begin{bmatrix} 1 & 1 \\ 1 & -1 \end{bmatrix}.$$

The coefficient $1/2$ rather than $1/\sqrt{2}$ is chosen because we will presently use probability vectors, which are normed by sums of absolute values rather than sums of squares.

4.3.2 B_2

Let us continue with B_2, likewise labelling its rows and columns by the 2-blocks to which they correspond. Zero matrix elements—for which there is no link in the de Bruijn diagram—will be suppressed in favor of dots.

	00	01	10	11
00	1	1	.	.
01	.	.	1	1
10	1	1	.	.
11	.	.	1	1

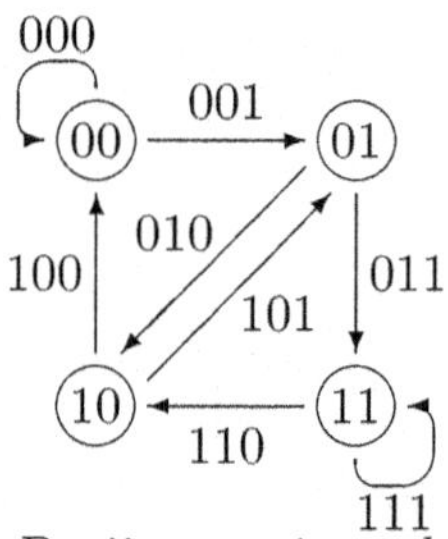

Figure 4.2: Two stage binary de Bruijn matrix and diagram.

By a simple calculation,

$$B_2^3 = 2B_2^2,$$

which is a minimal equation rather than a characteristic equation since it is not of degree 4. Its roots are $\lambda = 2, 0, 0$. B_2 is evidently singular because of the repeated rows, but on closer inspection we find that it shows the Jordan canonical form with respect to a 2×2 block belonging to eigenvalue 0.

The matrix of principal vectors reducing B_2 to canonical form could be

$$\frac{1}{4}\begin{bmatrix} 1 & 1 & 1 & 1 \\ 1 & -1 & -1 & 1 \\ 1 & 1 & -1 & -1 \\ 1 & -1 & 1 & -1 \end{bmatrix}.$$

The first three columns are eigenvectors, the last a principal vector.

Although characteristics of the de Bruijn matrices are already evident in B_2, it is worth going on to examine B_3 for good measure.

4.3.3 B_3

This time the minimal equation is

$$B_3^4 = 2B_3^3.$$

	000	001	010	011	100	101	110	111
000	1	1	.	.	.	.	.	.
001	.	.	1	1	.	.	.	.
010	.	.	.	.	1	1	.	.
011	.	.	.	.	.	.	1	1
100	1	1	.	.	.	.	.	.
101	.	.	1	1	.	.	.	.
110	.	.	.	.	1	1	.	.
111	.	.	.	.	.	.	1	1

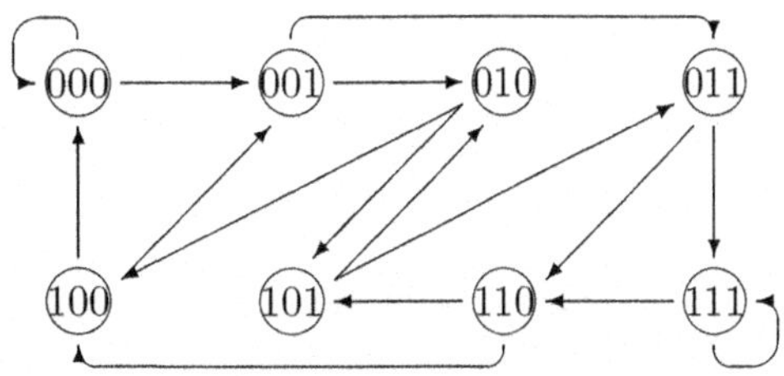

Figure 4.3: Three stage binary de Bruijn matrix and diagram.

The general result is that $B_{k,n}$ is a $k^n \times k^n$ matrix whose minimal equation is

$$B_{k,n}^{n+1} = kB_{k,n}^{n},$$

and which exhibits the Jordan canonical form for $k > 2$.

This result shows that it takes at most n links to get from one node to another in an n-stage de Bruijn diagram; in terms of shift register sequences this is precisely the shift required to completely replace one sequence by another. If the node labels have internal symmetry, a smaller shift and hence a shorter path through the diagram may exist.

Although the form of the minimal equation can be conjectured readily enough by inspection, an instructive proof can be given, whose elements will be useful later on. We begin with two factorizations of the de Bruijn matrix, motivated by the observation that it would be block diagonal if only its rows were rearranged.

4.3.4 B_n as product and sum

$$B_3 = \begin{bmatrix} 1 & . & . & . & . & . & . & . \\ . & . & 1 & . & . & . & . & . \\ . & . & . & . & 1 & . & . & . \\ . & . & . & . & . & . & 1 & . \\ . & 1 & . & . & . & . & . & . \\ . & . & . & 1 & . & . & . & . \\ . & . & . & . & . & 1 & . & . \\ . & . & . & . & . & . & . & 1 \end{bmatrix} \times \begin{bmatrix} 1 & 1 & . & . & . & . & . & . \\ 1 & 1 & . & . & . & . & . & . \\ . & . & 1 & 1 & . & . & . & . \\ . & . & 1 & 1 & . & . & . & . \\ . & . & . & . & 1 & 1 & . & . \\ . & . & . & . & 1 & 1 & . & . \\ . & . & . & . & . & . & 1 & 1 \\ . & . & . & . & . & . & 1 & 1 \end{bmatrix}$$

Expressing this equation in the form $B_n = P_n U_n$ implicitly defines the orthogonal permutation matrix P and a block diagonal U. These matrices satisfy the equations $P_n^n = I$, $U_n^2 = 2U_n$, where I is the unit matrix. Note that U_n is a Kronecker product, $U_n = I_{n-1} \otimes U_1$, and that the following two identities hold

$$B_n = P_n(I_{n-1} \otimes U_1)$$

$$B_n = (U_1 \otimes I_{n-1})P_n.$$

Due to the symmetry of the matrix U, a slightly different factorization is possible, in terms of a permutation matrix which we could call Q_n.

$$B_3 = \begin{bmatrix} \cdot & 1 & \cdot & \cdot & \cdot & \cdot & \cdot & \cdot \\ \cdot & \cdot & \cdot & 1 & \cdot & \cdot & \cdot & \cdot \\ \cdot & \cdot & \cdot & \cdot & \cdot & 1 & \cdot & \cdot \\ \cdot & \cdot & \cdot & \cdot & \cdot & \cdot & \cdot & 1 \\ 1 & \cdot & \cdot & \cdot & \cdot & \cdot & \cdot & \cdot \\ \cdot & \cdot & 1 & \cdot & \cdot & \cdot & \cdot & \cdot \\ \cdot & \cdot & \cdot & \cdot & 1 & \cdot & \cdot & \cdot \\ \cdot & \cdot & \cdot & \cdot & \cdot & \cdot & 1 & \cdot \end{bmatrix} \times \begin{bmatrix} 1 & 1 & \cdot & \cdot & \cdot & \cdot & \cdot & \cdot \\ 1 & 1 & \cdot & \cdot & \cdot & \cdot & \cdot & \cdot \\ \cdot & \cdot & 1 & 1 & \cdot & \cdot & \cdot & \cdot \\ \cdot & \cdot & 1 & 1 & \cdot & \cdot & \cdot & \cdot \\ \cdot & \cdot & \cdot & \cdot & 1 & 1 & \cdot & \cdot \\ \cdot & \cdot & \cdot & \cdot & 1 & 1 & \cdot & \cdot \\ \cdot & \cdot & \cdot & \cdot & \cdot & \cdot & 1 & 1 \\ \cdot & \cdot & \cdot & \cdot & \cdot & \cdot & 1 & 1 \end{bmatrix}$$

Q_n satisfies the equation $Q_n^{2n} = I$, but P and Q do not commute. It is interesting to note that

$$B_n = P_n + Q_n.$$

Thus there are both factorizations and sum decompositions for the de Bruijn matrix, all readily obtained in a way that indicates that many more representations are possible, although the others would not be as symmetrical as the ones shown.

If we adopted a more formal definition of P and Q we would find

$$[P_n]_{i,j} = \delta(2i, j) \bmod 2^n$$

$$[Q_n]_{i,j} = \delta(2i+1, j) \bmod 2^n$$

from which would follow

$$\begin{aligned} B_n^2 &= (P_n + Q_n)^2 \\ &= P_nP_n + P_nQ_n + Q_nP_n + Q_nQ_n \end{aligned}$$

Writing the four terms of this sum according to their matrix elements, we find (where δ is Kronecker's delta):

$$\begin{aligned} [P_n^2]_{i,j} &= \delta(4i, j) \bmod 2^n \\ [P_nQ_n]_{i,j} &= \delta(4i+2, j) \bmod 2^n \\ [Q_nP_n]_{i,j} &= \delta(4i+1, j) \bmod 2^n \\ [Q_n^2]_{i,j} &= \delta(4i+3, j) \bmod 2^n \end{aligned}$$

a representation which can readily be generalized for any power of B, and also for any number of states per cell, k. In particular, the form of both the minimal equation and the characteristic equation for the de Bruijn matrices follows. Note that the congruence in these equations is multiplicative, not additive, making 2^N congruent to 1, not 0.

The factorized form of the de Bruijn matrices can be used to obtain the determinants and inverses of their probabilistic generalization; since the strict de Bruijn matrices are singular their zero determinant can be derived in this way, but of course they have no inverses.

4.4 Periods and other properties

There are two ways that a de Bruijn diagram can be associated with the neighborhoods of a linear cellular automaton. The neighborhoods can be attributed to the nodes or to the links. The latter is the more compact representation; then the node between two links would describe the overlapping portion of the two neighborhoods. Any classification of neighborhoods immediately becomes a classification of the links. In searching for still lifes, we would discard all those links which did not stand for a neighborhood whose central cell evolved into itself.

If the remaining links formed chains, they would be chains which remained unchanged from one generation to the next. Unless they closed into loops however, they might dwindle away in each generation. If there were suitable loops, their lengths would determine the lengths of the ring automata in which they could be found. For example, chains of alternating ones and zeroes can only be found in rings of even length according to Rule 22, whereas the zero chain can exist in rings of any length.

Consequently loops often have to be extracted from larger fragments of de Bruijn diagrams, just as they have to be to get cycles from the evolutionary diagrams; in both cases the transients leading into the loops get discarded.

The formation of barriers between macrocells is one exception to the rule that only loops give interesting still lifes. The criterion for such a barrier is that the terminal nodes have a full complement of incoming or outgoing links according to the end of the chain which they occupy.

Another combination which often occurs is that a loop will be connected through a series of links to another loop, but that there is no return path. Thus some pattern may occupy the left side of an infinite chain, then undergo a transition into another pattern which occupies the right side of the chain. Such configurations were called *fuses* in Gardner's column and Wainwright's newsletter, and can exist in one dimension just as well as in two.

Nor are still lifes the only patterns which can be deduced from de Bruijn diagrams. To obtain all the patterns of period two, it is only necessary to iterate the evolutionary rule once and then look for still lifes. This would transform a (k, r) automaton into a $(k, 2r)$ automaton using a de Bruijn diagram of $4r$ stages, whose links would be classified by their behaviour after two generations of evolution. Similar considerations apply to periods of three or higher.

Any binary property at all of a neighborhood of cells can be reflected by its inclusion as a link in the de Bruijn diagram for neighborhoods of that length, and be extended to a larger region by pursuing chains throughout the diagram. Cells other than the central cell can repeat themselves, so that it is just as easy to search for shifting patterns as for periodic ones. Evolution into constant chains is likewise easy to test, and is helpful in determining the behaviour of backgounds, particulary in automata with many states per cell.

Figure 4.4 shows several examples; in fact configurations possessing any of the attributes mentioned in Section 4.2, and many more, can be read off by inspection, the $(2, 1)$ de Bruijn diagram being quite small.

Two less obvious properties can be mentioned. One is the search for *superluminal* configurations. Quiescent automata are ones in which there is a state q

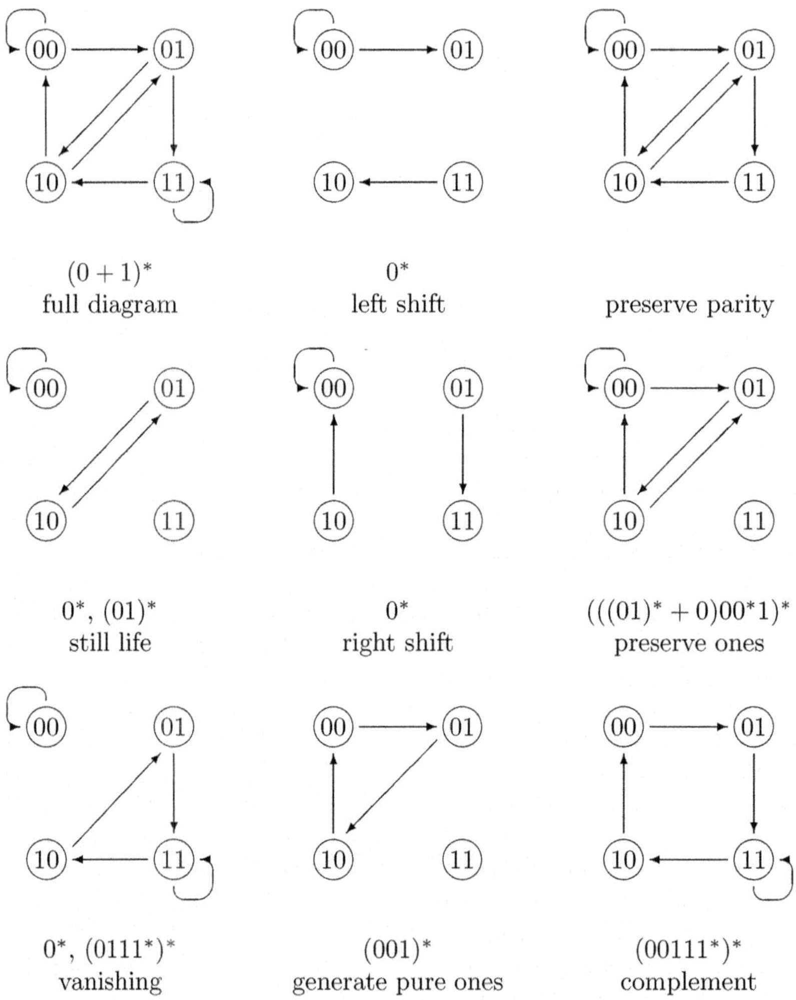

Figure 4.4: Simple properties of Rule 22 via de Bruijn diagrams.

such that $\varphi(q,q,q) = q$; when such a state exists, it is generally assigned the value zero. Actually, especially for automata which are not binary, there may be several quiescent states, in which case only one of them would be labelled zero. Quiescence simply means that a non-quiet cell cannot arise unless there is another non-quiet cell in its neighborhood, which is to say, "nearby." In quiescent automata information can propagate through a quiet region at a maximum velocity equal to one radius per time step; this velocity is picturesquely called the *velocity of light.*

A pattern can move faster than the velocity of light if it moves through a region which is densely enough populated by non-quiet cells. Superluminal configurations can always be found for any velocity, but they may turn out to be so trivial as to be uninteresting. For example, one may simultaneously be superluminal with a smaller velocity, or even be a higher harmonic of a subluminal configuration. The argument is that one can choose an arbitrary interval consisting of one super-light-second plus half the width of the neighborhood. Say its length is ℓ. Anything beyond this interval is fixed by the combination of evolution and shifting. Of course, the shifted image which emerges does not have to be part of the original interval, but there is not much else that it can be because there are only 2^ℓ intervals and $2^{\ell-2g}$ images after g generations. So, there has to be *some* interval which generates a recurrent pattern.

Idempotent rules lead to a particularly restrictive type of behaviour, but not all rules are idempotent. Still, one of the boolean questions which can be asked of second generation evolution is whether

$$\varphi(\varphi(a,b,c),\varphi(b,c,d),\varphi(c,d,e)) = \varphi(b,c,d)$$

and consequently whether a given rule admits any configurations which have an idempotent evolution even if all of them do not.

4.5 Superluminal configurations for Rule 22

Conceptually superluminal configurations may be somewhat easier to describe than those of lower velocity. For convenience, consider a .1(2) configuration, which is supposed to have a single phase which shifts by a distance of 2 in each generation. Its evolution would have the symbolic form

$$\begin{array}{ccccccccccc} . & . & . & . & . & a & b & c & d & e & f & . \\ . & . & a & b & c & d & e & f & . & . & . \end{array}$$

From this it is evident that a, b, and c are arbitrary, but that necessarily

$$\begin{aligned} d &= \varphi(a,b,c) \\ e &= \varphi(b,c,d) \\ &= \varphi(b,c,\varphi(a,b,c)) \\ f &= \varphi(c,d,e) \\ &= \varphi(c,\varphi(a,b,c),\varphi(b,c,\varphi(a,b,c))) \\ &\dots \end{aligned}$$

The point is that any seed sequence a, b, c determines the entire sequence of cells from some point on, but only eight different binary sequences of three cells are possible. The original seed need not repeat itself, but some seed must repeat within at least eight cells. Thus only eight strings need to be examined to ascertain all the possible .1(2) configurations; for Rule 22 we would find:

								configuration
0	0	0	0	.	.	.	.	0^*
0	0	1	1	0	0	1	.	$(0011)^*$
0	1	0	1	0	.	.	.	$(01)^*$
0	1	1	0	0	1	1	.	$(0011)^*$
1	0	0	1	1	0	0	.	$(0011)^*$
1	0	1	0	1	.	.	.	$(01)^*$
1	1	0	0	1	1	0	.	$(0011)^*$
1	1	1	0	0	1	1	0	$(0011)^*$

These configurations are not very interesting, because they all have simpler descriptions; but they all fulfill the technical description, and are moreover the only configurations which do so.

If one wanted to find .1(3) configurations, the seed would have length 4, and in general a $.p(d)$ configuration would have a seed of length $p+d$ and the superluminal configuration would have length less than 2^{p+d} and have to be manifest within that many cycles of iteration of the above procedure.

The following superluminal configurations (single phase, shift no longer than 12) exist for Rule 22:

.1(4)	11.1	.1(8)	29.1
.1(5)	30.1	.1(9)	132.1
.1(6)	46.1	.1(10)	191.1
.1(7)	31.1	.1(11)	96.1
.1(8)	37.1	.1(12)	83.1

Subluminal configurations can be found by the same procedure, but since the displaced seed overlaps its own neighborhood, consistency requirements must be met. Thus superluminal configurations always exist even if they are degenerate or trivial, but subluminal configurations may fail the additional requirement. In any event, the de Bruijn diagram yields a more systematic procedure for all configurations, and the foregoing discussion is mainly useful an existence theorem.

4.6 Periods for Rule 22

Whereas the diagram for the cycles of a linear cellular automaton is a collection of "trees rooted in cycles," its periods are obtained from the de Bruijn diagram which consists of an enormous number of cycles, all interlacing. Still, the length of a ring with a given period is limited by the number of nodes in its de Bruijn diagram, which after g generations is k^{2rg}. For one-generation properties of a (2,1) automaton this number is 4. This means that any still life must be evident in a ring whose length is no longer than 4, and that if there is a still life in a longer ring, it must contain repeated segments whose length is no longer than 4.

There is only one loop of length 4 in a 2-stage binary de Bruijn diagram, formed by the sequence 0011; to require it to represent a still life makes the following demands on the transition rule:

$$\begin{array}{cccccccc} 111 & 110 & 101 & 100 & 011 & 010 & 001 & 000 \\ a & 1 & b & 0 & 1 & c & 0 & d \end{array}.$$

Thus four neighborhoods have their evolution prescribed and for the other four it is arbitrary. For the maximum variety, a, b, c, and d should be chosen opposite to the central cell, yielding Rule 105, but there are altogether sixteen rules for which the sequence ...001100110011... is a still life. Of course, there may be additional still lifes besides the one which was stipulated; for example if d were chosen to be zero in this example, any number of additional zeroes could be inserted into the sequence besides the two shown, and we would have Rule 104. Then ...0011000001100011... would also be a still life. At the points where additional zeroes can be inserted, there is no restriction on the exact number.

It is harder to survey the possible periods for a rule than the possible cycles because the lengths of the configurations that have to be checked grow twice as fast with increasing period as with increasing cycle length; thus if it is feasible to survey all cycles up to length 30, the corresponding limit for periods is 15. Furthermore, the diagram for cycles has only simple loops, whereas the period diagrams can have multiple loops, whose branching is somewhat harder to describe.

Shifting configurations are just as easy to determine as are the periodic configurations; let the notation $P.p(d)$ denote a configuration which has period P with p phases after which time it has shifted d cells to the left (a negative d signifies a right shift, which only need be catalogued separately for non-symmetric rules). Reflective configurations of the form $2p.p$ can be found among the $2p.2p(0)$ configurations, the calculation required to detect them being the same in either case.

Actually it is better not to specify to specify the period, since it will depend somewhat on the length N of the ring in which it occurs. That is, if d does not divide N, the pattern may have to make several circuits of the ring before repeating itself, and

$$P = p \times \text{lcm}(d, N).$$

The smaller subluminal periods for Rule 22 have been determined; they include the following up through five generations:

1 generation :

- $.1(0)$ - 0^*, $(01)^*$
- $.1(1)$ - 0^*

2 generations :

- $.2(0)$ - 0^*, $(01)^*$, $(0011)^*$
- $.2(1)$ - 0^*, $(0001101)^*$
- $.2(2)$ - 0^*, $(01)^*$

3 generations :

.3(0) - 0^*, $(01)^*$
.3(1) - 0^*, $(000010011\ 11)^*$
.3(2) - 0^*, Figure 4.5
.3(3) - 0^*

4 generations :

.4(0) - 0^*, $(1(0 + 00 + 000)10000(0 + 00 + 000))^*$ and evolutes
.4(1) - 0^*
.4(2) - 0^*, $(01)^*$, $(0001101)^*$
.4(3) - 0^*, Figure 4.6
.4(4) - 0^*, $(01)^*$, $(0011)^*$

5 generations :

.5(0) - 0^*, $(01)^*$, $(100001000\ 0(0 + 00 + 000))^*$ and evolutes
.5(1) - 0^*, $(0001101)^*$
.5(2) - 0^*, $(0000111100\ 1)^*$, $(0011)^*$, $(01)^*$
.5(3) - 0^*, $(0000011100\ 0010001001\ 1101111000\ 0000010)^*$
.5(4) - 0^*, $(0001110001\ 0001011101\ 1000000010\ 0)^*$, $(01)^*$
.5(5) - 0^*, $(0010011000\ 0101110001\ 0000011111)^*$

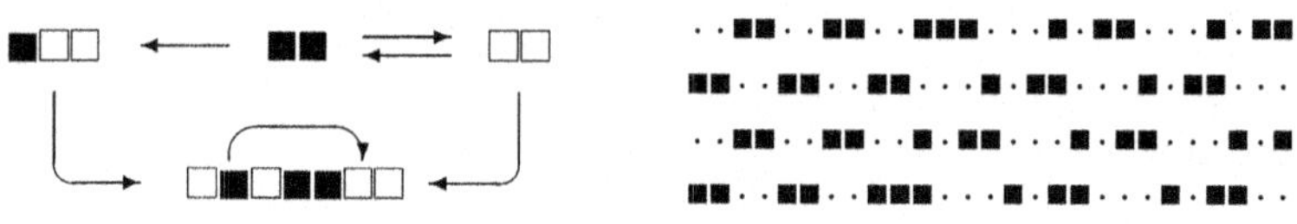

Figure 4.5: Configurations satisfying .3(2) for Rule 22.

The unshifted period six is interesting for a wide variety of cycles, all of whose lengths are multiples of ten, situated as vertical faces on the hexagonal prism of Figure 4.7. Of course the table continues indefinitely; shifting configurations of period six exist, not to mention longer periods. Taking into account additional shifting and longer periods tends to produce diagrams of increasing complexity, but three classes of diagram are already apparent in the results shown here.

First, there are simple cycles which evolve by shifting (including still lifes with zero shift); a single diagram serves all generations.

Next, there are diagrams, such as the .4(0) or .5(0) families, in which succeeding generations are described by distinct components of a disconnected diagram, although symmetric images of the first (and subsequent) generations may be encountered before the full period closes.

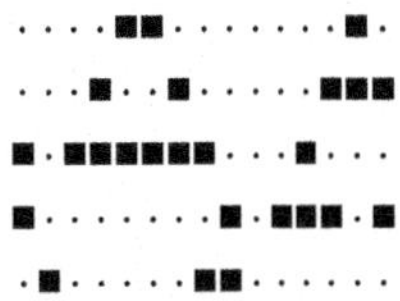

15.4(3r) left hand rhombus

29.1(8r) outer periphery

31.1(7ℓ) inner periphery

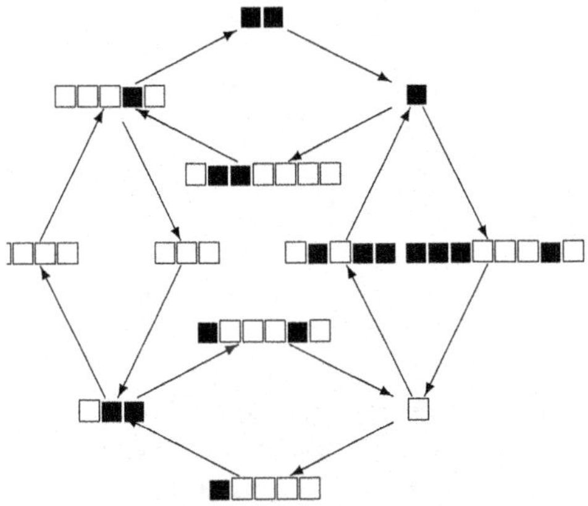

Figure 4.6: Configurations satisfying .4(3) for Rule 22.

The third class, typified by the .4(3) configurations of Figure 4.6 or the .6(0) configurations of Figure 4.7, combines several different phases within a single generation, producing a much more intricate single component diagram.

The .5(0) configurations arise from a single pair of cells embedded in a torus; two gaps separate them of which one always has length 4, the other may have length 5, 6, or 7. Altogether the cycle length may be 11, 12, or 13; longer configurations arise from joining the triplets in sequence.

By contrast, the .4(0) configurations also feature a live cell pair, but this time the short gap can have length 1, 2, or 3 while the long gap's length can be 5, 6, or 7; this time an ennead whose members can be sequenced.

In summary, the de Bruijn diagram determines sequences of neighborhoods, and therefore cyclic or infinite configurations which meet some requirement which can be placed on its links. Cyclic configurations follow from the loops inherent in the diagram, but aperiodic configurations also exist; these can arise in two ways.

If there are two loops with a one-way connection between them, such as can be seen in Figure 4.5 describing .3(2) configurations, a transition between the two loops can be made at some point. The result in some cases has been called a "fuse," a left hand field arises which is different from the right hand field; the boundary between them may be moving or static. To make a good fuse, the field from which the boundary retreats should be quiescent.

If two loops are mutually connected, the transition back and forth can be made many times; imagine that the choice is made to correspond to the binary expansion

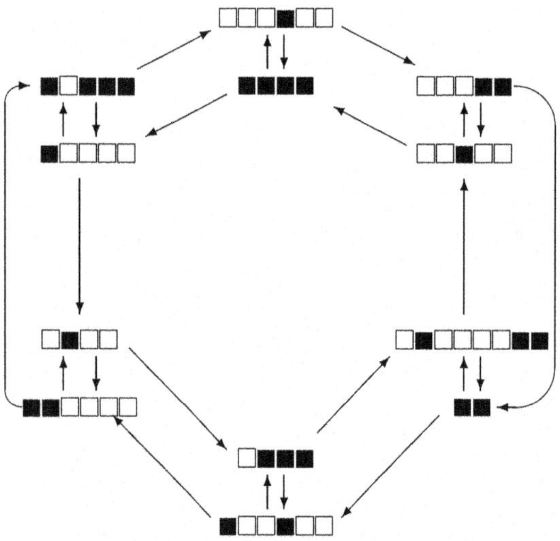

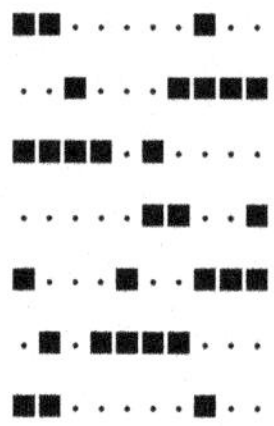

Shortest 10.6(0), period 6.

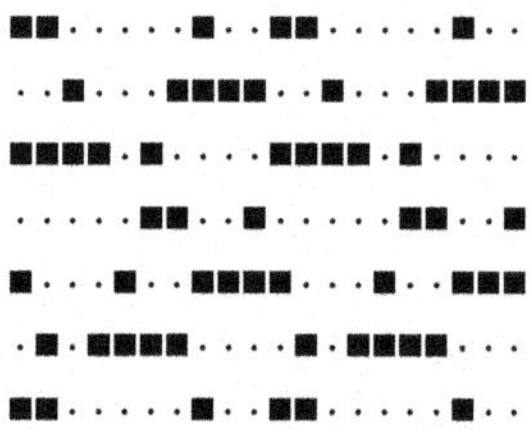

First 20.6(0),
repeating 10.6(0) twice.

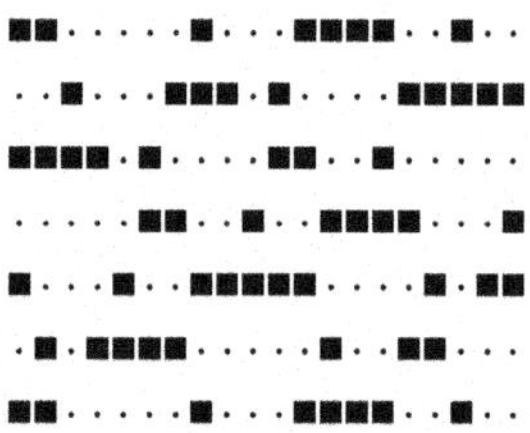

Second 20.6(0),
annexing adjacent face.

Figure 4.7: Configurations satisfying .6(0) for Rule 22.

of an irrational number to obtain an aperiodic configuration. The greater the variety of loops in the diagram, the greater the variety of configurations, but a connected pair is enough for aperiodcity.

4.7 The gap theorems

In the course of obtaining all the cycles of length up to $N = 34$ for Rule 22, rings of cells were examined in such an order that those with more leading zeroes were examined first. Moreover, rings were discarded if, in the course of evolution, a ring appeared which might have been used earlier as a starting point, or was equivalent to such a ring by some symmetry operation. In effect, that meant if a ring evolved which contained a string with more zeroes than there were leading zeroes in the original ring, the analysis was halted and the next case considered.

It was quickly observed and proved that when a single leading zero was encountered, only one cycle was going to be discovered — the still life if the length of the ring was even. Likewise, two leading zeroes would only lead to the ring of period two, and then only when the length of the ring was a multiple of four. If we say that a sequence of m zeroes (presumably bounded by ones at both ends) is a *gap* of length m, then the conditions in the following table were observed

gap	increases to	unless
1	2	$(01)^*$
2	3	$(0011)^*$
3	4	$(0001011)^*$
4	5	$(00100001111)^*$

A conjecture that the only exceptions to gap 5 were the configurations of period 4 and period 6 was made, but proved intractable to verify. Essentially the conclusion was drawn that there would always be some phase of a cycle with a larger gap than a certain value unless the ring were of a particular form, generally of low period, in which the given gap was conserved.

These observations had practical importance because the search for cycles could be stopped when the leading gap got to be small enough. The search time doubled each time the leading gap was decreased by one, so that it was worthwhile to stop with the largest possible leading gap.

By the use of the de Bruijn diagram, the low order gap theorems can be derived; the procedure is applicable to any rule in any automaton. It is convenient to change the point of view of the theorems slightly. Rather than worry about "gaps increasing," the question could be: What strings of cells *do not* evolve into gap m? This property is just one of many predicates which could be used to select subdiagrams from a de Bruijn diagram. For example, which strings of length three do not evolve into zero? Or, which strings of length four do not evolve into 00?

Since a small gap is contained in a larger one (gaps without worrying about boundaries), it must mean that only strictly smaller gaps can occur when a given gap is excluded from the next generation. Rings from which such gaps are excluded can only be made up from surviving loops in the de Bruijn diagram, and hence are even limited in their maximum length.

The simple exclusion of gaps from the second generation does not mean that the gaps were not present in the first generation, or that they might not appear in the third or subsequent generation. So, drop them from the first generation as well, leaving a finite set of candidates whose periods must be determined and which must be followed through their full cycle of evolution to see whether overly large gaps occur at some stage.

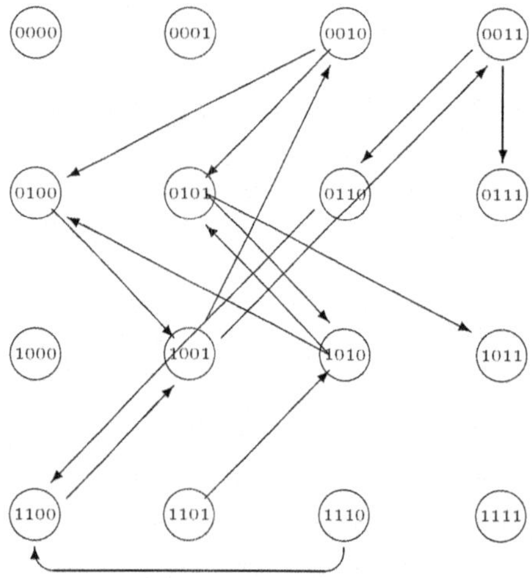

(2,2) de Bruijn diagram

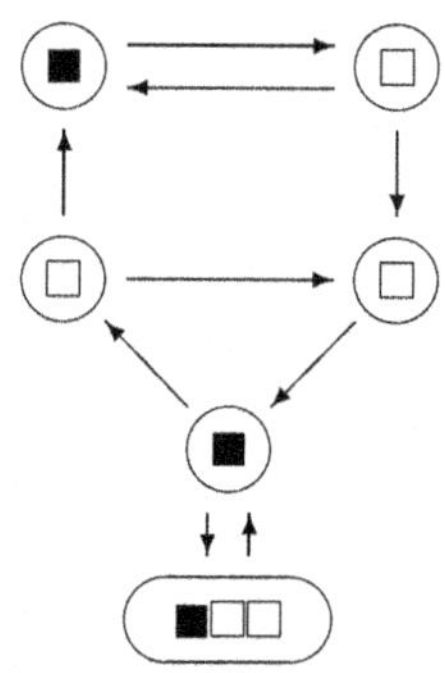

consolidated version

Figure 4.8: (2,2) de Bruijn diagram excluding 000 in first and second generations.

Figure 4.8 shows the process applied to Rule 22 for gap 3; using the table of Section 3.8 all links are excluded which contain three consecutive zeroes in either the first or the second generation; consolidating the results shows that either an additional generation should have been considered (resulting in a larger starting diagram) or some of the known counterimages of three zeroes should also have been dropped (a rather *ad hoc* procedure).

The elimination of a gap from a sequence is similar to the construction of Cantor's set, wherein the digit '1' is excluded from triadic decimals constructed from the digits $\{0, 1, 2\}$; but clearly additional and fancier exclusions are possible. The difference lies in extending the exclusion to several, and possibly all, generations; the latter requires skill, persistence, and luck.

Carrying a diagram through additional generations generally reduces the number of configurations which are admissible; the triangular bridge lying between the cycles $(01)^*$ and $(0011)^*$ disappears in the next generation, leaving results which are provably applicable to all future generations. The process works well because these cycles lack ancestors other than themselves; but it is another matter to arrive at gap 5 which can be found in all Rule 22's period four configurations, all of which have large numbers of ancestors with small gaps.

Useful results have been obtained for practically important cases with small gaps, but the theorems are slightly more complicated if aperiodic configurations are included: there are some interesting "black hole" states. Also to be borne in mind is the fact that for symmetrical rules such as Rule 22, symmetric images of exceptions are also exceptions, as we have tacitly assumed.

gap	nonincreasing configurations	delay
0	0^*	0
1	$(01)^*$	1
2	$(0011)^*$	3
3	$(0011)^*(0001011)^*$	5
3	$(0001011)^*(1100)^*$	5
3	$(1101000)^{3n}(1100)^{7n}0(0001011)^{3n}$	$21n$
4	$(11110000100)^*$	4
4	$(00100001111)^*$	4
4	$(1101000)^*(1100)^*11(0001011)^*$	

Chapter 5

The Garden of Eden

The de Bruijn diagram is particularly useful for the purpose of extending the properties of a neighborhood in a cellular automaton to the properties of chains of cells because they give a particularly graphical way to deal with the overlapping of neighborhoods. Links in the diagram correspond to neighborhoods, nodes to the intervals by which they overlap, and paths through the diagram either to extended neighborhoods or to the sequence of cells into which they evolve.

The de Bruijn diagram differs from the evolutionary diagram in one important respect - it consists wholly of cycles and there are as many links leading into a node as emerging from it; a number which is uniform for each node. Of course this fine balance may be upset when a subdiagram of the de Bruijn diagram is chosen, and one of the important problems may consist in locating the loops which have survived the pruning which created the subdiagram.

5.1 The subset construction

There is another way in which an imbalance may occur. Let us suppose that each link is labelled by the state into which its neighborhood evolves. A path made of such links has evolved from some state of the automaton, namely the one formed by reading off the sequence of overlapping neighborhoods rather than the sequence of evolved cells. We may invert this process. Suppose that we are given an arbitrary sequence of states. If we can locate it somewhere in the diagram, then we have found at least one extended neighborhood which is its ancestor. Sometimes we can locate the sequence along several paths; but it might also occur that such a path is nowhere to be found.

Although the number of links emanating from each node is the same, it is up to the particular rule what the mixture of labels will be. For example, Rule 0 would show a label of zero on *all* links, with a one nowhere to be found. Which is correct; it is a rule under which no ones can evolve. It is a dangerous situation if there is an imbalance at a certain node, so that some state does not label a link there and some other state labels more than one link. If one is ever obliged to use that node in forming a chain of evolved cells, the combination in which the missing cell is the next one required will be impossible. It may or may not happen that the desired sequence can be can be located somewhere else in the diagram, so we need an accurate determination of just which sequences can be folded up and fitted into the diagram.

Moore's subset construction [166] fulfills this requirement very nicely. The basic idea is that if we were systematic, we would start with the first cell of our evolved sequence and look through the diagram taking note of all the links matching that cell. They would emanate from a subset of nodes, so we could make a new diagram in which that subset was a node. The nodes to which the acceptable links lead form a new subset, which we write down and connect to the first subset by a single link with the common label. Turning to the second cell of the evolved sequence, we would examine only those nodes in the second subset node to see which new subset we might form and link with the second cell as label. It would be pointless to examine more nodes in the de Bruijn diagram, because they wouldn't even make the first cell in the evolution come out right.

The final result would be a network of subsets linked and labelled with the same labels used in the de Bruijn diagram. There is no reason not to complete the diagram showing other subsets and other links, but there are exponentially many subsets and they may not all be required if the ancestor of only one chain is sought. The difference is that the nodes and links would now cover all the possibilities in a systematic fashion rather than after the alternative laborious search. In effect the search has already been made, but in an orderly way. Note that in the complete subset diagram, every node has exactly one emerging link for each of the k states of an individual cell; thus one can never reach an impasse seeking a route through the subset diagram.

What one may find instead is that one has arrived at the empty set, which is simply a polite way of saying that the search has been fruitless.

The subset diagram serves a dual purpose. By avoiding the empty set, it is possible to obtain all the possible configurations of the second generation; since the diagram is finite and sequences of cells can be aribritrarily long, the diagram necessarily contains loops, just as does the de Bruijn diagram. By seeking out the empty set, the class of sequences is found which do not have ancestors, and which therefore belong to the Garden of Eden. By examining the final approaches to the empty set, but mainly by rejecting loops, sequences may be found which spoil any other sequence in which they appear, and thus are the basic excluded words for their rule.

5.2 Excluded states for Rule 22

As an example, consider Rule 22. For brevity, label the nodes A, C, D, F. They are linked by zeroes and ones according to the following table:

node	0 leads to	1 leads to
A	A	C
C	D	F
D	C	A
F	D, F	(*none*)

The links are not evenly distributed between zeroes and ones; indeed there are five zeroes and three ones so there just couldn't be a balance. In this case the discrepancy shows up at node F, which will surely cause trouble.

Figure 5.1 shows the connections between the subsets of the initial subset {A C D F}, and the resulting subset diagram. Note that the penultimate row, containing the unit classes, almost reflects the de Bruijn diagram, except where the imbalance in links results in a link descending to the empty set or rising to a higher level (which can only be to the row of doublets, for a binary rule).

subset	0 leads to	1 leads to
$\{ACDF\}$	$\{ACDF\}$	$\{ACD\}$
$\{ACD\}$	$\{ACF\}$	$\{ACD\}$
$\{ACF\}$	$\{ADF\}$	$\{CD\}$
$\{ADF\}$	$\{ACDF\}$	$\{AC\}$
$\{CD\}$	$\{CF\}$	$\{AD\}$
$\{AC\}$	$\{AF\}$	$\{CD\}$
$\{CF\}$	$\{DF\}$	$\{D\}$
$\{AD\}$	$\{AC\}$	$\{AC\}$
$\{AF\}$	$\{ADF\}$	$\{C\}$
$\{DF\}$	$\{CDF\}$	$\{A\}$
$\{D\}$	$\{C\}$	$\{A\}$
$\{C\}$	$\{F\}$	$\{D\}$
$\{CDF\}$	$\{CDF\}$	$\{AD\}$
$\{A\}$	$\{A\}$	$\{C\}$
$\{F\}$	$\{DF\}$	$\{\}$
$\{\}$	$\{\}$	$\{\}$

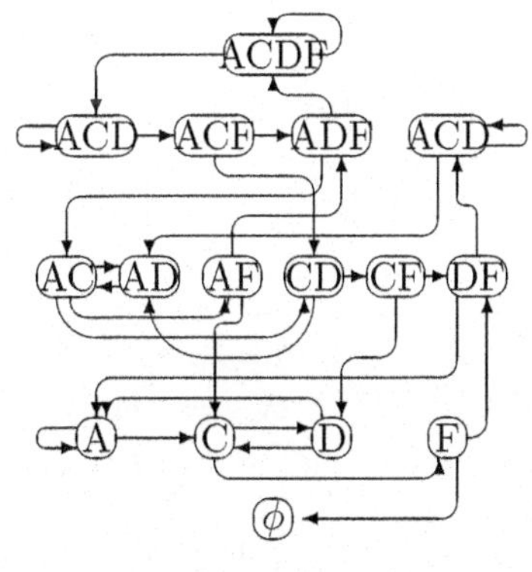

Figure 5.1: The subset diagram (exit links) for Rule 22.

It seems that all sixteen subsets have to be used to complete the analysis. It also appears that both 10101001 and its reverse, 10010101, lead directly to the null set, so they are both excluded words; the shortest ones that can be found for Rule 22.

5.3 Symbolic equations

Diagrams can be represented by connection matrices, through which many properties of the diagram can be deduced more readily than from the diagram itself. There is another representation, in terms of regular expressions, which is often useful. To see how this works, let us try to express the diagram through a set of symbolic equations. Let nodes be represented by capital letters, links by small letters. Then the two-stage de Bruijn diagram for Rule 22 can be written in the form

$$\begin{aligned} A &= A0 + D1 + \lambda \\ C &= A1 + D0 + \lambda \\ D &= C1 + F0 + \lambda \\ F &= C0 + F0 + \lambda \end{aligned}$$

In such an equation, "products" represent concatination and are usually written without any explicit operator sign, and "sums" represent alternative choices and may be read as "or." The symbol λ represents a null string and acts like a one in products. It is included in the equations to signify that one can get to a node by doing nothing; that is, by following a null link.

A system of equations is "solved" in much the same way that a system of algebraic equations would be solved. The distributive law holds, and any variable (the nodes) may be substituted by an equivalent expression. The process of elimination would eventually leave a single variable defined in terms of constants, following which the whole chain of substitutions could be unravelled to obtain the values of all the variables.

There is just one technical point, however, which concerns the procedure to be followed when the same variable occurs on both sides of an equation, as in

$$X = Xa + b.$$

An attempt to write

$$X(\lambda - a) = b$$

and thence

$$\begin{aligned} X &= b(\lambda - a)^{-1} \\ &= b(\lambda + a + aa + aaa + \ldots) \end{aligned}$$

is very suggestive but hardly justified in a system which lacks any trace of subtraction, division, infinite series, convergence, and the like.

The correct procedure is to set about a series of substitutions

$$\begin{aligned} X &= Xa + b \\ &= (Xa + b)a + b \\ &= ((Xa + b)a + b)a + b \\ &\ldots \end{aligned}$$

and observe the general form of the result

$$X = Xa \ldots aaa + b(\ldots aaa + aa + a + \lambda).$$

It is the first term which refuses to disappear; in the context of arithmetic, multiplication, and a numerical a less than 1, it could be argued that its magnitude eventually becomes arbitrarily small and hence it can safely be neglected. Here the argument might run along the lines, that if the symbolic a is not the null chain, λ, then the first term becomes arbitrarily long and can safely be ignored if we are seeking *finite* solutions to the symbolic equation.

The case $a = \lambda$ is special, because it really doesn't define anything; any value of X whatsoever, including b, will work. On the other hand, if b is empty and a is not null, the solution must be empty because otherwise we would be requiring X to be equal to something strictly longer than itself. That, however, is a different matter.

5.4 Arden's lemma

If this point of view is adopted, then we have three rules of inference.

- two expressions equal to a third are equal to each other
- an expression may be substituted for an equal expression
- the solution to $X = Xa + b, a \neq \lambda$ is $X = ba^*$

The notation a^* stands for the continued alternative $\lambda + a + aa + aaa + \ldots$, and is read "a star." Any symbolic expression constructed by the aid of concatination, alternative selection, and the star operation, is called a *regular expression.* Regular expressions are ideally suited to describing paths through a diagram. and conversely, any regular expression has a diagrammatic representation. Conway's monograph [48] is a concise reference for regular expressions and their properties.

If we apply these principles to the de Bruijn diagram for Rule 22 above, we find the following chain of substitutions

$$\begin{aligned} A &= (D1 + \lambda)0^* \\ F &= (C0 + \lambda)0^* \\ C &= D(10^*1 + 0) + 0^*1 + \lambda \\ D &= C(1 + 00^*0) + 0^*0 + \lambda \\ C &= C(1 + 00^*0)(0 + 10^*1) + (00^* + \lambda)(10^*1 + 0) + \lambda)^*1 + \lambda \end{aligned}$$

Finally we have

$$C = ((1 + 00^*0)(0 + 10^*1)) * ((00^* + \lambda)(10^*1 + 0) + \lambda)^*1 + \lambda)$$

of which the leading term is significant and the remainder represent transients.

Although we have set up the equations for the evolution of one generation of a $(2,1)$ automaton for the special case of Rule 22, in fact the equations have the same form for all rules. Given the general evolution table

111	110	101	100	011	010	001	000
h	g	f	e	d	c	b	a

the equations are, in symbolic form,

$$\begin{aligned} A &= Aa + Dh + \lambda \\ C &= Ab + Dd + \lambda \\ D &= Cc + Fg + \lambda \\ F &= Ce + Ff + \lambda \end{aligned}$$

with the general solution

$$\begin{aligned}
A &= Dha^* + a^* \\
F &= Cef^* + f^* \\
C &= ((f^*g + \lambda)(d + ha^*b) + a^*b + \lambda)((c + ef^*g)(d + ha^*b))^* \\
D &= ((a^*b + \lambda)(c + ef^*g) + f^*g + \lambda)((d + ha^*b)(c + ef^*g))^*
\end{aligned}$$

For symmetric rules we would have $b = h$ and $e = g$. If 0 were a quiescent state, we would have $a = 0$.

For infinite or cyclic chains the interesting part of these solutions would be the starred right hand expression common to the four nodes. In essence the remainder of the expression tells how to get to the designated node from any other, while the main term tells how to keep returning to that node. For purposes of reference we could tabulate the results for all the symmetric, quiescent rules for $(2,1)$ automata. These are just Wolfram's "legal" rules.

5.5 The use of symbolic equations

Unfortunately the scheme which solves systems of symbolic linear equations tends to multiply the length of the solution by a constant factor, equal to the number of alternatives, for each node that has to be resolved by the star formula. Since the symbolic coefficients do not tend to simplify in general, symbolic solutions become impractical for all but the very smallest diagrams. We might also note that the table above describes a considerable variety of automata, and yet they are uniformly described by formulae of the same general appearance. Many of them can be further simplified, in spite of the fact that most cannot. For example the expression describing Rule 0 transforms into 000^*; the descriptor of Rule 4, for which 11 is an excluded word, simplifies to $(100^* + 0000^*)^*$.

Some additional remarks ought to be made concerning symbolic equations. The equations which we have shown are *entrance equations*, in the sense that $C = A1 + D0 + \lambda$ says that you enter node C via link 1 from node A, via link 0 from node D or you do nothing. But it is also true that $C = 1D + 0F + \lambda$, wherein you leave node C via link 1 and continue from D, or else leave via link 0 and continue from F. The complete set of conditions would constitute a description of the de Bruijn diagram via *exit equations,* and can be equally solved by regular expressions.

Using the same example as before, the exit equations for Rule 22 would take the form

$$\begin{aligned}
A &= 0A + 1C + \lambda \\
C &= 1D + 0F + \lambda \\
D &= 1A + 0C + \lambda \\
F &= 0D + 0F + \lambda
\end{aligned}$$

For many purposes, the entrance equations and the exit equations of a diagram give equivalent descriptions. However, the path running between the full set and the empty set which corresponds to the excluded words for a rule ought to be be defined by entrance equations rather than exit equations. After all, interest lies in how an impasse is arrived at, not in what to do once it has occurred.

We have not explained why λ has been incorporated as an alternative in each equation. Were it omitted, the equations would be shorter and thereby easier to solve. However the final equation for Rule 22 would then read $C = C(1+00^*0)(0+10^*1)$ whose solution according to Arden's lemma would be the empty set—that is, no solution. After all, the equation requires that C be equal to something strictly longer than itself. Nevertheless the "infinite recursion" $C = ((1+00^*0)(0+10^*1))^*$ is just the part of the solution we extracted to represent a cyclic or infinite state for this automaton.

There is evidently a philosophical issue involved, concerning the differences between our conceptions of finite and infinite. However, it is not a particularly difficult philosophy. Had we been more selective in the introduction of the λ's, we would have paid more attention to the fact that they represent a starting point or a stopping point, according to the handedness of the equations. By including a λ in each of the entrance equations, we have implied that any node in the diagram could be a starting point, and thus we should be prepared to see that the cyclic part of the equation for each node is preceded by the transients leading into it from the other nodes.

If we are looking for finite paths using entrance equations, only the initial nodes should be provided with λ's and the equations should be solved for the terminal nodes; not only will the equations be simpler, but the results will be more accurate. For exit equations, the situation is reversed; the terminal nodes get their λ's in preparation for solving for the initial states.

If we are genuinely interested in infinite solutions, we can entertain alternatives to Arden's lemma, but we should beware that the result may depend upon the node which we have chosen for the final equation; it will only express a solution involving that node. It often happens that a diagram contains a loop which makes a transition to a second loop, but that there is no return path to the first loop. In de Bruijn diagrams for cellular automata, such a configuration corresponds to one of Wainwright's fuses. If the equations are solved for a node in the first loop, the existence of the second loop will not be evident.

The algebra of regular expressions is not as direct as one would like, and often questions involving regular expressions are most easily resolved by constructing an equivalent diagram, transforming the diagram, and converting the result back into a regular expression. For example, regular expressions form a Boolean algebra, but the definition of a regular expression only involves unions—the regular expression sum. Intersections, which represent words conforming to several regular expression descriptions, and complements, which are words which in no way conform to the description, are readily defined in terms of diagrams, but not by operations directly on the regular expression itself. For example, excluded words are obtained through the subset construction, even though they are then readily described by regular expressions.

A related problem is the one of determining the equality of two regular expressions, and is the reason that there is no canonical form for regular expressions.

Equality can be determined by constructing equivalent diagrams, reducing them to their simplest form and comparing the results; however there is no unique regular expression corresponding to any given diagram.

Another result for which there is no simple calculus is to obtain the regular expression which results from the evolution through one or more generations of a given regular expression. The difficulty clearly lies in accounting for all the overlapping which has to be taken into account where sums are involved. It is easier to form the composite rule corresponding to the desired evolution and solve the equations converting the de Bruijn diagram of the composite rule into a regular expression.

Although each generation of evolution of an automaton can be described by a regular expression, the tendency is for the expression to become more and more complicated with each increasing generation. Likewise the description of the words excluded from each generation becomes increasingly complex as the exclusions from the previous generation have to be taken into account. One may naturally wonder what kinds of limit there may be to all this increasing complexity, and indeed whether the limit may be in some cases simpler than all the confusion which preceded it.

Generally speaking, the limit of a sequence of regular expressions need not be a regular expression. One has only to recall that balanced parentheses are not described by a regular expression, although parentheses nested to any finite depth can be listed explicitly and thus described by a regular expression. Here, then, is a limit which is conceptually far simpler than its approximations, but for which there are approximations more than adequate for practical purposes.

There is something qualitatively different between keeping a huge list of acceptable parenthesis sequences to check parentheses, and a huge list of binary numbers to check the regular expression $(0+1)^*$; the latter is simply unnecessary because a simpler procedure suffices. In our present situation we might find that we have to let a cellular automaton run for a very long time to find out something that another approach might discover much more quickly. Nevertheless, it is still interesting to know what the possibilities are, or when such a situation might be occurring.

Certain transformations can be carried out on cellular automata themselves which can be used to clarify their limiting behaviour. These transformations mostly go in the direction of showing how cellular automata can be used to simulate Turing machines, from whence many known results concerning computability and decidability can be passed over to cellular automata. These results in turn affect the possible types of limiting behaviour which may be encountered.

5.6 Systems of symbolic equations

In Conway's monograph [48] it is shown that the premise of Arden's lemma, that the symbolic equation $X = aX + b$ can be solved by $X = a^*b$, is just as applicable to systems of equations if they are written in matrix notation. Thus if we define the following vectors and matrices

$$\mathbf{X} = \begin{bmatrix} P \\ Q \end{bmatrix}, \quad \mathbf{b} = \begin{bmatrix} u \\ v \end{bmatrix}$$

$$\mathbf{a} = \begin{bmatrix} a & b \\ c & d \end{bmatrix},$$

we can write one single matrix equation

$$\mathbf{X} = \mathbf{aX} + \mathbf{b}$$

whose solution would be

$$\mathbf{X} = \mathbf{a}^*\mathbf{b}.$$

Of course, such a solution is not of much value unless there is an effective way to calculate the "star" of a matrix. Fortunately the matrix elements of the star are simple regular expressions of the elements of the matrix being starred, expressing in a concise form the results of the chain of substitutions that would otherwise be carried out explicitly each time a system of symbolic equations was to be solved. Indeed, the easiest way to establish the point is to go ahead and solve the equations: Given

$$\begin{aligned} P &= aP + bQ + u \\ Q &= cP + dQ + v \end{aligned}$$

we deduce in succession

$$\begin{aligned} P &= a^*bQ + a^*u \\ Q &= d^*cP + d^*v \\ P &= a^*bd^*cP + a^*bd^*v + a^*u \\ &= (a^*bd^*c)^*(a^*bd^*v + a^*u) \\ Q &= d^*ca^*bQ + d^*ca^*u + d^*v \\ &= (d^*ca^*b)^*(d^*ca^*u + d^*v) \end{aligned}$$

Disentangling u and v from the final equations for P and Q yields a formula for $\mathbf{a}^*$

$$\mathbf{a}^* = \begin{bmatrix} (a^*bd^*c)^*a^* & (a^*bd^*c)^*a^*bd^* \\ (d^*ca^*b)^*d^*ca^* & (d^*ca^*b)^*d^* \end{bmatrix}.$$

Although these results have been stated for a 2×2 matrix, the fact that the matrix elements can be matrices themselves means that the results are valid in terms of submatrices. By repeated partitioning, in principle a system of any size can be reached. In practice, each partition multiplies the complexity of the symbolic expressions involved, quickly reaching unmanageable proportions. Alternative forms for these matrix elements can be derived using some of the identities satisfied by regular expressions, but the difficulty is a fundamental one—the formulas required are intrinsically complicated.

5.7 Ancestorless states for Rule 18

Rule 18, whose statistical properties have been extensively studied by Grassberger, is very similar to Rule 22, except for the fact that single cells die out rather than remaining to affect further evolution. The consequence is that isolated cells form growth clusters whose interiors die out, just as for Rule 22. However, the fate of colliding frontiers from two different clusters depends on the parity with which they collide, and as a result the evolution of Rule 18 is largely characterized by the conflict of neighboring domains of opposite parity. Occasionally a domain will shrink and disappear, allowing its two neighbors to coalesce. Grassberger was interested in the random walk involved.

The following script shows the subset machine for Rule 18.

```
(2,1) Rule #18
Like LIFE but single cells die
(00*(0+10*1))*

Transitions from original machine

Node A: Link 0:(A ) Link 1:(C )
Node C: Link 0:(D F ) Link 1:(% )
Node D: Link 0:(C ) Link 1:(A )
Node F: Link 0:(D F ) Link 1:(% )

Transitions from subset machine

Node ADF: Link 0:(ACDF ) Link 1:(AC )
Node CDF: Link 0:(CDF ) Link 1:(A )
Node A: Link 0:(A ) Link 1:(C )
Node DF: Link 0:(CDF ) Link 1:(A )
Node : Link 0:( ) Link 1:( )
Node C: Link 0:(DF ) Link 1:( )
Node AC: Link 0:(ADF ) Link 1:(C )
Node ACDF: Link 0:(ACDF ) Link 1:(AC )
```

The next script shows the symbolic equation set of the subset machine for Rule 18. (The nodes have been relabelled according to $a = \{ACDF\}$, $b = \{AC\}$, $c = \{ADF\}$, $d = \{C\}$, $e = \{DF\}$, $f = \{CDF\}$, $g = \{A\}$, $h = \{\}$)

```
a=a.0+c.0+%
b=a.1+c.1
c=b.0
d=b.1+g.1
```

```
e=d.0
f=e.0+f.0
g=e.1+f.1+g.0
h=d.1
```

Since there are relatively few equations, we can write their symbolic coefficient matrix fairly easily. By partitioning it into submatrices certain characteristics of the equations are emphasized; namely that certain groups of nodes are selfcontained. Links lead into the group but there are no links leading out again. Figure 5.2 shows the coefficient matrix (assuming the nodes to form a column vector) and the part of the subset diagram derived from the full subset.

	a	b	c	d	e	f	g	h
a	0	1	.	.	.	.	.	.
b	.	.	0	1	.	.	.	.
c	0	1	.	.	.	.	.	.
d	.	.	.	.	0	.	.	1
e	.	.	.	.	.	0	1	.
f	.	.	.	.	.	0	1	.
g	.	.	.	1	.	.	0	.
h	.	.	.	.	.	.	.	.

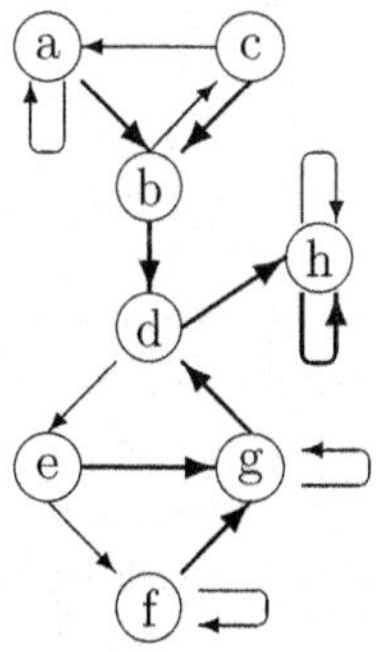

Figure 5.2: Subset diagram for Rule 18.

The solutions for these equations can be obtained in a way which emphasizes the triangular nature of the coefficient matrix:

```
b=(0*1)(000*1+01)*
d=b.1((01+000*1)0*1)*
h=d.1
g=d.(01+000*1)0*
f=d.000*
e=d.0
a=b.000*+0*
c=b.0
```

The coefficients of b and d can be simplified slightly. The new coefficient $0^*1(00^*1)^*$ means that (reading from left to right) after meeting the first 1, we can remain on the highest level only if 1's do not come in joined pairs. Having met the first such pair, we may now remain on the second level as long as the sequence 00^*10^*1 repeats itself, which it will do unless *three* 1's come along in sequence, or else a pair of pairs enclosing a single isolated 1. Descent to level three ensues, finally leaving a sequence without an ancestor. Thus three ones in sequence is a sure mark of a Garden of Eden sequence, but the other condition is much less specific, as it does not refer to some sequence of fixed length.

5.8 Factors

Regular expressions form an algebraic system in the same way that many other structures, such as groups, semigroups, rings and fields; they probably resemble semigroups more than any of the others. Consequently there are standard aspects of structure theory which may be brought to bear, including the study of equivalence relations, order relations, and mappings which are consistent with the regular operations of sum, concatination and star. Conway's monograph touches on all these points, without explicitly mentioning them. His theory of "factors" is related to the decomposition of an algebraic structure into ideals, with implications for the establishment of a canonical form for regular expressions.

In the case of Garden of Eden configurations, it is noticed that the mere presence of certain strings in a sequence of cells will prevent the whole sequence from having an ancestor; for instance 10101001 is sufficient to block the formation of ancestors for Rule 22. Tracing back from the empty set in the subset diagram, it is seen that all the strings lacking ancestors have to end with 01. A question which can be asked is whether the lack of ancestors is always a local matter, for which definite strings can be assigned responsibility, or whether difficulties can build up more slowly, with the possibility of being reversed before they become inescapable. The presence of loops in the paths leading from the universal set to the empty set allow for a wide variety of ancestorless chains, included among which there is a finite collection of loopless paths, whose maximum length is (exponentially) bounded.

Rule	a	b	c	d	e	f	g	h	$((c+ef^*e)(d+h0^*h))^*$
0	0	0	0	0	0	0	0	0	$((0+00^*0)(0+00^*0))^*$
4	0	0	1	0	0	0	0	0	$((1+00^*0)(0+00^*0))^*$
18	0	1	0	0	0	0	0	1	$((0+00^*0)(0+10^*1))^*$
22	0	1	1	0	0	0	0	1	$((1+00^*0)(0+10^*1))^*$
32	0	0	0	1	0	0	0	0	$((0+00^*0)(1+00^*0))^*$
36	0	0	1	1	0	0	0	0	$((1+00^*0)(1+00^*0))^*$
50	0	1	0	1	0	0	0	1	$((0+00^*0)(1+10^*1))^*$
54	0	1	1	1	0	0	0	1	$((1+00^*0)(1+10^*1))^*$
72	0	0	0	0	1	0	1	0	$((0+10^*1)(0+00^*0))^*$
76	0	0	1	0	1	0	1	0	$((1+10^*1)(0+00^*0))^*$
90	0	1	0	0	1	0	1	1	$((0+10^*1)(0+10^*1))^*$
94	0	1	1	0	1	0	1	1	$((1+10^*1)(0+10^*1))^*$
104	0	0	0	1	1	0	1	0	$((0+10^*1)(1+00^*0))^*$
108	0	0	1	1	1	0	1	0	$((1+10^*1)(1+00^*0))^*$
122	0	1	0	1	1	0	1	1	$((0+10^*1)(1+10^*1))^*$
126	0	1	1	1	1	0	1	1	$((1+10^*1)(1+10^*1))^*$
128	0	0	0	0	0	1	0	0	$((0+01^*0)(0+00^*0))^*$
132	0	0	1	0	0	1	0	0	$((1+01^*0)(0+00^*0))^*$
146	0	1	0	0	0	1	0	1	$((0+01^*0)(0+10^*1))^*$
150	0	1	1	0	0	1	0	1	$((1+01^*0)(0+10^*1))^*$
160	0	0	0	1	0	1	0	0	$((0+01^*0)(1+00^*0))^*$
164	0	0	1	1	0	1	0	0	$((1+01^*0)(1+00^*0))^*$
178	0	1	0	1	0	1	0	1	$((0+01^*0)(1+10^*1))^*$
182	0	1	1	1	0	1	0	1	$((1+01^*0)(1+10^*1))^*$
200	0	0	0	0	1	1	1	0	$((0+10^*1)(0+00^*0))^*$
204	0	0	1	0	1	1	1	0	$((1+10^*1)(0+00^*0))^*$
218	0	1	0	0	1	1	1	1	$((0+10^*1)(0+10^*1))^*$
222	0	1	1	0	1	1	1	1	$((1+10^*1)(0+10^*1))^*$
232	0	0	0	1	1	1	1	0	$((0+0f^*1)(1+00^*0))^*$
236	0	0	1	1	1	1	1	0	$((1+0f^*1)(1+00^*0))^*$
250	0	1	0	1	1	1	1	1	$((0+10^*1)(1+10^*1))^*$
254	0	1	1	1	1	1	1	1	$((1+10^*1)(1+10^*1))^*$

Table 5.1: Paths through the de Bruijn diagrams of Wolfram's "legal" Rules.

Chapter 6

The calculus of regular expressions

After having watched the evolution of a variety of cellular automata and developed a certain amount of theory regarding their behaviour, it is not only possible to begin to predict the behaviour of a previously unfamiliar automaton, but one also begins to feel that certain desired effects could be obtained if only certain changes were made in the rules of evolution. This is the beginning of automaton engineering, wherein automata are designed to order; at least within the limits of certain constraints. To do this job well, it is necessary to continue analyzing the structure of automata, particularly along the lines of formal structure theory, possible mappings between automata, and the possible embedment of one automaton in another.

6.1 Derivatives

Regular expressions are the natural formalism arising from an attempt to devise solutions to the symbolic equations encountered in trying to trace out all the possible paths through a network. They comprise one of three productive viewpoints; the other two being, first, the diagrammatic or graphical representation of the network, and second, a relatively numerical description of the network via its connectivity matrix.

The symbolic equations defining regular expressions are soon seen to be the simplest examples of a whole hierarchy of symbolic equations, both implicit and explicit, with various restrictions on the precise form which the system is permitted to assume. Just because they lie at the bottom of the hierarchy, it is worthwhile to make an exhaustive study of their properties, which will surely be encountered again and again in any more comprehensive theory.

Having obtained regular expressions as the description of the paths through a diagram, the converse question arises as to whether there is a diagram for an arbitrarily given regular expression. If such there were, it would be a diagram relating how one regular expression differed from another by its initial letter, in the case the regular expression were solving exit equations, or by its terminal letter, in the case of solutions to entrance equations. Since the two cases are symmetric, it suffices to study one of them—which might as well the case of prefixes.

One of the charming aspects of this theory is its formal resemblance to the differential calculus of analysis. J. Brzozowski developed this point of view, explicitly incorporating the notation of partial derivatives, in an article [36] published in the *Journal* of the Association for Computing Machinery in 1961. The basic concept is the following: Let $\Sigma = \{a, b, c, \ldots\}$ be a set of letters, Σ^* be the set of all finite sequences of letters from Σ, and Ω be some other subset of Σ. Perhaps it could be the set of words represented by a given regular expression, but this is not necessary for the definition.

The derivative of Ω with respect to a is defined to be the set

$$\frac{\partial \Omega}{\partial a} = \{\omega \in \Sigma^* | a\omega \in \Omega\}$$

In other words, it is the set of tails of the words in Ω which begin with the letter a.

Sets are the derivatives of sets; since regular expressions define sets it remains only to verify that their derivatives are also described by regular expressions, to be able to talk directly about the derivatives of regular expressions. The easiest way to attack the problem is to resort to the axiomatic definition of regular expressions, thereby obtaining a list of rules defining the derivatives of composites in terms of the derivatives of their constituents. It is at this point that the formal resemblance to differential calculus becomes apparent, evidently being thereby responsible for the terminology.

A functional notation is convenient; for example $f(a, b, c, \ldots)$ might be a regular (or other) expression built up explicitly from variables $a, b, c, \ldots$, all belonging to Σ, using specific operations or further functions similarly defined. At one point it is required to know whether the null word λ could be represented by f; the easiest way to find out is to evaluate f with zero (empty set, not null word; they're different) arguments. Let us designate this particular value of f by $[f]$:

$$[f(a, b, c, \ldots)] = f(\phi, \phi, \phi, \ldots)$$

Then, given Σ as above, λ null, and ϕ zero, we have:

- definition of a regular expression

  1. λ is a regular expression
  2. ϕ is a regular expression
  3. $a \in \Sigma$ is a regular expression
  4. if x and y are regular expressions, so is xy
  5. if x and y are regular expressions, so is $x + y$
  6. if x is a regular expression, so is x^*

- regular expression whose arguments are zero

  1. $[\lambda] = \lambda$

2. $[\phi] = \phi$
3. $[a] = \phi$
4. $[xy] = [x][y]$
5. $[x + y] = [x] + [y]$
6. $[x^*] = \lambda$

- derivative of a regular expression

1. $\frac{\partial \lambda}{\partial a} = \phi$
2. $\frac{\partial \phi}{\partial a} = \phi$
3. $\frac{\partial a}{\partial a} = \lambda$
4. $\frac{\partial b}{\partial a} = \phi \qquad (a \neq b)$
5. $\frac{\partial xy}{\partial a} = \frac{\partial x}{\partial a} y + [x] \frac{\partial y}{\partial a}$
6. $\frac{\partial (x+y)}{\partial a} = \frac{\partial x}{\partial a} + \frac{\partial y}{\partial a}$
7. $\frac{\partial x^*}{\partial a} = \frac{\partial x}{\partial a} x^*$

Note that $[x]$ is required in the "product" rule because we must devote our attention to the second "factor" if the first is the null word; otherwise the first "factor" is long enough to yield a derivative without regard to what follows it.

- canonical form

$$f(a, b, c, \ldots) = [f] + a\frac{\partial f}{\partial a} + b\frac{\partial f}{\partial b} + c\frac{\partial f}{\partial c} + \cdots$$

The evident way to get a diagram corresponding to a regular expression f is to start with a node labelled f, and to connect it to a node labelled $\frac{\partial f}{\partial a}$ via an outgoing link labelled a, to a node labelled $\frac{\partial f}{\partial b}$ via an outgoing link labelled b, and so on. The canonical form guarantees that we have the correct set of exit equations; the term $[f]$ determines whether we have a terminal node or not.

To incorporate additional nodes into the diagram, it is necessary to take further derivatives, which can be done recursively by applying the definitions once again. Our only real doubt is whether or not there is a finite total number of derivatives, so that the diagram will be finite. Higher derivatives correspond to words which begin with a specified sequence of letters rather than just a single letter; sometimes this is written as a single derivative with respect to the longer sequence.

6.2 Ideals and factors

One of the goals of any theory of the structure of an algebraic system is to decompose the system into standard combinations of standard components; in the case of vector spaces the result is to express any vector as a linear combination of a certain number of basis vectors. However, a vector space represents the ultimate in uniformity; the typical decomposition in other systems may be neither as complete nor as uniform, but there are usually some vestiges of both a basis and of coefficients.

When the system has both a multiplication and an addition, these terms bearing their usual connotations with respect to things like the commutative and the distributive laws, the principal object of interest is the behaviour of products. If there is a subset, such as Ω, typically closed with respect to addition, and an entire set Σ, we may classify products according to which of these two sets contains them, particularly with respect to the source of the factors. The most notorious example of this relationship is the fact that the product of any factor with zero results in zero; the idea is to generalize the properties of zero.

The following table expresses the possible relationships:

left factor	right factor	product	name
$a \in \Sigma$	$w \in \Omega$	$aw \in \Omega$	left ideal
$w \in \Omega$	$a \in \Sigma$	$wa \in \Omega$	right ideal
$x \in \Omega$	$y \in \Omega$	$xy \in \Omega$	subwhatever

Any subset which is simultaneously a left ideal and a right ideal is simply called an ideal; the set consisting entirely of zero is the prototypical ideal, but so also is the entire set Σ.

In the case of regular expressions, Conway has explored these concepts directly, without making any reference either to ideals or to the general theory of semigroups from which the results might possibly be taken. In doing so he exploits a natural order which exists for regular expressions, or any system having the regular expression operators; namely

$$x \leq y. \equiv .\exists z \ni x + z = y$$

zero is clearly a minimum element for this order; the goal is to factorize any regular expression E as a product of other regular expressions, but this may not be possible; thus one must work up to the concept in stages. The first is to define a *trial factorization* through the equation

$$F.G \ldots H \ldots J.K \leq E$$

and then define successively

- another trial factorization $F'.G' \ldots H' \ldots J'K'$ is *dominant* if $F \leq F', G \leq G', \ldots H \leq H', \ldots, J \leq J', K \leq K'$
- any term is *maximal* if it cannot be increased while keeping the product less than E
- a *factorization* is a trial factorization in which every term is maximal
- any term of a factorization is a *factor*, further characterized as *left* or *right* according to whether it occupies that position in some factorization

Amongst others, the canonical form of a regular expression provides trial factorizations whose terms are letters and derivatives. By taking further derivatives, trial factorizations may be developed with arbitrary words as leading terms. The point of starting with trial factorizations is the expectation that they can be promoted to factorizations by systematically increasing each of their terms.

In doing so we depend upon the fact that sums provide the route to upper bounds. Thus if $P \leq E$ and $Q \leq E$, $P+Q \leq E$, follows readily from the definition of inequality; as long as we do not insist that the result necessarily be a regular expression, the same conclusion also holds for arbitrary sums. Consequently the maximal element less than E is simply the sum of *all* the elements less than E.

A product is linear in each of its terms individually, so replacing G by summing the set

$$G' = \{\sigma \in \Sigma^* | F.\sigma \ldots H \ldots J.K \leq E\}$$

gives a new trial factorization dominating the first. Moreover the replacement does not spoil the maximality of any of the other terms; if a maximal term were increased, the distributive law shows that the expression which previously offended the inequality would still be present. Consequently trials can be refined one term at a time until they are finally factorizations.

Without further restriction, a factorization of E need not be unique; the trial factorizations even less so. One such restriction is that all terms but one be maximal; there is only one way the remaining term can be refined, and that must produce the term itself if it were already maximal. The concept of factor, defined as a maximal term, depends upon the environment in which it is encountered, but the environment can be reduced to a product of two terms for either a left or a right factor, or three terms for an interior factor, by using the associative law to consolidate the remaining terms.

Finally it is possible to conclude that, whatever the number of left factors, they are paired with unique right factors; thus they can be given a common index set reflecting this correspondence; let us call them L_i and R_i; they all satisfy $L_iR_i \leq E$, each of which is dominant. It is a temptation to combine them with each other. Thus we further define $E_{i,j}$ as the maximal solution to the trial factorization

$$L_i E_{i,j} R_j \leq E$$

dominating $L_i.\phi.R_j$

Several special cases come to mind. Since $\lambda.E = E$, there must be an index ℓ for which $\lambda \leq L_\ell$ and $R_\ell = E$; likewise an index r for which $\lambda \leq R_r$ and $L_r = E$. Finally, since $L_\ell.E.R_r = E$ we conclude $E_{\ell,r} = E$.

As for the L's and R's, note that by the definition of inequality, $L_iRi+X = E$, so that $L_\ell L_i R_i + L_\ell X = L_\ell E = E$, making $L_\ell L_i R_i$ a trial factorization with maximal terms—a factorization. Consequently $L_i = E_{\ell,i}$, and similarly $R_i = E_{i,r}$. Thus the $E_{i,j}$'s can be arranged into a square matrix, wherein the left factors form a certain row, the right factors a certain column, and E lies at their intersection. The $E_{i,j}$'s in general behave like elementary matrices, satisfying the following rules:

- $\lambda \leq E_{i,i}$ [non-zero diagonal]
- $E_{i,j}E_{j,k} \leq E_{i,k}$ [matrix basis]

- $AB \leq E_{i,k} \Leftrightarrow A \leq E_{i,j} \& B \leq E_{j,k}$
- $F_1 F_2 \ldots F_n \leq E \Leftrightarrow F_1 \leq E_{\ell,i} \& c$

One consequence of these rules is that by extending inequality elementwise to matrices, $E \times E \leq E$, which implies that

$$E^* = E.$$

Interesting as it may be, this discussion has so far been purely theoretical, depending on maximizing through the formation of arbitrary sums. Turning to the canonical representation and symbolic derivatives it is possible to have more concrete results. Note that a right factor of E is a maximal R satisfying $LR \leq E$; in terms of words belonging to L this means

$$R = \bigcap_{w \in L} \frac{\partial E}{\partial w};$$

the pairing between L and R is quite evident from the structure of this formula. By building up the parallel theory of right derivatives, a similar relationship can be created to express L in terms of R. If it is found inconvenient to work with intersections, de Morgan's rules for regular expressions can be invoked to obtain

$$\bar{R} = \sum_{w \in L} \frac{\partial \bar{E}}{\partial w},$$

using the notation that $\bar{X}$ is the complement of X, which may itself require some effort to obtain.

6.3 Rule 18

The generation evolving from the regular expression $(0+1)^*$ according to Rule 18 is $((0+00^*0)(0+10^*1))^*$, which can be simplified slightly to $(00^*(0+10^*1))^*$. If we calculate derivatives strictly according to the rules, simplifying only to remove λ's (% in the following script) and terms beginning with ϕ's (#'s), we obtain for the first few derivatives

```
(00*(0+10*1))*/0= (0*(0+10*1))(00*(0+10*1))*
(00*(0+10*1))*/1= #
(00*(0+10*1))*/00= (0*(0+10*1)+%)(00*(0+10*1))*
(00*(0+10*1))*/01= (0*1)(00*(0+10*1))*
(00*(0+10*1))*/10= #
(00*(0+10*1))*/11= #
(00*(0+10*1))*/000= (0*(0+10*1)+%)(00*(0+10*1))*+(0*(0+10*1))(00*(0+10*1))*
(00*(0+10*1))*/001= (0*1)(00*(0+10*1))*
(00*(0+10*1))*/010= (0*1)(00*(0+10*1))*
(00*(0+10*1))*/011= (00*(0+10*1))*
(00*(0+10*1))*/100= #
(00*(0+10*1))*/101= #
(00*(0+10*1))*/110= #
(00*(0+10*1))*/111= #
```

If we introduce the definitions

$$\begin{aligned} e &= (00^*(0+10^*1))^* \\ a &= 0^*(0+10^*1) \\ b &= 0^*1 \end{aligned}$$

we may draw the conclusion that there are only five distinct derivatives, one of which is zero, and one of which is a sum. Thus the following table is sufficient to deduce all the remaining derivatives.

$$\begin{aligned} \frac{\partial e}{\partial 0} &= ae \\ \frac{\partial e}{\partial 1} &= \phi \\ \frac{\partial ae}{\partial 0} &= (a+\lambda)e \\ \frac{\partial ae}{\partial 1} &= be \\ \frac{\partial be}{\partial 0} &= be \\ \frac{\partial be}{\partial 1} &= e \end{aligned}$$

The five distinct derivatives are also the nodes in a diagram describing the second generation of Rule 18; it is an exit diagram, whose links are shown in Figure 6.1. Comparison with Figure 5.2 shows them to be equivalent, although slightly different.

node	(λ,ϕ)	0 leads to	1 leads to
e	λ	ae	ϕ
ae	ϕ	$(a+\lambda)e$	be
$(a+\lambda)e$	λ	$(a+\lambda)e$	be
be	ϕ	be	e

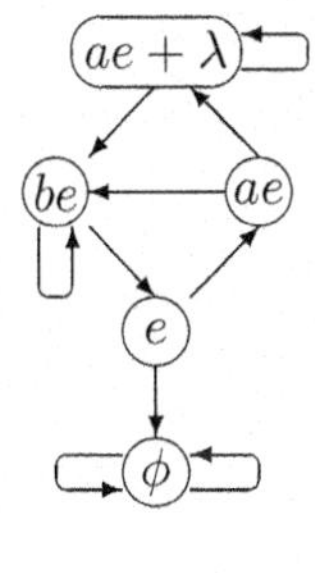

Figure 6.1: Exit diagram for Rule 18.

The right factors will be formed from intersections of these left derivatives.

It is possible to work out the right derivatives of this same regular expression; One might think that the results should be the same because Rule 18 is a symmetrical rule. However, the regular expression describing it is not, and consequently the remainder of the analysis must procede accordingly. We can use the same program to calculate the derivatives, however, just by reversing the order of the factors:

```
((0+10*1)0*0)*/0= (0*0)((0+10*1)0*0)*
```

```
((0+10*1)0*0)*/1= ((0*1)0*0)((0+10*1)0*0)*
((0+10*1)0*0)*/00= (0*0+%)((0+10*1)0*0)*
((0+10*1)0*0)*/01= #
((0+10*1)0*0)*/10= ((0*1)0*0)((0+10*1)0*0)*
((0+10*1)0*0)*/11= (0*0)((0+10*1)0*0)*
((0+10*1)0*0)*/000= (0*0+%)((0+10*1)0*0)**+(0*0)((0+10*1)0*0)*
((0+10*1)0*0)*/001= ((0*1)0*0)((0+10*1)0*0)*
((0+10*1)0*0)*/010= #
((0+10*1)0*0)*/011= #
((0+10*1)0*0)*/100= ((0*1)0*0)((0+10*1)0*0)*
((0+10*1)0*0)*/101= (0*0)((0+10*1)0*0)*
((0+10*1)0*0)*/110= (0*0+%)((0+10*1)0*0)*
((0+10*1)0*0)*/111= #
```

This time we define

$$\begin{aligned} e &= (00^*(0+10^*1))^* \\ f &= 00^* \\ g &= 00^*(10^*), \end{aligned}$$

obtain the derivatives

$$\begin{aligned} \left(\frac{\partial e}{\partial 0}\right)_r &= ef \\ \left(\frac{\partial e}{\partial 1}\right)_r &= eg \\ \left(\frac{\partial ef}{\partial 0}\right)_r &= e(f+\lambda) \\ \left(\frac{\partial ef}{\partial 1}\right)_r &= \phi \\ \left(\frac{\partial eg}{\partial 0}\right)_r &= eg \\ \left(\frac{\partial eg}{\partial 1}\right)_r &= ef \end{aligned}$$

and the entrance diagram

node	(λ, ϕ)	0 comes from	1 comes from
e	λ	ef	eg
ef	ϕ	$e(f+\lambda)$	ϕ
$e(f+\lambda)$	λ	$e(f+\lambda)$	eg
eg	ϕ	eg	ef

While different from the exit diagram, it evidently serves the same purpose. To use an entrance diagram one builds up an expression from right to left reading *backwards* along the arrows until an initial node is reached.

6.4 Factors for Rule 18

Now that we have all the derivatives for the second generation of Rule 18, it is possible to calculate its factors using the representation that they are intersections of the derivatives. The easiest way to calculate an intersection is to represent each intersectand via its diagram, collect corresponding nodes into pairs, and work out the regular expression representing the terminal nodes in this new diagram. A terminal pair is one which is terminal for both of its components, signified by the presence of λ's in their own diagrams.

Accordingly the exit diagram for the intersections of the left derivatives can be gotten from a table of links, which can be written in a compact form if certain preparations are made. Define $x = a + \lambda$; then deduce the links for e and its derivatives by writing them in canonical form, and finally display all the pairwise intersections in a matrix.

	e	ae	xe	be
e	$\begin{pmatrix}\lambda + 0ae + 1\phi \\ \lambda + 0ae + 1\phi\end{pmatrix}$	$\begin{pmatrix}\lambda + 0ae + 1\phi \\ \phi + 0xe + 1be\end{pmatrix}$	$\begin{pmatrix}\lambda + 0ae + 1\phi \\ \lambda + 0xe + 1be\end{pmatrix}$	$\begin{pmatrix}\lambda + 0ae + 1\phi \\ \phi + 0be + 1e\end{pmatrix}$
ae	$\begin{pmatrix}\phi + 0xe + 1be \\ \lambda + 0ae + 1\phi\end{pmatrix}$	$\begin{pmatrix}\phi + 0xe + 1be \\ \phi + 0xe + 1be\end{pmatrix}$	$\begin{pmatrix}\phi + 0xe + 1be \\ \lambda + 0xe + 1be\end{pmatrix}$	$\begin{pmatrix}\phi + 0xe + 1be \\ \phi + 0be + 1e\end{pmatrix}$
xe	$\begin{pmatrix}\lambda + 0xe + 1be \\ \lambda + 0ae + 1\phi\end{pmatrix}$	$\begin{pmatrix}\lambda + 0xe + 1be \\ \phi + 0xe + 1be\end{pmatrix}$	$\begin{pmatrix}\lambda + 0xe + 1be \\ \lambda + 0xe + 1be\end{pmatrix}$	$\begin{pmatrix}\lambda + 0xe + 1be \\ \phi + 0be + 1e\end{pmatrix}$
be	$\begin{pmatrix}\phi + 0be + 1e \\ \lambda + 0ae + 1\phi\end{pmatrix}$	$\begin{pmatrix}\phi + 0be + 1e \\ \phi + 0xe + 1be\end{pmatrix}$	$\begin{pmatrix}\phi + 0be + 1e \\ \lambda + 0xe + 1be\end{pmatrix}$	$\begin{pmatrix}\phi + 0be + 1e \\ \phi + 0be + 1e\end{pmatrix}$

To read this array, suppose that the intersection $ae \wedge be$ is to be calculated. Then from the row ae, column be we see from (ϕ, ϕ) that the node is not terminal, that 0 leads from (ae, be) to (xe, be) and that 1 leads from (xe, be) to (be, e). If further links are traced onward from these, a diagram with eight nodes is obtained, but none of them is a terminal state. Thus this intersection is empty, $ae \wedge be = \phi$, a conclusion which is confirmed by observing that ae always contains an even number of 1's, while the number of 1's in be is always odd. Since the node (be, e) occurs in the same diagram with (ae, e) we can also conclude that that intersection is empty.

It is evident from the definition of xe that $e \leq xe$. The diagram for (ae, e) reads

$$\begin{aligned}
(ae, e) &= (\phi, \lambda) + 0.(xe, ae) + 1.(be, \phi) \\
(xe, ae) &= (\lambda, \lambda) + 0.(xe, xe) + 1.(be, be) \\
(xe, xe) &= (\lambda, \lambda) + 0.(xe, xe) + 1.(be, be) \\
(be, be) &= (\phi, \phi) + 0.(be, be) + 1.(e, e) \\
(e, e) &= (\lambda, \lambda) + \cdots
\end{aligned}$$

These equations can be solved by inspection to obtain

$$\begin{aligned}
ae \wedge e &= 0 + 00xe + 01be \quad (= 0ae), \\
ae \wedge xe &= 0xe + 1be \quad (= ae).
\end{aligned}$$

The only triple intersection not deducible from the foregoing results is

$$e \wedge ae \wedge xe = 00xe + 01be \qquad (= 0ae),$$

finally giving us all the possible right factors. By examining canonical forms for the derivatives these results can be somewhat simplified to the forms shown in parentheses. As was to be expected, they all involve incomplete segments of the basic unit in e. All told, including the null intersection Σ^*, they are seven in number—ϕ, Σ^*, ae, be, xe ($= ae + \lambda$), $0ae$, and e ($= 0ae + \lambda$).

By repeating essentially the same calculation with $y = e(f + \lambda)$, the left factors are found to be ϕ, Σ^*, ef, eg, ey ($= ef + \lambda$), $ef0$, and e ($= ef0 + \lambda$).

In general, the n-fold intersections can be deduced from a table of n-tuples, and in each case a diagram can be made up and solved to determine the regular expression corresponding to each intersection. The only slight doubt that might remain is whether the diagram could be reduced, due to the unsuspected equality of some of its nodes. Sometimes this is evident from inspection; as when two different nodes solve the same equation, or when the expression is fairly short with a clear interpretation. Otherwise it might be necesary to resort to testing the diagram for equivalence classes.

Having developed this theory to a fair degree of elaboration, the question naturally arises as to its applicability. For studying infinite or cyclic systems formed from symmetric rules, it does not seem likely that there will be a useful distinction between left factors and right factors; they seem more related to starting up or ending a finite expression, as the analysis which we have made for Rule 18 shows. They are also likely to be of importance for rules which admit fuses or gliders.

On the other hand, it cannot be denied that the combination of Arden's lemma and the calculus of regular expressions yields a very concise and elegant way to describe diagrams symbolically.

6.5 A geometrical representation

Regular expressions represent sequences of symbols—letters taken from some alphabet. When the sequence is relatively short, the sequence is readily perceived and its properties may be examined. The notation is especially nice for summarizing cycles which can be repeated an arbitrary number of times, since a single star signifies this infinite collection of alternatives. When the regular expression is complicated, and particularly when it subsumes several distinct cycles at the same time, it rapidly becomes more complicated, both to write down completely and to comprehend once it has been written.

Although difficult of typography, the diagram representing a regular expression is preferable for expressions of moderate complexity, although eventually the many crossing lines and increased density of nodes take their toll, and the figure can no longer be comprehended.

Another representation has some advantages, although it is typographically even more difficult to deal with. To begin with, suppose that the symbols are chosen to be consecutive numbers, just as we are accustomed to assuming when dealing with automata. If a decimal point were to be placed in front of a finite sequence, it could be interpreted as a number with a positional representation in base k if there were k symbols running from 0 to $k - 1$. Finite sequences would be finite decimals, infinite sequences could be interpreted after the fashion of real numbers.

The principal difference is that in base 10 .10000... and .09999... are ordinarily considered to be the same real number; but have to be understood to be distinct regular expressions—10^* in the first case and 09^* in the second.

The advantage of this representation is that every regular expression, no matter how complicated, is a point in the interval $0 \leq x \leq 1$, and so is represented in a finite region of space. The counterbalancing penalty is that this region can be filled more densely than is convenient, often too densely for a clear presentation, since the individual points are always fairly bulky and can never be presented visually with the mathematical precision of occupying zero area.

We are interested in regular expressions describing configuations of cellular automata, but these are sequences infinite in both directions. To capture both directions in such a string, break it at some point, then write the left hand side backwards. Inserting a figurative decimal point as before now produces two intervals of unit length; the natural thing is to treat them as two coordinates generating a unit square. The vertical, or y-coordinate represents the left half of the sequence, the horizontal or x-coordinate represents the right half.

Shifting, or changing the decision as to where to break the sequence, will naturally change the point which represents the sequence; in any event it is a geometric operation whose general nature can be described. Indeed this is a fundamental operation on sequences: a right shift splits the unit square into k horizontal strips and reassembles them into k vertical strips after stretching them in one direction and compressing them in the other.

The numerical value associated with each sequence defines a distance, and consequently the distance between two sequences is just the sequence of differences. At this point the difference between this representation and that of the real numbers manifests itself; although different terms are included in the sum in the normal way, there are no carries from one term to another and so there is no confusion that a number may have two different representations as a repeating decimal.

In mathematical terms, suppose that there are two sequences

$$a_{-n}, a_{-n+1} \ldots, a_{-1}, a_0, a_1, \ldots, a_{n-1}, a_n$$

$$b_{-n}, b_{-n+1} \ldots, b_{-1}, b_0, b_1, \ldots, b_{n-1}, b_n$$

their distance is defined as

$$\|\mathbf{a} - \mathbf{b}\| = \sum_{i=-n}^{n} \frac{|a_i - b_i|}{k^{|i|}}.$$

The distance is small, or the sequences are close, when their central regions agree; the closer they are, the longer the matching core must be.

To have an idea of how this representation works, let us consider Rule 18 for a moment. We have seen that the sequence 111, amongst others, is excluded from the second generation. If we divide the interval $0 \leq x \leq 1$ into eighths, any sequence $.xyz\ldots$ beginning with .111 will occupy some position in the last eighth of this interval. This interval will include sequences of the form .1111..., but sequences of the form .0111... will ocuppy the last eighth of the interval $0 \leq x \leq \frac{1}{2}$. Recursively, the last eighth of any binary subdivision of the unit interval is excluded. Similar comments apply to sequences extending to the left, so that as the unit square represents various doubly infinite sequences the excluded sequences will come to resemble a Scotch plaid.

Additionaly it is necessary to consider sequences whose central point lies amongst the three excluded 1's. For example 1.11 will refer to points within the upper half of the square, but in the rightmost quarter; 11.1 to the upper quarter but right half. These regions have to be added to the plaid, as well as the further subdivisions implied by further decimals. Thus the same basic pattern is repeated recursively in each quadrant, 16-ant, and so on of the unit square.

Sequences of three 1's are not the only configuration excluded from the second generation, since it is also impossible to encounter a single 1 in a sequence bounded by pairs of 1's. Taking into account all of the excluded sequences simply gives an even finer structure to the plaid. In general the plaid is a two dimensional version of the classical Cantor set, which is the residue remaining after the middle thirds of the unit interval and surviving subintervals have been removed. In arithmetic terms, the Cantor set is the set of all decimals in the unit interval which contain no 1's when they have been displayed with respect to the base 3; which is one of the simplest possible examples of an excluded sequence.

The plaid can be used as the starting point for further discussions. For example the third generation must occupy a subset of the permitted portion of the second generation plaid, and so on as the automaton evolves through further generations; it is even meaningful to think of a limiting set which consists of just those points, if any, which remain after an infinite number of generations. The limiting set will include all the points representing cycles of whatever period, as well as gliders, fuses, and a variety of structures which may have no finite or cyclical representation.

Metrizing regular expressions, and Σ^* in general, has the advantage of creating a topological space, to which the whole machinery of analysis may be applied. As a compact topological space, even stronger conclusions, such as the necessary existence of certain limits, can often be drawn. An extensive literature concerns the application of these ideas to probability and symbolical dynamical systems; but there is no reason that it cannot be broadened to include regular expressions as well. In particular, the interaction between probability measures and linear cellular automata can be studied through this model, to the mutual enlightenment of both disciplines.

Chapter 7

Probabilistic de Bruijn matrix

The reduced evolutionary matrix and the de Bruijn diagram have probabilistic versions, in which the zeroes and ones which enable links in the diagram are replaced by probabilities that the links are to be used. This does more than express the likelihood that one thing or another will occur; it allows some quantitative comparisons to be made.

Since the reduced evolution matrix enumerates the numbers of $(n+2r)$-block ancestors of n-blocks, the probabilistic evolution matrix can be used estimate the likelihood that the ancestors actually occur, and thus to develop self-consistent estimates for their probabilities. Probabilistic de Bruijn matrices are useful for studying correlations between cells or strings of cells situated at a distance from one another because its matrix elements could describe the probability that one n-block will overlap the next; powers of the matrix would relate blocks through a chain of overlaps.

7.1 Block probabilities

The rules of probability are widely known and frequently employed in a wide variety of contexts; nevertheless it is a probability which assumes such things as independent events and the lack of correlation. Under those conditions probabilities of alternatives are summed and probabilities of coincidences are multiplied. Sometimes it seems reasonable that the justifying hypotheses are valid, other times it is simply assumed so with the hope that the calculations will still be reliable.

When the evolution of cellular automata is treated probabilistically, it does not seem reasonable that the evolution of neighboring cells is independent, inasmuch as they share a common portion of their neighborhood and evolve according to deterministic rules. Nevertheless the rules are complicated, and the overlapping segment of the neighborhoods enters differently into the evolution of each cell, so the coupling between the two evolutions might not be all that strong.

Going ahead with naive probabilistic calculations for cellular automata produces results which are neither very good nor very bad. Tendencies are evident and the general conclusions seem to be reliable, but the results are clearly defective as soon as any accuracy is required.

One way to get better results is to work out the rules for more complicated combinations of probabilities, as is routinely done to get the probabilities for the values of all kinds of mathematical functions in terms of the probability distributions of their arguments and the fundamental arithmetic operations. This approach does not work well for cellular automata because of the complicated way that the rule of evolution would have to be expressed as an arithmetic function of a real variable.

A more tractable approach seems to be to work with probabilities for sets of cells rather than with individual cells; in particular, with the linear sequences of cells from which linear cellular automata are already composed. Thus correlations between cells are seen in the differences in the probabilities of different pairs of cells, which do not necessarily have to be the products of probabilities for the individual cells. Further correlations can be deduced by studying the probabilities of triples of cells, quadruples, and so on. The extreme logical limit to this approach would be to assign probabilities to each and every possible sequence, no matter how long.

Sequences of cells will be called *blocks,* to emphasize their linear arrangement; the term n-block will be used when its exact length is required. Some blocks form parts of others; indeed the best way to describe $(n+1)$-blocks is by systematically extending n-blocks in all possible ways; since they are linear, the extension can be made to the left or to the right. Insertions could be made in the middle, but that will be regarded as a compound extension.

The assignment of probabilities to blocks ought to respect the process of extension. Since there are more extensions than bases, each of them should have a smaller probability than its base; moreover the probability of the base could be expected to be the sum of the probabilities of all its extensions. In this way a constant amount of probability will be subdivided more and more finely as the blocks grow longer.

When formalized, these restraints are called the Kolmogorov consistency conditions. For linear sequences there are two such conditions, according to the handedness of the extension. Thus, if abc were a 3-block, Σ, the collection of cell states, the requirements are

$$\begin{aligned} p(abc) &= \sum_{x\in\Sigma} p(xabc) \\ p(abc) &= \sum_{x\in\Sigma} p(abcx). \end{aligned}$$

By repeated application of the consistency conditions the probabilities of all m-blocks for $m \leq n$ are determined once all the n-block probabilities are known. But because there are *two* consistency conditions, the choice of n-block probabilities is not completely arbitrary. For example, in a binary alphabet, 1 extends to the left to become 01 or 11, to the right to become 10 or 11. This implies that $p(01)+p(11) = p(10)+p(11)$ or in short that $p(01) = p(10)$, which is a restraint on the assignments of probabilities to 2-blocks.

7.2 Kolmogorov conditions in matrix form

Block probabilities for linear chains can be defined and their properties described by an extremely elegant matrix formulation. Binary sequences already illustrate the principles quite clearly, as we can see by writing the equations defining block probabilities up to length two. Generally it is understood that $p(\lambda) = 1$.

$$\begin{aligned} p(\lambda) &= p(0) + p(1) \\ p(0) &= p(00) + p(01) \\ p(1) &= p(10) + p(11) \\ p(0) &= p(00) + p(01) \\ p(00) &= p(000) + p(001) \\ p(01) &= p(010) + p(011) \\ p(10) &= p(100) + p(101) \\ p(11) &= p(110) + p(111) \\ &\ldots \end{aligned}$$

Concentrating on the last group of equations, four 2-block probabilities are written as linear combinations of eight 3-block probabilities:

$$\begin{bmatrix} p(00) \\ p(01) \\ p(10) \\ p(11) \end{bmatrix} = \begin{bmatrix} 1 & 1 & . & . & . & . & . & . \\ . & . & 1 & 1 & . & . & . & . \\ . & . & . & . & 1 & 1 & . & . \\ . & . & . & . & . & . & 1 & 1 \end{bmatrix} \begin{bmatrix} p(000) \\ p(001) \\ p(010) \\ p(011) \\ p(001) \\ p(001) \\ p(001) \\ p(001) \end{bmatrix}.$$

If $U = (1, 1)$ is a row vector, I a unit matrix, the matrix in the above equation can be written as a tensor product, $I \otimes U$.

However, there is another way that the same system of equations can be written, which is:

$$\begin{bmatrix} p(00) \\ p(01) \\ p(10) \\ p(11) \end{bmatrix} = \begin{bmatrix} p(000) & p(001) & . & . & . & . & . & . \\ . & . & p(010) & p(011) & . & . & . & . \\ . & . & . & . & p(100) & p(101) & . & . \\ . & . & . & . & . & . & p(110) & p(111) \end{bmatrix} \begin{bmatrix} 1 \\ 1 \\ 1 \\ 1 \\ 1 \\ 1 \\ 1 \\ 1 \end{bmatrix}.$$

Next, we partition the rectangular matrix of probabilities into two square matrices A and B, and partition the column of eight 1's correspondingly into two identical columns. Then, in terms of submatrices, the right hand side of this equation takes the form $AV + BV = (A + B)V$.

Finally the equations relating 2-blocks to 3-blocks become

$$\begin{bmatrix} p(00) \\ p(01) \\ p(10) \\ p(11) \end{bmatrix} = \begin{bmatrix} p(000) & p(001) & \cdot & \cdot \\ \cdot & \cdot & p(010) & p(011) \\ p(100) & p(101) & \cdot & \cdot \\ \cdot & \cdot & p(110) & p(111) \end{bmatrix} \begin{bmatrix} 1 \\ 1 \\ 1 \\ 1 \end{bmatrix}.$$

In which the matrix $(A + B)$ will be recognized as having the form of a three-stage de Bruijn matrix, but with probabilities as non-zero elements.

The interesting point is that this entire derivation can be repeated, with slight variations, for left extension rather than right extension, to obtain the 2-block probabilities as sums of 3-block probabilities. The same "de Bruijn" matrix will make its appearance if we write the 2-block probabilities as a row vector rather than a column vector:

$$\begin{bmatrix} p(00) & p(01) & p(10) & p(11) \end{bmatrix} = \begin{bmatrix} 1 & 1 & 1 & 1 \end{bmatrix} \begin{bmatrix} p(000) & p(001) & \cdot & \cdot \\ \cdot & \cdot & p(010) & p(011) \\ p(100) & p(101) & \cdot & \cdot \\ \cdot & \cdot & p(110) & p(111) \end{bmatrix}.$$

In this form, the Kolmogorov consistency conditions require that the row sums of the "de Bruijn" matrix match the column sums. The reason that the adjective 'de Bruijn' has been enclosed in quotes is that this is not quite the matrix which we intend to honor by such a description; the term is reserved for a slightly different matrix for which the consistency conditions will be met.

7.3 Probabilistic de Bruijn matrix

A de Bruijn diagram is a basic structure which can be applied in various ways. It describes all the possible ways that two sequences of symbols, or windows, can overlap if they are displaced with respect to each other; just the relationship between the neighborhoods of two successive cells in a linear cellular automaton. Selecting a subset of the diagram by retaining or discarding selected links discloses states of a given period, gliders, and other information. An intermediate approach would assign probabilities to the links rather than selecting them. As an extreme case, a probability of zero would exclude a link, a probability of one would surely include it, giving other values in the same interval intermediate interpretations.

The connectivity matrix of a graph is readily adapted to show the probabilities of the links; one has only to define

$$M_{i,j} = \begin{cases} p(i \to j) & i \to j \\ 0 & \text{otherwise} \end{cases}$$

The rules of matrix multiplication coincide with the rules for compound probabilities if it is desired to calculate the probability of forming chains of several links. Even the de Bruijn matrix itself is readily converted into a probability matrix by replacing its non-zero elements by the value $1/k$, making the choice of all paths equally likely.

Just as probabilities can be associated with the links of a de Bruijn diagram, they also can be associated with the nodes; to conserve the probabilistic interpretation of matrix multiplication the nodal probabilities might be used to form a vector, with matrix equations describing the linkages between nodes. A vector of probabilities would have non-negative real components—with a unit sum, if it happens that all the different components comprise a complete set of alternatives.

Any matrix intended to preserve each and every such column vector must necessarily have columns of unit sum, as can be seen by testing the matrix on unit vectors. Such a matrix is called a *stochastic* matrix. Likewise the preservation of unit row sums would require the matrix to have unit row sums. Sometimes the conditions for rows and columns must be met simultaneously, for which the adjective *doubly stochastic* would be used.

In terms of the de Bruijn diagram, if the incoming probabilities at a node were required to have a unit sum, the probabilistic de Bruijn matrix would have to be column-stochastic. This property would refer to the rows if the outgoing probabilities were constrained. Expressing either constraint in matrix form guarantees a unit eigenvalue, with a probability vector as the equilibrium eigenvector of opposite handedness. These and certain other conclusions characterize stochastic matrices, all of which information can be found in any one of several recent textbooks[89, 21, 206] which can be consulted for descriptions and proofs of the results.

If the links in an n-stage de Bruijn matrix are replaced by their corresponding n-block probabilities, the row or column probabilities do not sum to 1, but rather to the $(n-1)$-block probabilities; this suggests dividing each column by its column sum to get a stochastic matrix; alternatively the rows could be divided by the row sums. Generally these are two distinct choices and result in two different de Bruijn matrices. For right extension, we define

$$\begin{aligned} p(ab \to bc) &= \frac{p(abc)}{\sum_x p(xbc)} \\ &= \frac{p(abc)}{p(bc)}; \end{aligned}$$

for left extension,

$$\begin{aligned} p(ab \leftarrow bc) &= \frac{p(abc)}{\sum_x p(abx)} \\ &= \frac{p(abc)}{p(ab)}. \end{aligned}$$

Among other things, these denomnators compensate for the overlapping between blocks when de Bruijn matrices are multiplied.

A stochastic de Bruijn matrix (of either handedness) can be used to generate a set of n-block probabilities which will satisfy the Kolmogorov consistency conditions. Suppose that the rows are indexed by the $(n-1)$-block sequences, so that writing one of the non-zero matrix elements as $m(axb)$ identifies it as belonging to the axth row and the xbth column. Suppose further that the matrix is column stochastic. Then the row $(1, 1, \ldots, 1)$ is an eigenvector belonging to unit eigenvalue, and we may call $p(xb)$ the components of the corresponding column eigenvector. By rewriting

$$p(ax) = \sum_b m(axb)p(xb)$$

in the form

$$p(ax) = \sum_b (m(axb)p(xb)) \times 1,$$

we see that the elements $p(ax)$ are row sums of a new matrix whose elements are defined implicitly by the equation

$$m(axb) = \frac{p(axb)}{p(xb)}.$$

Since we have only renamed the elements $m(axb)$, the matrix M is still column stochastic; since $p(axb)$ has the same denominator throughout a given column we have

$$1 = \sum_a \frac{p(axb)}{p(xb)};$$

when rewritten in the form

$$p(xb) = \sum_a p(axb)$$

we have the other half of the consistency condition.

7.4 Some properties of n-block probabilities

The stochastic de Bruijn matrices suggest one way to obtain a set of n-block probabilities satisfying the Kolmogorov consistency conditions, at least for $(n-1)$-block probabilities in terms of n-block probabilities. All that is required is to fill in the matrix elements of the de Bruijn matrix with positive numbers, taking care not to exceed, say, the unit column sum. The choice is relatively arbitrary until the selection of the final element of the column, which must be chosen to complete the sum. For k symbols, this means that only k^{n-1} of the k^n non-zero elements lack any freedom of choice; the remaining $(k-1)k^{n-1}$ elements are then parameters to be chosen.

Once this is done, the equilibrium eigenvector is to be determined and each of the columns of the stochastic matrix is to be multiplied by the corresponding component, to produce the n-block probabilities. The components of the eigenvector are the $(n-1)$-block probabilities; the probabilities for smaller blocks follow from summing longer blocks.

It is interesting to observe that once consistency has been established for $(n-1)$-block probabilities relative to n-block probabilities, the consistency of all smaller blocks follows. Starting with n-blocks, the inductive step consists of verifying consistency for $(n-2)$-blocks:

$$\begin{aligned}
p(a) &= \sum_x p(ax) \\
&= \sum_x \sum_y p(axy) \\
&= \sum_x \sum_y p(yax) \\
&= \sum_y \sum_x p(yax) \\
&= \sum_y p(ya),
\end{aligned}$$

and to arrive at successively shorter blocks. In similar fashion it follows that the probability of any block up to length n is simply the sum of the probabilities of all the possible extensions which can be derived by the same route; for example,

$$p(0) = \sum_x \sum_y p(x0y).$$

Given the interrelation between n-blocks and all the shorter blocks, some consideration can be given to using the probabilities of short blocks as parameters. It is not possible to use *only* the probabilities of short blocks as parameters; were it otherwise there would be no need to to build up a theory of block probabilities. Nevertheless it is possible to attempt the maximum usage of the probabilities of short blocks. For example, $p(1)$, $p(2)$, ..., $p(k-1)$ can all be chosen before the value of $p(0)$ is fixed in order to make the complete sum equal to 1—a total of $k-1$ parameters. Or, of a total of k parameters, one–$p(0)$–has been excluded.

Then, among the 2-blocks, $p(x1)$, $p(x2)$, ..., $p(x(k-1))$ can be chosen before $p(i0)$ is needed to complete a sum of $p(x)$. Alternatively, we can consider that $p(x0)$ is excluded as a parameter, as is $p(0x)$ when the left extensions are considered. Going on to 3-blocks, we see that all the probabilities $p(0xy)$ or $p(xy0)$ are excluded. Continuing through n-blocks, the total number of parameters comprising probabilities in which 0 is neither an initial nor a terminal symbol is

$$(k-1) + (k-1)(k-1) + (k-1)k(k-1) + \cdots + (k-1)k^{n-2}(k-1)$$

which can be summed to give the value $(k-1)k^{n-1}$; just the same as the number of parameters required to make a stochastic de Bruijn matrix.

Even when full use is made of the short blocks to define parameters the fraction of parameters which are n-block probabilities is

$$\frac{(k-1)k^{n-2}(k-1)}{(k-1)k^{n-1}},$$

which reduces to $(k-1)/k$; 1/2 for a binary sequence, larger and approaching 100% for others.

7.5 Some simple examples

The generic form of a 2×2 matrix with unit column sum is

$$\begin{bmatrix} a & b \\ 1-a & 1-b \end{bmatrix},$$

whose eigenvalues are

$$\lambda = 1, a - b;$$

whose matrix of left eigenvectors is

$$\begin{bmatrix} 1 & 1 \\ 1-a & b \end{bmatrix},$$

and whose matrix of right eigenvectors is

$$\begin{bmatrix} b/(1+b-a) & 1 \\ (1-a)/(1+b-a) & -1 \end{bmatrix}.$$

The eigenvalue 1, whose eigenvector has components proportional to the off diagonal elements of the matrix, represents an equilibrium probability. The second eigenvalue lies in the range $-1 \leq \lambda \leq 1$, with an eigenvector suitable for probability differences. Indeed it is the factor by which the disequilibrium decreases in each generation; if it is zero equilibrium is reached in one step, otherwise there will be an exponential approach to equilibrium. Depending on the sign, there may be oscillations about equilibrium, or a uniform approach to equilibrium. However, if the second eigenvalue is as large as one, there will either be a degeneracy by which any probability is in equilibrium, or the probabilities for zeroes and ones will be exchanged with each other and alternate forever after.

A doubly stochastic one-stage de Bruijn diagram would have a probability matrix

$$\mathcal{P}_1 = \begin{bmatrix} p & q \\ q & p \end{bmatrix}$$

with $q = 1 - p$, whose eigenvalues would be $\lambda = 1, p - q$; being symmetric, the matrix of both left and right eigenvectors would be

$$\frac{1}{2}\begin{bmatrix} 1 & 1 \\ 1 & -1 \end{bmatrix}.$$

The probability matrix of a two stage diagram is a little more complicated.

$$\mathcal{P}_2 = \begin{bmatrix} p_1 & q_2 & \cdot & \cdot \\ \cdot & \cdot & p_3 & q_4 \\ q_1 & p_2 & \cdot & \cdot \\ \cdot & \cdot & q_3 & p_4 \end{bmatrix}$$

with $q_i = 1 - p_i$, and the characteristic polynomial

$$(\lambda^4 - \lambda^3(p_1 + p_4) - \lambda^2(p_1p_4 - p_2p_3) + \lambda(p_2d_{34} + p_3d_{12}) - d_{12}d_34$$

with two determinants defined by

$$\begin{aligned} d_{12} &= (p_1p_2 - q_1q_2) \\ d_{34} &= (p_3p_4 - q_3q_4). \end{aligned}$$

The determinant of the de Bruijn matrix is the product of two smaller determinants corresponding to the evident blocks in the de Bruijn matrix; in this we have a special case of a quite general result. The p's and q's were defined as they are with the thought of equating the subscript pairs (1,2) and (3,4) to get a doubly stochastic matrix, but the best interpretation of the vanishing of the small determinants is that the biases of the probabilities in their submatrices are equal.

The constant term in the characteristic equation will be zero, giving a single zero root, if either determinant is zero. The coefficient of λ will also be zero if both are zero, producing a double zero root. In order to get three equal roots and thus open up the possibility of the Jordan normal form, requires some similarity between the two submatrices in addition.

7.6 Determinant and inverse

The representations of the de Bruijn matrix as sums and products need very little change to apply to the probabilistic versions as well. The factorizations give immediate formulas for determinants; when they do not vanish, the block diagonal form of of one of the factors, together with the fact that the other is a permutation matrix quickly reveals a form for the inverse matrix. Taking as an example the matrix of the last section, we have

$$\mathcal{P}_2^{-1} = \begin{bmatrix} \frac{p_2}{d_{12}} & \cdot & \frac{-q_2}{d_{12}} & \cdot \\ \frac{-q_1}{d_{12}} & \cdot & \frac{p_1}{d_{12}} & \cdot \\ \cdot & \frac{p_4}{d_{34}} & \cdot & \frac{-q_4}{d_{34}} \\ \cdot & \frac{-q_3}{d_{34}} & \cdot & \frac{p_3}{d_{34}} \end{bmatrix}.$$

The Gerschgorin limit of this matrix requires its maximum eigenvalue to be less than the greatest reciprocal determinant $1/d_{12}$ or $1/d_{34}$ (all with respect to absolute values). Generalizing, this means that *every* eigenvalue of $\mathcal{P}_n$ must be greater than the lesser of these determinants. Of course, if one of them is zero, we know that the lower bound is reached and $\mathcal{P}_n$ is singular.

In the other direction, unless some p_i or q_i is zero, every matrix element of $\mathcal{P}_n^n$ will be strictly positive, although in the probabilistic case no two of them have to be equal. This ensures that the maximum eigenvalue, $\lambda = 1$, will be unique. Conversely, if some probability or coprobability vanishes, the possibility of degeneracy exists (but is not obligatory).

7.7 Chacteristic equation

Unfortunately information about the other eigenvalues, particularly the second largest of them, is not so readily available—either directly or from the factored forms. However, formulas for the coefficients of the characteristic equation can be derived from the symbolic form of the de Bruijn matrix elements; their expression is even more elegant if a new rule of composition is introduced for regular expresions, particularly for sequences of symbols.

7.7.1 Merged product

Let us suppose a and b are letters, and that x is a sequence, possibly null and possibly empty. We define the merged product of the sequences s and t, denoted $s \vee t$, by

$$s \vee t = \begin{cases} axb & s = ax \text{ and } t = xb \\ \phi & \text{otherwise} \end{cases}$$

likewise, define the overlap, $s \wedge t$, by

$$s \wedge t = \begin{cases} x & s = ax \text{ and } t = xb \\ \phi & \text{otherwise} \end{cases}$$

These are just the combinations of indices required to work with the de Bruijn matrices. For example, the $2^n \times 2^n$ column-stochastic matrix is defined in terms of n- and $(n+1)$-block probabilities by the formula

$$[\mathcal{C}_n]_{ax,yb} = \begin{cases} \frac{p(ax \vee yb)}{p(ax)} & p(ax) \neq 0 \\ 0 & \text{otherwise} \end{cases}$$

Correspondingly the row-stochastic matrix is defined by

$$[\mathcal{R}_n]_{ax,yb} = \begin{cases} \frac{p(ax \vee yb)}{p(xb)} & p(xb) \neq 0 \\ 0 & \text{otherwise} \end{cases}$$

Unless stated otherwise, we will assume that a probabilistic de Bruijn matrix is column-stochastic. Since it is not always convenient to show the matrix elements as a quotient, but it is essential to know which of them intrinsically vanish, let us write

$$[\mathcal{C}_n]_{ax,yb} = m(ax, yb)$$

and note that it vanishes unless $x = y$, or alternatively when $ax \vee yb = \phi$.

7.7.2 Trace

The trace of $\mathcal{C}_n$, which is the coefficient of λ^{n-1} in the characteristic equation, satisfies

$$\begin{aligned} Tr(\mathcal{C}) &= \sum_{ax} m(ax, ax) \\ &= \sum_{i|ax=xi} m(ax, xi) \\ &= \sum_{i} m(i^*, i^*). \end{aligned}$$

The essential element in all of these calculations is to factor the column index in such a way that it can match a row index and yet produce a non-zero merged product.

7.7.3 Second coefficient

$$\begin{aligned} a_2 &= \sum_{ax \leq by} (m(ax, ax)m(by, by) - m(ax, by)m(by, ax)) \\ &= \sum_{i|ax=xi} \sum_{j|by=yj} m(ax, xi)m(by, yj) - \sum_{i|by=xi} \sum_{j|ax=yj} m(ax, xi)m(by, yj) \\ &= \sum_{i \leq j} m(i^*, i^*)m(j^*, j^*) - \sum_{i \leq j} m((ij)^*)m((ji)^*). \end{aligned}$$

In essence, this sum runs over all cycles of length 2 in the de Bruijn diagram. The solution to the requirements that $by = xi$ and $ax = yj$ is that $y = (ab)^*$ insofar as this notation can be respected on account of the fact that y might not have even length.

7.7.4 Principal minors

Striking out rows and columns intersecting on a diagonal will always leave some surviving elementary determinants, until so many have been stricken out that it is no longer possible to form a complete determinant. The coefficient of the elementary determinants will depend on how much of the remaining elementary determinant can be salvaged.

In general, the determinant of the de Bruijn $|m(ix, yj)|$ factorizes, so that

$$|m(ix, yj)| = \prod_{x,x} |m(ix, xj)|,$$

where the factors are $k \times k$ determinants obtained by varying the indices i and j with fixed x. To get the minor of the hz, hz diagonal element, we need to consider two cases. If this pivot does not belong to any subdeterminant, it can be ignored.

If it does, we drop the z, z term from the product, replacing it by the internal minor from within the omitted factor. Thus

$$a_1 = \sum_{hz=wh} cofactor(m(hz, hz)) \prod_{x \neq z} |m(ix, xj)|.$$

7.7.5 Determinant

Although we already have the determinant of the probabilistic de Bruijn matrices from their factored forms, it is worth observing that it can be calculated directly by following the symbolic procedure.

$$|m(ix, yj)| = \prod_{x,x} |m(ix, xj)|$$

7.8 Correlations

One application of the probabilistic $(n+1)$-stage de Bruijn matrix is to finding the probability of encountering one n-block in the vicinity of another; depending upon the direction, the basic probabilities are obtained by multiplying a vector of n-block probabilities by the matrix of the corresponding handedness to get the probabilities of a shifted block. Powers of the matrix would then yield blocks shifted to greater and greater distances, giving some importance to knowing about the behaviour of such powers. The degeneracy of the largest eigenvalue and the size of the second largest eigenvalue are the quantities which determine the limiting behaviour of powers and how rapidly their limits are reached.

If the largest eigenvalue is non-degenerate, which will be assured if none of the links has zero probability, then both the dominant eigenvalue and its normalized eigenvector will be unique. The second largest eigenvalue governs the exponential approach to equilibrium, which will be the more rapid the smaller the eigenvalue. In general a slow approach to equilibrium is associated with a large bias. Unfortunately the commonest estimators of eigenvalues tend to estimate either the largest eigenvalue, or else the smallest by first inverting the matrix, but not any of the others.

There is a certain informal expectation that if a block is long enough, probabilities will not be much affected by dropping the final cell, especially if comparative probabilities are considered. Illustrating this concept with a fairly short sequence of three cells, we might expect to find

$$\frac{p(abc)}{p(ab)} = \frac{p(bc}{p(b)}.$$

There is no real justification for such an assumption, but it gives an alternative viewpoint to the equivalent equation

$$p(abc) = \frac{p(ab)p(bc)}{p(b)},$$

which says that we should multiply the probabilities for the two sequences *ab* and *bc* because the presence of both is required to form the sequence *abc*, but that the probability of *b* should be divided out because it is common to both sequences and thus counted twice when they are joined.

As yet, the source of the block probabilities which enter into the formation of an actual probabilistic de Bruijn matrix has not been discussed; only the fact that that they ought to satisfy the Kolmogorov consistency conditions. One way to obtain them would be to to look for probabilities which were self consistent with respect to evolution; the context in which to do this is the probabilistic version of the reduced evolution matrix.

Chapter 8

Probabilistic evolution matrix

Empirically it has neen noted that automata go through three phases of evolution. If a ring is started out with an arbitrary initial configuration the randomness decreases fairly rapidly, followed by a longer period with a fairly constant density of states, and a final period which will be reached after a time depending on the length of the ring, in which the evolution is completely periodic.

Paying closer attention to the intermediate phase, it is seen that the density of states is not constant but that it fluctuates, and that there may be short range correlations between the states. Nevertheless, these densities, correlations, and even standard deviations can be calculated and they seem to agree fairly well with experience.

8.1 Regularities and anomalies

There are both temporal and spatial correlations; we have seen how the latter can be estimated by the probabilistic version of the de Bruijn diagram. It is interesting that the most serious logical discrepancy in probabilistic estimates arises from the excluded states which can be determined from the subset construction. Exclusion amounts to assigning a probability of zero, but zero probabilities cannot arise in probabilistic calculations. which involve only sums and products of positive quantities. Thus calculations based on the probabilities of individual cells will not suffice, and strings of cells must be taken into account from the outset.

Since there are arbitrarily long strings which are excluded for the first time (which is to say that none of their shorter segments is excluded, but that they are) it would seem that no theory based on the extrapolation of probabilities from finite strings could be mathematically exact. Nevertheless it can be hoped that there is a degree of approximation which is sufficient for practical purposes, but still not so complicated as to be beyond reasonable access.

For example, 10101001 is an excluded word for Rule 22, but it would have a probability of $1/256 = 0.004$ if zeroes and ones were considered equally probable.

This is not the worst distortion which Rule 22 suffers, because the distribution of frequencies for other short sequences of is by no means uniform, much less a simple function of the number of zeroes and ones they contain. For example, segments of the form $(01)^*$ are generally hard to come by, inasmuch as that is a sequence which it is its only ancestor. Most rules have their own excluded words, some of them even shorter than the eight letters which form the shortest excluded words for Rule 22.

In its probabilistic version, the de Bruijn diagram serves to predict the probabilities of the possible sequences comprising a shifting window in a long chain of states, and thus the correlations between such a sequence and a similar one occuring somewhere else in the chain. The conclusion afforded by the Frobenius-Perron theory is that there will always be an equilibrium vector of probabilities, that under certain circumstances will be unique. The particular form of the de Bruijn matrix shows that the lack of any extreme bias is sufficient to make the equilibrium unique; also that any zero biases will result in zero eigenvalues. We even know that the equilibrium probabilities are the n-block probabilities if the de Bruijn matrix is derived from $(n+1)$-block probabilities.

Although the disequilibrium eigenvalues and eigenvectors do depend in detail upon the transition probabilities of each particular diagram, the general conclusion is that the greater the bias the slower the approach to equilibrium, the node with the least bias establishing an upper limit to the speed. Zero bias results in an immediate equilibrium which may be accompanied by degenerate eigenvectors, while extreme bias offers the only possibility of alternate equilibria.

Whatever may be the spatial correlations of the moment, it is time evolution which drives the statistics of a linear automaton, continually challenging any correlations or lack thereof which may be found in the spatial distribution of the cells. Consequently we should look for probabilities, either of single cells or correlated probabilities of sequences of cells, which are consistent with the rule of evolution of the automaton.

8.2 Mean field theory

There are two approaches to this computation in the literature. One is called *mean field theory* and begins by assigning probabilities to each of the k states of the automaton, and then calculates the probabilities in the next generation on the basis of the usual combinatorial rules of probability, assuming that the probabilities for each of the cells in a neighborhood are independent. It is then possible to solve for a set of self consistent probabilities for each state. The results of such calculations are generally plausible but do deviate significantly from empirical observations. The suspect element in the calculation is the assumption of independence.

To see how this works, let us once again recall the transitions defining Rule 22:

$$\begin{array}{cccccccc} 111 & 110 & 101 & 100 & 011 & 010 & 001 & 000 \\ 0 & 0 & 0 & 1 & 0 & 1 & 1 & 0 \end{array}.$$

Five neighborhoods evolve into zeroes, three into ones; thus one might predict 37.5% ones would be found each generation on the basis of the number of ancestral neighborhoods. This is a better estimate than saying that 50% of the cells ought to be ones because there are only two different values they can have, but we have no reason to believe that all neigborhoods are equally likely either.

Taking the probability of finding a one as p, its coprobability as q, we could estimate $p' = 3pq^2$ for the probability of finding a one in the following generation, based on the makeup of the three neighborhoods that evolve to one. Mean field theory takes the fixed point of this estimate as the equilibrium density of ones for this rule. The self-consistent values for p are 0 and $1 - 1/\sqrt{3}$, or approximately 42%.

A slightly more detailed approach to the same information would be to set up an evolution matrix, in which the probabilities of each of the cell values are components of a vector, while the elements of the matrix describe the probabilities that one value of the cell evolves into another.

$$\begin{bmatrix} q \\ p \end{bmatrix}' = \begin{bmatrix} p^2 + q^2 & p^2 + 2pq \\ 2pq & q^2 \end{bmatrix} \begin{bmatrix} q \\ p \end{bmatrix}.$$

The eigenvalues of this column stochastic matrix are $\lambda = 1, q(2q-1)$, A matrix of column eigenvectors is

$$U = \begin{bmatrix} p^2 + 2pq & 1 \\ 2pq & -1 \end{bmatrix}$$

Self-consistency is judged as before, with the same equation for $p' = p$; but now additional information about the rate of decay of disequilibrium is available. Disequilibrium is very long lived for the self-consistent value $p = 0$, and vanishes for $p = 50\%$.

8.3 More refined theories

Dresden and Wong [67] showed how to write the rule of evolution of Conway's Life in an algebraic from to which the rules for combining probability distribution functions could be applied. Schulman and Seiden [202] proceded to obtain an explicit form for the evolution of probability, finding the cummulants hard to deal with, but nevertheless worked out an approximation and applied it to Life.

Wilbur, Lipman, and Shamma [242] decided instead to work with the probabilities of chains or blocks of cells rather than individual cells, including a survey of the self-consistent probabilities of triples of cells according to Wolfram's thirty two "legal" (2, 1) automata in their article. They obtain self-consistency from estimating the probabilities of the different ancestors in the reduced evolution matrix. They use the the same chains they are studying to estimate the probabilities of the extensions required as ancestors, treating the extensions as though they were a Markov process.

That is, starting with n-block probabilities, they use the quotient

$$p_\ell(a, x) = \frac{p(ix)}{\sum_i p(ix)}$$

as the probability that the $(n-1)$-block chain x can be extended to the n-block chain ax; in terms of the notation we have already introduced, they assume $p_\ell(ax) =$

$p(ax)/p(x)$. A similar result, $p_r(xa) = p(xa)/p(x)$ is supposed to hold for right extensions. Their working hypothesis is that the same relation serves to extend an n-block probability to an $(n+1)$-block probability; making two extensions, one on each side, to get the ancestor of a given block, their equations for self-consistent block probabilities read

$$p(z) = \sum_{\varphi(abxcd)=z} \frac{p(abx)}{p(bx)} \frac{p(xcd)}{p(xc)} p(bxc).$$

All terms refer to n-block probabilities or their sums.

8.4 Local structure theory

Perhaps the most extensive pursuit of this theme is to be found in Gutowitz, Victor, and Knight's [107] *local structure theory.* It differs from the approach of Wilbur *et.al.* by their use of the theory of probability measures to justify the derivation of their equations for the probabilities of the blocks, even though the final equations are the same. Both groups of authors survey several classes of rules, comparing the theoretical results with empirically observed frequencies. There is general agreement that the probability of a block in one generation should equal the probability of its ancestor in the previous generation; differences arise both from the philosophies and the actual techniques used to obtain the required probability.

To derive their equations, Gutowitz *et.al.* require $2r+1$ consecutive overlapping n-blocks to build up the $n+2r$-block ancestor of a given n-block from other n-blocks. The choice of a notation in which to express the equations is important; we shall explore certain alternatives. To begin with, suppose that B and X are words, the remaining symbols single letters. The equations for self consistent probabilities then read

$$p(ABC) = \sum_{VWXYZ \in \varphi^{-1}(ABC)} \frac{p(VWX)p(WXY)p(XYZ)}{p(WX)p(XY)}.$$

As Gutowitz *et.al.* demonstrated, these equations reduce to the mean field equations for 1-block probabilities, making their local field theory a plausible extension of mean field theory. Since the 1-block, or mean field theory, denominators reduce to the constant value $p(\lambda) = 1$, the mean field theory equations are polynomial equations, wheras the general local structure theory equations involve rational fractions. As a result it is much harder to establish the existence, uniqueness, or stability of their solutions.

Nevertheless, the local structure theory approach makes contact with a very general framework within probability theory, and also presents the basic equations in a symmetrical form admitting a greater variety of interpretations. In particular we can recover the point of view of Wilbur *et.al.* , interpreting the equations as a set of linear equations whose matrix of coefficients depends on the same solutions which it defines. The nonlinearity of the equations is thus of a very special form, reminiscent of the Hartree equations or the Hartee-Fock equations of quantum mechanics, which are generally solved by an iterative process.

Numerically, the equations of Gutowitz *et.al.* can be solved exclusively by iteration. Diagonalizing the coefficient matrix might accelerate convergence, but at the cost of the time expended on diagonalization. It is of greater importance that the form of the equations reveals particular properties of the solution and its convergence that are not otherwise evident.

The ostensible variables are not manifest in the denominators of the equations, but they are readily obtained by using the Kolmogorov consistency conditions. Since the terms of the numerators are associated with links in the de Bruijn diagram, and the denominators with nodes, the whimsically minded might fancy that they see a resemblance to the propagators and Feynman diagrams of field theory. In any event we are going to take up a slightly different interpretation.

8.5 Hartree-Fock approach

The Hartree-Fock-like assumption is to define

$$\mu_V = \frac{p(VWX)}{p(WX)}$$

$$\nu_Z = \frac{p(XYZ)}{p(XY)}$$

both these quantities are positive numbers less than 1; if any numerator is zero, the entire fraction is taken to be zero, even in those cases where the denominator might also be zero.

In fact μ and ν are probabilities which can be used in various ways, such as constructing the Markov matrix for spatial correlations; in an ℓ-stage de Bruijn diagram they give the relative probabilities for the different entering or emerging symbol during a shift.

The self-consistency equations can now be written

$$p(ABC) = \sum_{WXY}\sum_{V,Z}\delta(VWXYZ, \varphi^{-1}(ABC))\mu_V\nu_Z p(WXY)$$

which presents the appearance of being a system of linear equations if one overlooks the fact that the μ's and ν's are not constants, but depend upon the very same unknown probabilities for which one is solving. The function δ is a set-theoretic Kronecker delta, 1 when its arguments coincide, zero when they do not.

The inner sum defines a matrix (writing all indices as arguments, not subscripts):

$$\mathcal{H}(ABC, WXY) = \sum_{V,Z}\delta(VWXYZ, \varphi^{-1}(ABC))\mu(V)\nu(Z),$$

mapping one vector of probabilities for the ℓ-blocks into another.

For symmetry we have linearized the equations of Gutowitz *et.al.* by taking $p(WXY)$ as the vector component, but we could just as easily have taken one of the other numerator terms; the Kolmogorov conditions are flexible enough to permit μ and ν to retain the same form in either case. Should the occasion arise to do so, we could distinguish the three different definitions of $\mathcal{H}$ by the adjectives left, cental, or right.

To implement the theory of Gutowitz *et.al.* , recall the reduced evolution matrix E_n. For n-chains of cells i and j, it is defined by

$$[E_n]_{i,j} = \begin{cases} 1 & i \in \varphi^{-1}(j) \\ 0 & \text{otherwise} \end{cases} .$$

The probabilistic version of E is just $\mathcal{H}$ defined above, at least for one particular way of estimating the probabilities of ancestors. In any event, many essential properties of $\mathcal{H}$ are determined by E, since both matrices have the same block diagonal structure, no matter whether the nonzero matrix elements are determined self-consistently from the nonlinear local structure theory or otherwise. E_n, which is always fairly crowded for low values of n, becomes sparser and sparser as n increases. One hopes that it tends toward a stable form which could be described analytically; and which might also describe the full matrix $\mathcal{H}_n$.

8.6 Kolmogorov consistency conditions

The motivation for working with block probabilities rather than cell probabilities is the hope that a better agreement with empirical observations can be obtained, supposing that up to some point longer blocks can better account for correlations between cells than short blocks can. Nevertheless, the empirical quantity which is usually calculated is the probability of individual cells. Sometimes variances or pair probabilities may also be calculated, but it is the density of cells which is the primary concern.

$$\begin{bmatrix} p(00) \\ p(01) \\ p(10) \\ p(11) \end{bmatrix} = \begin{bmatrix} 1000001100010111 \\ 0100000000101000 \\ 0001010010000000 \\ 0010100001000000 \end{bmatrix} \begin{bmatrix} p(0000) \\ p(0001) \\ p(0010) \\ p(0011) \\ p(0100) \\ p(0101) \\ p(0110) \\ p(0111) \\ p(1000) \\ p(1001) \\ p(1010) \\ p(1011) \\ p(1100) \\ p(1101) \\ p(1110) \\ p(1111) \\ p(1100) \\ p(1101) \\ p(1110) \\ p(1111) \end{bmatrix}$$

Figure 8.1: Two-block probabilities summed from four-block probabilities.

Cell density can be inferred from block probabilities by summing up the probabilities of blocks in various ways, all of which are guaranteed to give consistent results by virtue of the Kolmogorov consistency conditions. It is thus worthwhile to investigate whether the iterative solution of the local field theory equations conserves the consistency conditions. We have shown that if the definition of $(n-1)$-block probabilities in terms of n-block probabilities is consistent, the definitions for all shorter blocks will also be consistent. Thus it suffices to show that one cycle of iteration conserves consistency at the the highest level—the one which is anyway involved in defining the denominators for $\mathcal{H}$.

Wilbur *et.al.* included a proof in their article, so the only practical question remaining is one of stability—whether numerical errors arising during the course of an iterative solution of the equations could prejudice the consistency of the results.

It is in any event instructive to write the field equations in a very extended explicit matrix form, shown here for 2-blocks evolving by Rule 22. We begin with equations (set aside in Figure 8.1 due to their bulk) expressing the probability of a block as the sum of the probabilities of its possible ancestors.

Either Wilbur *et.al.* 's theory or Gutowitz *et.al.* 's equations written in Hartree-Fock form estimate the probability of each 4-block ancestor in terms of the probabilities of the 1- and 2-block segments into which the ancestor can be decomposed. The matrix form of this estimate, in which the precursor of the de Bruijn format is quite apparent, is shown in Figure 8.2.

$$\begin{bmatrix} p(0000) \\ p(0001) \\ p(0010) \\ p(0011) \\ p(0100) \\ p(0101) \\ p(0110) \\ p(0111) \\ p(1000) \\ p(1001) \\ p(1010) \\ p(1011) \\ p(1100) \\ p(1101) \\ p(1110) \\ p(1111) \\ p(1100) \\ p(1101) \\ p(1110) \\ p(1111) \end{bmatrix} = \begin{bmatrix} \frac{p(00)}{p(0)}\frac{p(00)}{p(0)} & \cdot & \cdot & \cdot \\ \frac{p(00)}{p(0)}\frac{p(01)}{p(0)} & \cdot & \cdot & \cdot \\ \cdot & \frac{p(00)}{p(0)}\frac{p(10)}{p(1)} & \cdot & \cdot \\ \cdot & \frac{p(00)}{p(0)}\frac{p(11)}{p(1)} & \cdot & \cdot \\ \cdot & \cdot & \frac{p(01)}{p(1)}\frac{p(00)}{p(0)} & \cdot \\ \cdot & \cdot & \frac{p(01)}{p(1)}\frac{p(00)}{p(0)} & \cdot \\ \cdot & \cdot & \cdot & \frac{p(01)}{p(1)}\frac{p(10)}{p(1)} \\ \cdot & \cdot & \cdot & \frac{p(01)}{p(1)}\frac{p(11)}{p(1)} \\ \frac{p(10)}{p(0)}\frac{p(00)}{p(0)} & \cdot & \cdot & \cdot \\ \frac{p(10)}{p(0)}\frac{p(01)}{p(0)} & \cdot & \cdot & \cdot \\ \cdot & \frac{p(10)}{p(0)}\frac{p(10)}{p(1)} & \cdot & \cdot \\ \cdot & \frac{p(10)}{p(0)}\frac{p(11)}{p(1)} & \cdot & \cdot \\ \cdot & \cdot & \frac{p(11)}{p(1)}\frac{p(00)}{p(0)} & \cdot \\ \cdot & \cdot & \frac{p(11)}{p(1)}\frac{p(00)}{p(0)} & \cdot \\ \cdot & \cdot & \cdot & \frac{p(11)}{p(0)}\frac{p(10)}{p(1)} \\ \cdot & \cdot & \cdot & \frac{p(11)}{p(0)}\frac{p(11)}{p(1)} \end{bmatrix} \begin{bmatrix} p(00) \\ p(01) \\ p(10) \\ p(11) \end{bmatrix}.$$

Figure 8.2: Four-block probabilities estimated from two-block probabilities.

Consolidating these two equations produces a 4×4 matrix of coefficients

$$\mathcal{H}_2 = \left[\begin{array}{c|cc|cc} & 00 & 01 & 10 & 11 \\ \hline 00 & \frac{p(00)}{p(0)}\frac{p(00)}{p(0)} & \frac{p(10)}{p(0)}\frac{p(11)}{p(1)} & \frac{p(11)}{p(1)}\frac{p(01)}{p(0)} & 1 \\ 01 & \frac{p(00)}{p(0)}\frac{p(01)}{p(0)} & \frac{p(10)}{p(0)}\frac{p(10)}{p(1)} & \frac{p(11)}{p(1)}\frac{p(00)}{p(0)} & 0 \\ \hline 10 & \frac{p(10)}{p(0)}\frac{p(00)}{p(0)} & \frac{p(00)}{p(0)}\frac{p(11)}{p(1)} & \frac{p(01)}{p(1)}\frac{p(01)}{p(0)} & 0 \\ 11 & \frac{p(10)}{p(0)}\frac{p(01)}{p(0)} & \frac{p(00)}{p(0)}\frac{p(10)}{p(1)} & \frac{p(01)}{p(1)}\frac{p(00)}{p(0)} & 0 \end{array}\right],$$

in which it can be verified by inspection that $p(0)$ is consistently defined by either $p(0) = p(00) + p(01)$ or by $p(0) = p(00) + p(10)$; or that equivalently $p(10) = p(01)$ on the right hand side implies the same relation for the new values on the left hand side of the equation.

It is easier to give the general proof in symbolic form, having first written the field equations in terms of the merged product:

$$p(Q) = \sum_{\varphi(X \vee Y \vee Z) = Q} \frac{p(X)p(Y)p(Z)}{p(X \wedge Y)p(Y \wedge Z)}$$

If it is intended to write $Q = AB$ so as to obtain

$$p(B) = \sum_A p(AB)$$

we need to know that

$$\varphi^{-1}(AB) = \varphi^{-1}(A) \vee \varphi^{-1}(B),$$

if $\varphi^{-1}(A)$ is incorporated into X we introduce W

$$X = \varphi^{-1}(A) \vee W$$

and note that

$$\sum_A \varphi^{-1}(A) \vee W = X \wedge Y,$$

so that

$$p(B) = \sum_{\varphi(Y \vee Z) = B} \frac{p(Y)p(Z)}{p(Y \wedge Z)}.$$

Since an entirely symmetrical expression results when Q is factored into $Q = BC$, comparing the two establishes the consistency of $p(B)$ after iteration, supposing that it was consistent before. The essential points in the proof are, first, that running through all initial letters in the sum for $p(B)$ guarantees a traversal of all the initial letters in the ancestors, and second, that the denominators in the probabilities are cancelled by the sums in the numerators as a consequence of the consistency hypothesis.

8.7 The vector subset diagram

If $\alpha(x)$ is the number of ancestors of x, we have

$$0 \leq \alpha(xy) \leq \alpha(x)\alpha(y),$$

because the $2r$ cells at which the ancestors of x overlap the ancestors of y when x and y are joined must coincide. Moreover, there can be no ancestor of xy which does not begin with an ancestor of x, nor which does not terminate with an ancestor of y.

The overlapping of neighborhoods which must be taken into account in calculating ancestors can be summarized very concisely by the relation

$$\varphi^{-1}(ax) = \varphi^{-1}(a) \vee \varphi^{-1}(x)$$

when a is a letter and x is a word (or the reverse). If both are words a similar relation holds, but the merged product was not defined with quite the generality required to restrict the overlap involved.

Since the determination of the ancestors of a given word plays an important role in all aspects of the calculation of evolutions, it is fortunate that the subset diagram used to locate excluded words can be generalized slightly to a form which will provide both the number of counterimages of a word and the counterimages themselves.

The subset diagram was derived from the $2r$-stage de Bruijn diagram whose links were labelled for the neighborhoods of a (k, r) automaton because the links also correspond to the cells that evolve from the neighborhoods which the links represent. Thus following out a path according to the evolved cells automatically yields the ancestor involved; the subset construction was invoked because of a need to obtain *all* the possible paths, and moreover to obtain them systematically.

Links were the only information recorded in the subset diagram because interest in the diagram was limited to knowing which combinations, or subsets, of nodes could be linked to get a given evolution. However, if each node is assigned a 2^{2r}-ple (quadruple for a $(k, 1)$ automaton), the exact linkage of neighborhoods could be shown, not just the mere indication that such a linkage exists.

If we use Rule 126 as an example, we need the table of transitions

111	110	101	100	011	010	001	000
0	1	1	1	1	1	1	0

In other words all neighborhoods evolve into ones with the excpetion of 111 and the quiescent neighborhood 000.

The de Bruijn diagram for this rule is

node	0 leads to	1 leads to
A	A	C
C	ϕ	D, F
D	ϕ	A, C
F	F	ϕ

while the vector subset diagram will have the form

subset	0 leads to	1 leads to
$\{ACDF\}$	(A, ϕ, ϕ, F)	$(D, A+D, C+F, C)$
$\{AF\}$	(A, ϕ, ϕ, F)	(ϕ, A, F, ϕ)
$\{CD\}$	(ϕ, ϕ, ϕ, ϕ)	(D, D, C, C)
$\{\}$	(ϕ, ϕ, ϕ, ϕ)	(ϕ, ϕ, ϕ, ϕ)

The shortest excluded word for this rule is 010; generally any sequence ending in 010 is excluded. To determine the subset to which a quadruple leads we have only to determine its non-ϕ components. To obtain the number of ancestors for any sequence, say 1101, we begin with the quadruple (1,1,1,1), and apply the formula within the quadruple belonging to each digit in turn. Finally we sum the components.

	node	count	
	$ACDF$	$(1,1,1,1)$	$(00, 01, 10, 11)$
1	$ACDF$	$(1,2,2,1)$	$(100, 001+101, 010+110, 011)$
1	$ACDF$	$(2,3,3,2)$	$(0100+1100, 1001+0101+1101, 0010+1010+0110, 0011+1011)$
0	AF	$(2,0,0,2)$	$(01000+11000, \phi, \phi, 00111+10111)$
1	CD	$(0,2,2,0)$	$(\phi, 010001+110001, 001110+101110, \phi)$

Thus we conclude that there are 4 ancestors of 1101 according to Rule 126. If we wanted the actual ancestors, we should have performed the arithmetic symbolically, as shown in the last column of the table.

8.8 Estimating the number of ancestors

Instead of working with formula bearing quadruples, we could have represented the links in the vector subset diagram by square matrices; most of them would be singular, but calculations invloving them could be carried out entirely in terms of matrix algebra. Moreover, it would be possible to ask selective questions about ancestors by examining individual matrix elements of the product; a multiple product is already indexed by the $2r$ letters with which the ancestor commences and the $2r$ letters with which it ends.

Even more useful is to construct the connection matrix of the vector subset diagram; links to the empty set can be omitted because of the triangular form of the full matrix. For the example of Rule 126, this matrix has the form

$$
\begin{array}{cccc|cccc|cccc}
\hline
\cdot & 1 & \cdot & \cdot & 1 & \cdot & \cdot & \cdot & \cdot & \cdot & \cdot & \cdot \\
\cdot & \cdot & 1 & 1 & \cdot & \cdot & \cdot & \cdot & \cdot & \cdot & \cdot & \cdot \\
1 & 1 & \cdot & \cdot & \cdot & \cdot & \cdot & \cdot & \cdot & \cdot & \cdot & \cdot \\
\cdot & \cdot & 1 & \cdot & \cdot & \cdot & \cdot & 1 & \cdot & \cdot & \cdot & \cdot \\
\hline
\cdot & \cdot & \cdot & \cdot & 1 & \cdot & \cdot & \cdot & \cdot & 1 & \cdot & \cdot \\
\cdot & \cdot & \cdot & \cdot & \cdot & \cdot & \cdot & \cdot & \cdot & \cdot & \cdot & \cdot \\
\cdot & \cdot & \cdot & \cdot & \cdot & \cdot & \cdot & \cdot & \cdot & \cdot & \cdot & \cdot \\
\cdot & \cdot & \cdot & \cdot & \cdot & \cdot & \cdot & 1 & \cdot & \cdot & 1 & \cdot \\
\hline
\cdot & \cdot & \cdot & \cdot & \cdot & \cdot & \cdot & \cdot & \cdot & \cdot & \cdot & \cdot \\
\cdot & \cdot & 1 & 1 & \cdot & \cdot & \cdot & \cdot & \cdot & \cdot & \cdot & \cdot \\
1 & 1 & \cdot & \cdot & \cdot & \cdot & \cdot & \cdot & \cdot & \cdot & \cdot & \cdot \\
\cdot & \cdot & \cdot & \cdot & \cdot & \cdot & \cdot & \cdot & \cdot & \cdot & \cdot & \cdot \\
\hline
\end{array}
$$

This matrix contains four distinct non-zero submatrices

$$
\alpha = \begin{bmatrix} \cdot & 1 & \cdot & \cdot \\ \cdot & \cdot & 1 & 1 \\ 1 & 1 & \cdot & \cdot \\ \cdot & \cdot & 1 & \cdot \end{bmatrix}, \beta = \begin{bmatrix} 1 & \cdot & \cdot & \cdot \\ \cdot & \cdot & \cdot & \cdot \\ \cdot & \cdot & \cdot & \cdot \\ \cdot & \cdot & \cdot & 1 \end{bmatrix}, \gamma = \begin{bmatrix} \cdot & 1 & \cdot & \cdot \\ \cdot & \cdot & \cdot & \cdot \\ \cdot & \cdot & \cdot & \cdot \\ \cdot & \cdot & 1 & \cdot \end{bmatrix}, \delta = \begin{bmatrix} \cdot & \cdot & \cdot & \cdot \\ \cdot & \cdot & 1 & 1 \\ 1 & 1 & \cdot & \cdot \\ \cdot & \cdot & \cdot & \cdot \end{bmatrix}
$$

which are the matrix representations of the quadruples of the vector subset diagram. It is convenient to regard them as hypercomplex numbers formed from a basis consisting of the following eight 4×4 matrices

$$
\beta = \begin{bmatrix} 1 & \cdot & \cdot & \cdot \\ \cdot & \cdot & \cdot & \cdot \\ \cdot & \cdot & \cdot & \cdot \\ \cdot & \cdot & \cdot & 1 \end{bmatrix}, \gamma = \begin{bmatrix} \cdot & 1 & \cdot & \cdot \\ \cdot & \cdot & \cdot & \cdot \\ \cdot & \cdot & \cdot & \cdot \\ \cdot & \cdot & 1 & \cdot \end{bmatrix}, \epsilon = \begin{bmatrix} \cdot & \cdot & \cdot & \cdot \\ \cdot & \cdot & 1 & \cdot \\ \cdot & 1 & \cdot & \cdot \\ \cdot & \cdot & \cdot & \cdot \end{bmatrix}, \zeta = \begin{bmatrix} \cdot & \cdot & \cdot & \cdot \\ \cdot & \cdot & \cdot & 1 \\ 1 & \cdot & \cdot & \cdot \\ \cdot & \cdot & \cdot & \cdot \end{bmatrix},
$$

$$
\eta = \begin{bmatrix} \cdot & \cdot & 1 & \cdot \\ \cdot & \cdot & \cdot & \cdot \\ \cdot & \cdot & \cdot & \cdot \\ \cdot & 1 & \cdot & \cdot \end{bmatrix}, \theta = \begin{bmatrix} \cdot & \cdot & \cdot & 1 \\ \cdot & \cdot & \cdot & \cdot \\ \cdot & \cdot & \cdot & \cdot \\ 1 & \cdot & \cdot & \cdot \end{bmatrix}, \iota = \begin{bmatrix} \cdot & \cdot & \cdot & \cdot \\ \cdot & 1 & \cdot & \cdot \\ \cdot & \cdot & 1 & \cdot \\ \cdot & \cdot & \cdot & \cdot \end{bmatrix}, \kappa = \begin{bmatrix} \cdot & \cdot & \cdot & \cdot \\ 1 & \cdot & \cdot & \cdot \\ \cdot & \cdot & \cdot & 1 \\ \cdot & \cdot & \cdot & \cdot \end{bmatrix},
$$

according to the definitions

$$
\begin{aligned}
\alpha &= \gamma + \zeta + \epsilon \\
\delta &= \zeta + \epsilon.
\end{aligned}
$$

These basis elements obey the multiplication table

	β	θ	ι	ϵ	γ	η	ζ	κ
β	β	θ	ϕ	ϕ	γ	η	ϕ	ϕ
θ	θ	β	ϕ	ϕ	η	γ	ϕ	ϕ
ι	ϕ	ϕ	ι	ϵ	ϕ	ϕ	ζ	κ
ϵ	ϕ	ϕ	ϵ	ι	ϕ	ϕ	κ	ζ
γ	ϕ	ϕ	γ	η	ϕ	ϕ	θ	β
η	ϕ	ϕ	η	γ	ϕ	ϕ	β	θ
ζ	ζ	κ	ϕ	ϕ	ϵ	ι	ϕ	ϕ
κ	κ	ζ	ϕ	ϕ	ι	ϵ	ϕ	ϕ

In general it is probably easier to work with the connection matrix and its powers than to deal with the algebra of hypercomplex numbers. Sometimes special properties of the algebra manifest themselves, as they do here. Every basis matrix has an even element sum, and likewise an even trace. This shows that there must always be an even number of ancestors of whatsoever open chain, and also of every ring. Thus there can be no orphan periods among the rings, in which no transients lead into the ring.

The following table shows the numbers of ancestors, of both open sequences, and rings, of three cells. To count open sequences all the elements of the final product matrix are to be summed; for rings only the trace is involved because the the ring cannot close unless the loose ends coincide.

cells	000	001	010	011	100	101	110	111
matrix	$\beta\beta\beta$	$\beta\beta\gamma$	$\beta\gamma\phi$	$\beta\gamma\delta$	$\alpha\beta\beta$	$\alpha\beta\gamma$	$\alpha\alpha\beta$	$\alpha\alpha\alpha$
product	β	γ	ϕ	$\theta+\eta$	ζ	ϵ	$\theta+\kappa$	$\eta+\beta+\gamma+\kappa+2\iota+\zeta+\epsilon$
total	2	2	0	4	2	2	4	16
ring	2	0	0	0	0	0	0	6

8.9 Trivial solutions

It would be surprising if the equations for self consistency had unique solutions. In fact, even for one-block equations, the quiescent rules will always have a solution for the quiescent state. Generally there is another solution (except for Wolfram's classes I and II, which practically exclude additional solutions by definition.) If there is more than one solution it will be found that some of them are stable while others are unstable; in fact there is a whole theory surrounding the existence and properties of the fixed points of nonlinear equations.

Whatever may be the nature of the solutions of the local field theory equations in general, there are some solutions which can be foreseen. If the length of the block corresponds to the length of a cycle for a given automaton, then the members of the cycle can be assigned equal probabilities and other blocks can be assigned probability zero. We have two cases to consider—the blocks obtained by symmetry but belonging to the same phase of evolution, and the blocks belonging to different phases. All the phases of the same block will form a cyclic submatrix of the local structure matrix, while different symmetry images will generate additional diagonal blocks. In any event, a given block has exactly one counterimage within the sequence of evolution, and everything has been assigned equal probabilities.

There are some patterns of cyclic evolution which have no other ancestors than their immediate predecessors in the cycle of evolution—the still life of alternating zeroes and ones in Rule 22 for example, as well as the cycles of period seven amd eleven. Presumably they represent unstable fixed points in the parameter space of blocks. Other cycles may be the endpoint of various transients, and thus be somewhat more stable. Generally the self-consistent probabilities for blocks will not correspond to the actual densities of cells for any particular pattern of evolution, but rather will be formed from a composite of all of them.

It is also not excluded that there will be probabilities which are not self-consistent, but rather which alternate between a finite number of values. This behaviour tends to occur for rules which do not have quiescent states, so that alternation between two or more backgrounds can occur.

Chapter 9

Positive matrices

Starting with the evolution and de Bruijn matrices, and continuing with their probabilistic versions, we find that we are dealing with a specialized class of matrices all of whose elements are positive—or more accurately, non-negative. The first two of these classes of matrices is more restricted, inasmuch as their elements must be integers, but it is not especially easy to obtain much advantage from that particular characteristic. However, positive matrices—including those with integer elements—enjoy two properties which can be exploited to considerable advantage. Avoiding some limiting cases arising from unfavorable groupings of zero matrix elements, these properties are:

- there is a unique maximum eigenvalue, whose value is bracketed by the row sums of the matrix (as well as the column sums), and
- whose eigenvector can be normalized so that all its components are strictly positive. That is, all have the same sign and none are zero.

9.1 Gerschgorin's disks

The fundamental tool for estimating the size of eigenvalues is the eigenvalue equation itself, written out in terms of components of the eigenvectors. Thus, from the matrix equation $MX = \lambda X$ there follows

$$\lambda x_i = \sum_{j=1}^{n} m_{ij} x_j.$$

Since the zero vector is never considered to be an eigenvector, there is at least one component x_i which is not zero; let us select the largest among the non-zero components, move the corresponding diagonal element to the left hand side of the equation, and apply the triangle inequality. If there is any ambiguity in selecting the largest component, any one of them will do. Then

$$|\lambda - m_{ii}||x_i| \leq \sum_{j \neq i} |m_{ij}||x_j|.$$

The purpose of selecting the largest component and ensuring that it was non-zero was to allow division of this equation by $|x_i|$ to obtain the factors $|x_j|/|x_i|$ which are all less than, or equal to, 1 (even when $i = j$). Thus

$$|\lambda - m_{ii}| \leq \sum_{j \neq i} |m_{ij}||\frac{x_j}{x_i}|.$$

Since the inequality can only be enhanced by making the terms on the right hand side larger by dropping their small multipliers, we finally obtain the basic result

$$|\lambda - m_{ii}| \leq \sum_{j \neq i} |m_{ij}|,$$

which applies even for complex matrices or for real matrices with complex eigenvalues. The complex setting allows a picturesque description of the results with respect to the geometry of the plane. Regarding the column sum excluding the diagonal element as a radius and the diagonal element as a center, we have found a circle which surely contains the eigenvalue.

In general, without having actually found an eigenvector, which would be its largest component would be rarely be evident; but if all the different disks were joined together, no eigenvalue could escape the collection. There is even some choice of the precise disks to be used, according to whether the diagonal element is taken as part of the radius or as part of the center; the procedure shown yields the smallest disks but, at times a common center might be preferable. Of course, the smallest disks are centered on the diagonal elements.

This estimate is due to Gerschgorin, whose name is associated with the disks. Generally the disks all intersect, and there is no correspondence between eigenvalues and disks because of the uncertainty as to which would be the largest component of any given eigenvector. Sometimes there will be disjoint clusters of disks for which continuity arguments give each cluster its quota of eigenvalues. That is, if the diagonal elements are retained and the remainder multiplied by a small parameter, continuity of the eigenvalues with respect to the coefficients of the matrix will locate the eigenvalues within small disks surrounding the diagonal elements. As the parameter is increased, eigenvalues can only wander amongst disks which overlap, but the number within a given cluster must remain constant.

Using the origin as a common center for the Gerschgorin disks leads to expressing the bounds directly in terms of row sums:

$$|\lambda| \leq \sum_{j=1}^{n} |m_{ij}|;$$

since the inequality must hold for at least one row, we could always select the worse case with full confidence that it expresses a valid bound.

9.2 Eigenvalues on the boundary

The approximations used to derive the Gerschgorin limits would seem to be fairly generous. If any one of the eigenvalues actually lies on the circumference of a Gerschgorin disk, the implied equality for that one equation could presumably be used to draw additional conclusions; even close proximity to the boundary might possibly be a source of further information.

Consider the point in the derivation at which the ratios $|x_j|/|x_i|$ were replaced by 1, and recall that i was the index of the largest component of the eigenvector; this substitution would surely alter the sum unless the matrix element m_{ij} multiplying it were zero. If *all* the remaining offdiagonal matrix elements were zero, we could stop and think about reducing the matrix; otherwise we have forced a number of components of the eigenvector to be equal in absolute value. But this means that there are other equations in which this same largest component occurs, but with other indices.

The same argument forces equality for any further components related through non-zero matrix elements, and we arrive at the eventual conclusion that all the components of the eigenvector have the same absolute value, and that *all* the Gerschgorin disks intersect at least in the eigenvalue in question. The only exception would arise from a matrix whose diagram were not connected, so that either the class of matrices involved must be restricted, or the conclusion must be confined to the components belonging to connected parts of the matrix.

Bearing this restriction in mind, we conclude that the Gerschgorin bounds can only be realized for matrices having uniform row sums, and thus any nonuniformity is certain evidence that the limits cannot be reached. Nevertheless, we have only derived a necessary—not a sufficient—condition, and it may well happen that the eigenvalues of matrices with uniform row sums lie well within the Gerschgorin disks. Examining simple 2×2 matrices with elements of mixed signs will quickly confirm this statement. The lack of sufficiency results from not yet having taken into account the liberalizing influence of the triangle inequality on Gerschgorin's inequalities; but if all matrix elements were positive, the uniformity of the row sums would immediately establish the vector with unit components as an eigenvector and the row sum as its eigenvalue.

Given that our arguments are insensitive to scalar multiplication of the whole matrix by a complex factor, we cannot postulate that we are working with a positive matrix. However, if we note that a collection of vectors of fixed lengths cannot reach their maximum sum unless they are all parallel, we see that the only way that the natural eigenvalue equation and the one resulting from inserting absolute values can hold simultaneously is for all the terms to have a common phase.

We can begin by removing phase factors from the components of the eigenvector by performing a similarity transformation via diagonal matrices bearing the phase factors. The rows of the resulting matrix must then have constant phase, which is necessarily the phase of the eigenvalue. If the eigenvalue were not zero, this would be the phase of the whole matrix, and could be discarded by treating it as a constant factor.

Thus non-uniform row sums guarantee that the largest row sum is not an eigenvalue, but even uniformity will only make the common row sum an eigenvalue when the matrix is effectively a positive matrix.

Even though we did not assume that a possible boundary eigenvalue was the largest eigenvalue, that conclusion seems inescapable, inasmuch as we have seen that a connected matrix possessing a boundary eigenvalue is essentially positive, for which summing row elements with anything other than equal weights will necessarily produce a smaller sum and with it a smaller eigenvalue. Likewise, we seem to be led to the uniqueness of this eigenvalue since only one eigenvector can be associated with it.

9.3 Minimax principle

Using Dirac's notation for row and column vectors, the equations

$$\lambda \leq \frac{<x|M|y>}{<x|y>},$$

used in deriving the Gerschgorin disks would arise directly from the so-called Rayleigh quotient, were it not for the fact that an inequality can only be stated for real quantities. However the necessary absolute values entering the equations could be avoided, given an assurance that all the terms were real, as perhaps might be anticipated for the largest eigenvalue of a real matrix.

It is generally understood that eigenvalues and eigenvectors satisfy a variational principle; for symmetric or hermitean matrices this leads to bounds involving the Rayleigh coefficient or to the the Courant minimax principle. If we suppose that $<x+\delta x|$ and $|y+\delta y>$ are vectors deviating slightly from $<x|$ and $|y>$ respectively, and ignoring quantities of second order we know that

$$\begin{aligned}\frac{<x+\delta x|M|y+\delta y>}{<x+\delta x|y+\delta y>} &= \frac{<x|M|y>}{<x|y>} + \\ & <\delta x|\left(M|y> - \frac{<x|M|y>}{<x|y>}|y>\right) + \\ & \left(<x|M - <x|\frac{<x|M|y>}{<x|y>}\right)|\delta y> + \\ & \ldots\end{aligned}$$

Requiring the vanishing of the first order terms independently of $<\delta x|$ and $|\delta y>$ establishes the Rayleigh quotient as an eigenvalue, providing that $<x|$ and $|y>$ are its corresponding eigenvectors. However it would be preferable to work directly from explicit inequalities and not have to extrapolate from extremals to bounds.

Consequently, without yet specifying P and Q, define

$$r = \max_{y \varepsilon Q} \min_{x \varepsilon P} \frac{<x|M|y>}{<x|y>}.$$

$$s = \min_{y \varepsilon Q} \max_{x \varepsilon P} \frac{<x|M|y>}{<x|y>}.$$

If P were the set of coordinate vectors and Q were the set of vectors with positive components, the Rayleigh quotient would correspond to the right hand side of the Gerschgorin equations. Interchanging the roles of P and Q in these same equations would produce results for column sums equivalent to those which we would otherwise obtain for row sums.

With these choices of P and Q, it is easy to follow the rationale behind the definition of r; vector by vector, we decide which component suffers the least magnification, even if it is very, very small. At least we know that the vector as a whole will recieve at least that much magnification. Noone can blame us for finally choosing the vector with the greatest overall magnification, the value of which is r.

Nevertheless, the conservatism involved in singling out the component with the least magnification means that the entire vector will always be magnified slightly more then the worst case estimate unless in fact each and every component is magnified by the same amount—in which case we have an eigenvector on our hands. Likewise, supposing the magnification is not uniform for all components, we could construct another vector whose minimum magnification was slightly larger by simply reducing the size of the minimally amplified component relative to the others. However, this latter procedure would definitely fail for the vector of maximum minimal amplification, so it would just have to be an eigenvector.

Similar considerations apply to the definition of s, since we would be estimating the maximum magnification that M could possibly impart to a transformed vector, and examining the vector whose estimate was the least extravagant, only to discover that the estimate was quite valid throughout the whole vector, and so to conclude once again that we had an eigenvector and that s was its eigenvalue.

To be certain of our conclusions we need to be assured that zero matrix elements do not occur in inconvenient places, and also that the limits involved in assuming the existence of maxima or minima over sets of vectors exist; the reason that the Rayleigh quotients contain the denominators which they do is to allow us to confine our attention to vectors of fixed norm, and thus to compact sets. The question of the possible configurations of zeroes does indeed reward more careful consideration, because of the possibility of finding disconnected portions or cyclic structures within M. Nevertheless, the central role is played by those matrices whose elements are strictly positive.

9.4 Largest eigenvalue

Whereas it is clear that the definition of r in the last section yields the largest eigenvalue with a positive eigenvector, we would like to know that it is the largest possible eigenvalue, and that it is unique. This information can be obtained somewhat indirectly by noticing that any eigenvector lacking zero components defines an invertible diagonal matrix R, containing those same components on its diagonal. Let X be the eigenvector, U the vector with unit elements. Then $X = RU$, so that the eigenvalue equation

$$MX = \lambda X$$

is equivalent to

$$R^{-1}MR\,U = \lambda U$$

which means that λ is the common row sum of the matrix $R^{-1}MR$. Since the properties of boundary eigenvalues of Gerschgorin disks establish the desired result for the equivalent matrix $R^{-1}MR$ they establish it for M as well.

Since M does not necessarily have uniform row sums, we are now interested in relating λ to the row sums, the largest of which we know to be an upper bound to λ, yet unequal to it. Here the variational definition of r is useful; let us write $r(M)$ to acknowledge the dependence of r on M, and increase M in any way whatsoever, even by increasing one single matrix element. Then each quantity $<x|M|y>$ must strictly increase, and with it both the minima and maxima, leading to a strictly increased maximum eigenvalue.

At this point we could alter some of the matrix elements of M to obtain a new matrix A, whose row sums were uniformly the minimal row sum of M. It is important to do this exclusively by reducing matrix elements in M. Conversely, we could selectively increase the elements of M to obtain a matrix B whose row sums were uniformly the row sums of M, but we already have the result which this would imply.

Applying the theorem on boundary eigenvalues to the matrices A, M, and B, we obtain both upper and lower bounds to λ, defined as r above:

$$\min_i \sum_{j=1}^{n} m_{ij} \leq \lambda \leq \max_i \sum_{j=1}^{n} m_{ij}.$$

However both inequalities are strict unless the bounds coincide. Although an estimate of the gaps would be useful, often the assurance that they exist is sufficient.

9.5 Second largest eigenvalue

The largest eigenvalue of a positive matrix is unique and belongs to eigenvectors with positive components, both the row eigenvector and the column eigenvector. Thus any other eigenvectors must have mixed signs both in their real and possible imaginary parts. Knowledge of the relative magnitude of the second eigenvalue of a matrix is often convenient, for example in judging the rate of convergence of most procedures for diagonalizing the matrix. The traditional way to obtain this information is to subtract the contribution of the largest eigenvalue from the matrix, followed by an estimate of the largest eigenvalue of the remaining matrix. There will often be a pair of second largest eigenvalues, complex conjugates on account of belonging to a real matrix; if so they must be considered together.

Removal of the largest eigenvalue requires an explicit knowledge of both the largest eigenvalue and its eigenvectors. It is convenient to avoid this explicit knowledge by supposing the matrix to have been previously transformed to stochastic form; if it is already stochastic, then no preparation is required. The transformation to a column stochastic matrix is accomplished by incorporating the elements of its row eigenvector into the matrix itself by column divisions and row multiplications, finally dividing the entire matrix by its eigenvalue.

Even so, we would still need to know the components of the column eigenvector to achieve the traditional reduction. By substituting a more readily available vector for the unknown column eigenvector we can still obtain useful estimates. If the matrix were stochastic to start with, no additional knowledge would be required. A plausible choice, which would still leave matrix elements of uniform sign, would be to form a vector by taking the least element in each row, or alternatively, the greatest element in each row.

To recapitulate, suppose that M is a column stochastic matrix so that 1 is its largest eigenvalue, with U^T, the row of ones, as its eigenvector. We are seeking an upper bound for λ, which is another eigenvalue with column eigenvector X, and we have formed the column C by taking the least element from each row of M. Since

$$U^T X = 0$$

we know that both

$$MX = \lambda X$$

and

$$(M - CU^T)X = \lambda X.$$

Either of these two matrices can be used to obtain bounds for λ since it is an eigenvalue of each. Taking Gerschgorin disks from the column sums, we see that the column sums of M are unity and those of C are all equal and equal to the sum of the minimal elements of each row. Thus

$$|\lambda| \leq 1 - \sum_{i=1}^{n} \min_{j=1}^{n} m_{ij}.$$

If maximal elements were utilized, there would be a reversal of sign and

$$|\lambda| \leq \left(\sum_{i=1}^{n} \max_{j=1}^{n} m_{ij} \right) - 1,$$

obviously one would choose the more restrictive of the two bounds.

If there is a considerable discrepancy in the sizes and an unfavorable distribution of matrix elements of M, the bounds provided by these two inequalities may not be very useful; for example the probabilistic de Bruijn matrices generally contain zeroes in every row and column so that the minimal inequality is vacuous. Likewise, the concentration of nonzero elements makes them so large that the maximal inequalities give worse limits than the knowledge that the matrices are stochastic. Nevertheless, powers of the de Bruijn are better behaved, and are more suitable for estimates.

9.6 Averaging and convergence

The fact that the largest eigenvalue of a positive matrix is isolated means that powers of the matrix should converge to a multiple of the idempotent matrix formed from the corresponding eigenvectors. By normalizing the eigenvalue and one of the eigenvectors, as has already been done for stochastic matrices, the limiting form of high powers is even further simplified. The rate of convergence depends on the ratio between the largest and second largest eigenvalue, or directly on the second if the matrix is stochastic. Conversely, one way of estimating the second eigenvalue is to study the behaviour of matrix powers, preferably already having transformed the matrix to stochastic from.

This can be done by commencing with a positive vector $X^{(0)}$ and defining recursively

$$\begin{aligned} X^{(k+1)} &= MX^{(k)} \\ \lambda_k^+ &= \max_i \frac{x_i^{(k+1)}}{x_i^{(k)}} \\ \lambda_k^- &= \min_i \frac{x_i^{(k+1)}}{x_i^{(k)}}. \end{aligned}$$

Then it will be found that the largest eigenvalue satisfies

$$\lambda_0^- \le \lambda_1^- \le \ldots \le \lambda \le \ldots \le \lambda_1^+ \le \lambda_0^+,$$

which confines the eigenvalue to a succession of ever smaller intervals. The outermost, widest, interval follows from the minimax characterization of the largest eigenvalue. So do each of the others, by specializing the choice of the column vector in the same characterization, but the important part of this result is that the intervals are nested.

Nesting is a consequence of multiplying successive vectors by M, but it is sufficient to see how multiplication by M^2 relates to multiplication by M. Moreover it will be more understandable to write the definition of λ_k^+ in terms of

$$MX \le \lambda_k^+ X,$$

the vector inequality being understood as holding for each component, and λ_k^+ being understood as the greatest constant for which the inequality holds. Then we would have

$$\lambda MX \le M(MX) \le \lambda_k^+ MX \le (\lambda_k^+)^2 X,$$

the first inequalith holds since λ itself serves for a general vector such as MX. Given that $\lambda \le \lambda_k^+$, we conclude that $\lambda_{k+1}^+ \le \lambda_k^+$, since it is the factor which applies to this final equation.

There is a corresponding convergence of vectors which is interesting, especially since the arguments apply equally well to any product of stochastic matrices and not just to powers of a single matrix. For simplicity suppose that the family of matrices is column stochastic, which means that U^T, the row of 1's, is a uniform eigenvector for the family. Then the dominant part of such a matrix would be the vector product of some column X with the row U^T. The action of U^T its action in multiplying any column would be to "average" the components of the column, just as its action on a matrix would be to form a new row of column averages. Strictly, an average would also include division by the number of terms summed.

One of the outstanding properties of averaging is its tendency to reduce variance. In the present context this would mean that the product of a large number of positive stochastic matrices would tend to a matrix with uniform rows, independently of whether the same matrix or different matrices were being multiplied. In the particular case of powers of a single matrix, including an additional factor would not change the limit, which would have to be a matrix of eigencolumns.

In verifying the details of the averaging process it is found that certain estimates are required for the size of the matrix elements. In particular, permitting very small or zero elements would be nice for certain applications, but would either complicate or invalidate the proofs. Here we only present one of the preliminary theorems, which describes the reduction in range of the elements of a vector with real components when multiplied by a stochastic matrix.

Suppose that M is the matrix, that X is a real vector whose algebraically largest and smallest components are A and a respectively, and that the corresponding extreme elements of MX are B and b, respectively. We also require the smallest element of M, which we might call ε. For the purpose of argumentation we introduce another vector Y constructed from X by replacing all of its elements save one of the smallest by A. Starting from the evident inequality

$$X \leq Y$$

we obtain

$$MX \leq MY$$

and a requirement to estimate MY. Each component will have the form $\alpha a + (1 - \alpha)A$, which can be rewritten $A - \alpha(A - a)$. The value of α will vary from component to component, but it is always greater than ε, so

$$B \leq A - \varepsilon(A - a).$$

The easiest way to get an estimate for b is to repeat the same argument for $-X$, which reverses all comparisons. Combining the result,

$$-b \leq -a - \varepsilon(-a + A)$$

with the bound for B, we find

$$B - b \leq (1 - 2\varepsilon)(A - a).$$

By its nature, we have $\varepsilon \leq 1/2$, so the range of variation of the components of MX is necessarily less than the corresponding range for X.

If we could rely on a fixed lower limit to ε for an entire family of matrices, it is clear that X would gradually be reduced to a multiple of U if it were multiplied by increasingly larger numbers of matrices of the family.

9.7 Non-negative matrices

The presence of zeroes in an otherwise positive matrix complicates the derivation of its properties. The sporadic presence of zeroes does not greatly complicate the derivations nor change the results; on the contrary certain systematic patterns of zeroes can alter the conclusions considerably. Fortunately it is relatively easy to predict the influence of zeroes, and to state the correct conclusions. Displaying the graph associated with the matrix leads to one of the most concise descriptions that seems to be possible.

The essential point is that zero elements are tolerable in a matrix if they disappear from a power of the matrix. Since powers of a matrix have the same eigenvectors as the original, and powers of the original eigenvalues, the necessary details of proofs can be worked out for the matrix power and then attributed to the matrix itself. Such circumvention could fail in two evident ways: either the zeroes persist through all the powers, or they simply keep shifting around. Both circumstances are fairly evident from an inspection of the graph of the matrix.

In the former case, there are nodes in the graph which are not accessible from one another, so that the basic requirement is for a connected graph. If the nodes are grouped according to their accessibility, inaccessible nodes are reflected in a triangular structure of the matrix which can be handled by partitioning the matrix and applying the relevant theorems to the resulting submatrices. Thus the solution to the problem is to treat disconnected portions of the graph separately, just as it is well to separate out transient parts of the graph and treat them separately from the others.

Even if the graph is connected, it may result that links of only certain lengths may exist; consider for example a graph with two subsets of nodes such that the only direct links occurred between nodes of different sets, none between two nodes of the same set. This structure signifies a partitioned matrix whose diagonal submatrices are zero. Its square would have diagonal submatrices but zero antidiagonal submatrices, while its cube would once again have the original form. But no power at all would be free of blocks of zeroes. Fortunately this alternative has a particularly elegant solution.

For example, the example cited refers to a matrix of the form

$$M = \begin{bmatrix} O & A \\ B & O \end{bmatrix},$$

but there exists another matrix

$$J = \begin{bmatrix} I & O \\ O & -I \end{bmatrix},$$

for which $MJ = -JM$. If

$$X = \begin{bmatrix} U \\ V \end{bmatrix}$$

is an eigenvector of M belonging to λ, then JX is another belonging to the eigenvalue $-\lambda$, as the use of the anticommutation relation readily shows. Thus all nonzero eigenvalues occur in negative pairs, with eigenvectors partitioned so that corresponding components differ at most in sign, in appropriate places.

The squared matrix M^2 is block diagonal, of the form

$$M^2 = \begin{bmatrix} AB & O \\ O & BA \end{bmatrix},$$

but the diagonal components AB and $BA = A^{-1}(AB)A$ are equivalent as long as A is square and non-singular. If it is not, a subspace belonging to eigenvalue 0 can be partitioned off, and the conclusions applied to the complementary space.

Generally M would contain a longer superdiagonal of submatrices, higher roots of unity would occupy the diagonal of J, and the canonical form would require k equivalent copies of a matrix of more tractable form. All eigenvalues would occur in cycles, related by kth roots of unity.

Chapter 10

Zeta function

The evolution and de Bruijn matrices are sparse, both in their normal and in their probabilistic versions; that is a consequence of their origins as connectivity matrices of diagrams of one type or another. It is therefore natural to seek a description of such matrices in terms of the visual properties of the same diagrams, one of the most evident of which is the network of cycles or loops which they contain. It is even more satisfying if the description will accomodate the accumulation of a variety of matrices into families. A recent article of Cvitanovic [58] reveals the process, as applied to an analysis of strange attractors.

10.1 Counting loops

It is convenient to begin with matrices with integer elements, and even more precisely with the connection matrices for diagrams, such as those which form the evolution matrices or the de Bruijn matrices. There is a mutual relationship between matrices and diagrams. To begin with, a diagram can be described in various ways by matrices, one of the most natural being that in which the rows and columns of the matrix are indexed by the nodes of the diagram. The matrix elements are then zero or one, according to whether the node indexing the row is linked to the node indexing the column. Such a matrix need not be symmetric, inasmuch as the links may be directed and not run in both directions.

It is even possible to accomodate multiple links between nodes, by allowing the matrix elements to be positive integers rather then confining them to just the two values zero and one. This interpretation is useful because it allows powers of the connection matrix to be interpreted in terms of the number of composite paths joining two nodes, the power of the matrix determining the number of links involved. Whether or not one would want to suppose that negative matrix elements denote a link in the reverse direction depends upon whether one expects to cancel a path by retracing it or not, a complication which we shall avoid.

Conversely it is possible to represent a square matrix by a diagram, by setting up nodes to correspond to the dimensions of the matrix, and then introducing a link wherever there are non-zero matrix elements. Of course, if none of the matrix elements is zero, every node will then be linked to every other. Although the diagrammatic representation would be most interesting for disentangling the relationships between the elements of sparse matrices, complete mutual accessibility is nevertheless an important special case. In particular, given a finite diagram in which every node can be reached from any other node through a long enough chain of links, there will be a corresponding power of the connection matrix in which no zeroes remain.

Diagonal elements of the connectivity matrix represent nodes linked to themselves. It might seem at first sight that a node would naturally be linked to itself, but it should be borne in mind that links and nodes are two different things, and that even null links have to be shown explicitly. Diagonal elements of powers of the connection matrix are probably easier to understand; aside from possible null links, they would represent loops in which a node was connected to a succession of others, the final link nevertheless returning to the original node.

Generally, the ijth element of the kth power of the connectivity matrix M counts the number of paths with exactly k links leading from node i to node j. The trace, $Tr\{M^k\}$, gives the total number of paths leading back to their point of origin, but under favorable circumstances each path is counted once for each of the k nodes which it contains. This is the case for loops which do not retrace a part of themselves, but not otherwise. Thus

$$N(k) = \frac{1}{k} Tr\{M^k\}$$

is a function which would represent the number of *loops* of length k in the diagram whenever the non-redundancy criterion was met.

Sometimes useful information can be gained by constructing a power series whose coefficients count some quantity or other, particularly if the count can be related to a convolution of the coefficients in some other series. We might begin by forming the series

$$m(t) = \sum_{k=0}^{\infty} M^k t^k.$$

By either ignoring convergence, or hoping for the best, the series is seen to be equivalent to

$$m(t) = (I - Mt)^{-1},$$

which is a variant form of the resolvent of M. It can be used as a generating function for the number of random walks through the diagram of M, but a more interesting generating function results from counting loops rather than random walks. Thus consider the series z(t) defined by

$$z(t) = \sum_{k=1}^{\infty} \frac{1}{k} Tr\{M^k\} t^k.$$

whose coefficients count loops. Since the trace is a linear operator, we can apply it after summing the matrix series, which this time is seen to correspond to a logarithm—the indefinite integral of the series for $m(t)$. Thus we have

$$z(t) = Tr\{-\ln(I - Mt)\}.$$

Since the logarithm is an awkward function to compute for matrices, it is fortunate that there is an identity relating exponentials of traces to determinants of exponentials, namely

$$\exp(Tr\{X\}) = Det\{\exp(X)\},$$

which can be used to define the *zeta function*,

$$\zeta(t) = \exp z(t) = Det\{m(t)\}.$$

Because the loop counts are embedded in an exponential they must be evaluated by forming logarithmic derivatives of the zeta function at the origin, but this is a small price to pay for working with such a convenient quantity as the determinant of the "resolvent" (note that t is not in its accustomed place).

Certain properties of the zeta function can be deduced from its representation as a determinant. Although the determinant of a product is a product of determinants, the involvement of the variable t in the resolvent precludes resolvents having the same property. However a matrix will sometimes have a block diagonal structure allowing the determinant of the matrix to be related to the determinants of the blocks; an example would be the connection matrix for two independent diagrams.

10.2 Traces, $\chi(t)$, $\zeta(t)$

Although the zeta function was not always given that name, there is a classical relationship between it, the characteristic polynomial, and the traces of the powers of a matrix, which are simply restatements of certain properties of polynomials. By convention, the characteristic polynomial of the matrix M is defined to be the determinant

$$\chi(t) = |M - It|$$

which is equivalent to

$$\chi(t) = \prod_{i=1}^{n}(\lambda_i - t),$$

for eigenvalues λ_i. Written as a polynomial in powers of t,

$$\chi(t) = \sum_{i=0}^{n}(-1)^i a_i t^{n-i},$$

wherein the coefficients a_i can be expressed either as homogeneous products of the roots λ_i or as sums of principal minors, both according to well known formulas.

For the transformations which we intend to make, it is more convenient to use $1/t$ as a variable or, to avoid confusion, to introduce the function

$$\varphi(t) = |I - Mt|$$

whose factored form is

$$\varphi(t) = \prod_{i=1}^{n}(1 - \lambda_i t).$$

As a polynomial it differs from χ only in attaching the coefficients a_i to ascending rather than descending powers of t:

$$\varphi(t) = \sum_{i=0}^{n} (-1)^i a_i t^i.$$

Both χ and φ are finite polynomials, whose reciprocals could be represented as power series, even treating them formally and ignoring questions of convergence. We are interested in ζ as the reciprocal of φ,

$$\zeta(t) = \prod_{i=1}^{n} (1 - \lambda_i t)^{-1},$$

which can be represented as a single power series by representing each factor in the denominator as a geometric series, multiplying them term by term, and collecting coefficients.

In that case

$$\zeta(t) = \sum_{i=0}^{\infty} h_i t^i,$$

where h_i is the sum of all possible homogeneous products of λ_i; when $n = 2$, for instance, $h_2 = \lambda_1^2 + \lambda_1\lambda_2 + \lambda_2^2$.

Of more interest are the coefficients of $z(t) = \ln\zeta(t)$

$$z(t) = \sum_{i=0}^{\infty} m_i t^i,$$

for which

$$m_i = \sum_{j=1}^{n} \lambda_j^i;$$

in other words, the traces of the powers of M.

The practical situation is that the coefficients m_i are always fairly easy to obtain, while the quantities that are really desired are either the coefficients a_i or h_i. Comparison of the series involved in each of the definitions yields equations of convolution type relating the coefficients, typically in the form of single determinants. Recovering the eigenvalues from the coefficients is much harder since that is just the problem of finding the roots of a polynomial; all the same, the task is hard to avoid since it is likely that some or all of the eigenvalues will be sought for.

Newton's identities relate the power sums m_i to the coefficients a_i of the characteristic polynomial; in the present context they are obtained by differentiating the formula $\varphi(t) = \exp(-z(t))$ to obtain $\varphi'(t) = -\varphi(t)z'(t)$. Comparing coefficients yields the convolution (with $a_0 = 1$)

$$ia_i = \sum_{j=1}^{i} (-1)^{j-1} m_j a_{i-j},$$

which can be solved for either a or m in terms of the other.

For example, the system of equations

$$\begin{bmatrix} m_1 \\ m_2 \\ m_3 \\ m_4 \\ . \end{bmatrix} = \begin{bmatrix} 1 & . & . & . & . \\ m_1 & -2 & . & . & . \\ m_2 & -m_1 & 3 & . & . \\ m_3 & -m_2 & m_1 & -4 & . \\ . & . & . & . & . \end{bmatrix} \begin{bmatrix} a_1 \\ a_2 \\ a_3 \\ a_4 \\ . \end{bmatrix}$$

is readily solved for the a's, to obtain

$$a_i = \frac{1}{i!} \begin{vmatrix} m_1 & 1 & . & . & \ldots & . & . \\ m_2 & m_1 & 2 & . & \ldots & . & . \\ m_3 & m_2 & m_1 & 3 & \ldots & . & . \\ \ldots & & & & & & \\ m_{i-1} & m_{i-2} & m_{i-3} & m_{i-4} & \ldots & m_1 & i-1 \\ m_i & m_{i-1} & m_{i-2} & m_{i-3} & \ldots & m_2 & m_1 \end{vmatrix}.$$

For example, $a_2 = \frac{1}{2}(m_1^2 - m_2)$, $a_3 = \frac{1}{6}(m_1^3 + 2m_3 - 3m_1m_2$.

10.3 Infinite de Bruijn matrix

It is rare that we will find a matrix which meets the non-reduncancy criterion for *all* of its cycles, so consideration should be given to the contrary case. The situation is nicely illusrated for the de Bruijn matrices, which have cycles of all possible varieties while still retaining a certain amount of systematic order. The shift-register interpretation of a de Bruijn matrix makes it clear that an n-stage matrix will have 2^n distinct cycles; but those which possess translational symmetry will be redundant. For example, a cycle of 5 zeroes will show up only once in the fifth power of a two-stage matrix, whereas the string 11010 will show up five times and be counted as one single loop in evaluating $\frac{1}{5}Tr\{M^5\}$.

There would appear to be two alternatives: either insist that the name "zeta function" refer exclusively to the generating function which counts loops although it might not incorporate the resolvent, or retain the elegance of the resolvent even though it may not always count loops correctly. It would appear that the latter alternative is the one that is generally chosen; moreover that such a definition was implicit in classical treatments of the theory of equations even though such a name was never used explicitly.

Since we know the characteristic equation of a finite de Bruijn matrix, it is a straightforward matter to see that they all have the same zeta function

$$\zeta(t) = (1 - 2t)^{-1}$$

from which we obtain

$$z_k = \frac{2^k}{k}$$

as the "number of cycles" of length k, although we know that this is in no way a natural representation of the true state of affairs. At least for de Bruijn matrices, we can get some idea of the discrepancy; using the probabilistic de Bruijn matrix produces an even more general result.

Beginning with the two stage matrix whose elements incorporate both a probability and the parameter t ($t_{ij} = p_{ij}t$),

$$\begin{bmatrix} t_{00} & t_{01} \\ t_{10} & t_{11} \end{bmatrix},$$

we have

$$\begin{aligned} Tr\{B\} &= t_{00} + t_{11} \\ Tr\{B^2\} &= t_{00}^2 + t_{11}^2 + 2t_{01}t_{10} \\ Tr\{B^3\} &= t_{00}^3 + t_{11}^3 + 3(t_{00} + t_{11})t_{01}t_{10} \\ Tr\{B^4\} &= t_{00}^4 + t_{11}^4 + 2(t_{01}t_{10})^2 + \\ & \quad 4(t_{00}^2 + t_{00}t_{11} + t_{11}^2)t_{01}t_{10}. \end{aligned}$$

On the other hand if we work with the three stage matrix,

$$\begin{bmatrix} t_{000} & t_{001} & \cdot & \cdot \\ \cdot & \cdot & t_{010} & t_{011} \\ t_{100} & t_{101} & \cdot & \cdot \\ \cdot & \cdot & t_{110} & t_{111} \end{bmatrix},$$

we have

$$\begin{aligned} Tr\{B\} &= t_{000} + t_{111} \\ Tr\{B^2\} &= t_{000}^2 + t_{111}^2 + 2t_{010}t_{101} \\ Tr\{B^3\} &= t_{000}^3 + t_{111}^3 + 3t_{001}t_{010}t_{100} + 3t_{011}t_{110}t_{101} \\ Tr\{B^4\} &= t_{000}^4 + t_{111}^4 + 2(t_{010}t_{101})^2 + \\ & \quad 4t_{000}t_{001}t_{010}t_{100} + 4t_{001}t_{011}t_{110}t_{100} + \\ & \quad 4t_{011}t_{111}t_{110}t_{101}. \end{aligned}$$

The first thing to be observed is that the structure of these matrices is quite similar, thanks to the fact that the de Bruijn matrices form a very regular family. In fact the same cycles will reappear in all the trace formulas, the only difference being the length of the overlappable fragment represented in each subscript. For example, $Tr\{B\}$ will always be the sum of two contibutions, one for the string 0^* and the other for the string 1^*; likewise a contribution from these two strings and another from $(01)^*$ will always appear in $Tr\{B^2\}$.

Another observation, although it is just barely evident from the small number of terms that have been written, is that the series can be rearrranged to group the terms belonging to one particular cycle into its own logarithmic series. It is clear that the terms t_{00} and t_{11} in the first example can be so arranged. There is an indication that the term $t_{01}t_{10}$ would participate in another, since the factor 2 cancels opportunely wherever it occurs.

Recalling the shift register interpretation of de Bruijn diagrams, we see that every cycle which has no internal translational symmetry will show up k times in

$Tr\{B^k\}$ only to be divided by k in the term $\frac{1}{k}Tr\{B^k\}$. If there is an internal symmetry present, the unit cell will have length ℓ; its cycle will occur to the power k/ℓ, but will only be ℓ such terms in the trace. Note how $(t_{01}t_{10})^2$ has coefficient 2, not 4, in $Tr\{B^4\}$, or that t_{00}^4 only has coefficient 1.

The series for $z(t)$ belonging to the two-stage de Bruijn matrix could then be rearranged to take the form

$$
\begin{aligned}
z(t) &= t_{00} + \frac{1}{2}t_{00}^2 + \frac{1}{3}t_{00}^3 + \ldots \\
&+ t_{11} + \frac{1}{2}t_{11}^2 + \frac{1}{3}t_{11}^3 + \ldots \\
&+ t_{01}t_{10} + \frac{1}{2}(t_{01}t_{10})^2 + \frac{1}{3}t_{00}^3 + \ldots \\
&- \ln(1 - t_{00}t_{01}t_{10}) - \\
&- \ln(1 - t_{01}t_{11}t_{10}) - \\
&- \ldots
\end{aligned}
$$

In turn this produces a zeta function

$$
\begin{aligned}
\zeta(t)^{-1} &= (1 - t_{00})(1 - t_{11})(1 - t_{01}t_{10}) \times \\
&\quad (1 - t_{00}t_{01}t_{10})(1 - t_{01}t_{11}t_{10}) \times \\
&\quad \ldots
\end{aligned}
$$

This is a very beautiful and elegant formula, and also blatantly at odds with the fact that B has only two eigenvalues, its characteristic polynomial can only vanish for two values of t, and that its zeta function should have at most two poles. Clearly it is a formula which does not meet the requirements for a convergent infinite product.

10.4 Cluster expansion

Before feeling unduly discouraged about the discrepancy between the apparent number of poles of the zeta function and the dimension of its matrix, we should examine the zeta function more carefully. Since we know the exact zeta function for the normal de Bruijn matrix B, all of whose non-zero matrix elements have the form $t_{ij} = t$, we would want to examine

$$
\begin{aligned}
\zeta(t)^{-1} &= (1 - t)(1 - t)(1 - t^2)(1 - t^3)(1 - t^3) \times \ldots \\
&= (1 - 2t + t^2)(1 - t^2)(1 - t^3)(1 - t^3) \times \ldots \\
&= (1 - 2t + 2t^3 - t^4)(1 - t^3)(1 - t^3) \times \ldots \\
&= (1 - 2t + 3t^4 - 3t^6 + 2t^9 - t^{10}) \times \ldots \\
&= \ldots \\
&= (1 - 2t).
\end{aligned}
$$

The limit expression for the zeta function is exactly the one that was expected, although truncating the product representation or otherwise taking it too literally leads to the appearance of spurious poles. Although the algebra is slightly more complicated for the probabilistic matrix, multiplying out the factors of the zeta function reproduces the characteristic polynomial for the general case just as well as it does for the normal case. In principle it must be possible to reconstruct the factored form of the characteristic polynomial, but to do so directly is little more than an exercise in going around in circles. Instead we should utilize the fact that the zeta function has been described in terms of cycles of matrix elements.

The difference between primitive and composite cycles is an important one. A cycle is primitive when its translational symmetry is trivial. Composites are aggregates of primitive cycles, not necessarily connected to one another. Thus, for a two stage diagram $(t_{00})(t_{01}t_{10})$, joins into one long cycle with repeated nodes; but in a three stage diagram the analogous $(t_{000})(t_{010}t_{101})$ forms two disconnected cycles. Because their primitive cycles are selfcontained and independent, composites are commutative, but they never contain repeated factors.

Let c stand for a product of matrix elements forming a cycle, such as $t_{00}t_{01}t_{10}$ and let C be a composite cycle, such as $(t_{00})(t_{01}t_{10})$. Let $|C|$ denote the total length of the composite, but let 1 be the composite cycle of length 0. Finally, let $\mu(C)$ be the parity of the composite cycle, which is required to expand the product of differences.

If a three-valued definition of parity is used, making it vanish for composites with repeated factors; and taking it to be unity for the null composite, μ is the classical Möbius function redefined for cycles:

$$\mu(C) = \begin{cases} -1 & \text{odd parity} \\ 0 & \text{repeated factor} \\ 1 & \text{even parity} \end{cases} .$$

Then, according to the distributive law, the product

$$\zeta(t) = \prod_{\text{cycles}} (1 - c)$$

can be replaced by a sum

$$\zeta(t) = \sum_{\text{composites}} \mu(C)C.$$

If the composites are grouped by length, the sum presents once again the form of a polynomial:

$$\zeta(t) = \sum_{i=0}^{\infty} \sum_{|C|=i} \mu(C)C.$$

If the dimension of B were finite, we would expect all the terms of high degree in this sum to vanish. For example, the terms with $|C| = 3$ belonging to the two-stage de Bruijn matrix are

$$(t_{00})(t_{01}t_{10}) + (t_{11})(t_{01}t_{10}) - (t_{00}t_{01}t_{10}) - (t_{11}t_{01}t_{10}),$$

which evidently vanishes. The corresponding term for the three-stage matrix is

$$(t_{000})(t_{010}t_{101}) + (t_{111})(t_{010}t_{101}) - (t_{001}t_{010}t_{100}) - (t_{110}t_{101}t_{110}),$$

which need not.

Thus we have a scheme which is capable of displaying the structure of the characteristic polynomial of a matrix explicitly in terms of loops formed from the elements of the matrix. Nevertheless, the same results would also have followed from a careful analysis of the characteristic determinant, although the assignment of parity to the terms is phrased in a somewhat different form.

Using subscripts to identify the cycles by their lexicographically ordered repeating unit, the zeta function for binary de Bruijn matrices takes a form independent of the number of stages, namely

$$\begin{aligned} \zeta(t)^{-1} &= 1 - (c_0 + c_1) - (c_{01} - c_0 c_1) - \\ &- (c_{001} + c_{011} - c_0 c_{01} - c_{01} c_1) - \\ &- (c_{0001} + c_{0011} + c_{0111} - c_0 c_{001} - c_{011} c_1 - c_0 c_{01} c_1) - \\ &- \ldots . \end{aligned}$$

When it is no longer possible to find long cycles which are not composites of shorter cycles, it may be concluded that the series represents a matrix of a finite number of stages; otherwise there is no intrinsic limitation on the series.

The parameters upon which this series depends are no longer the probabilities of individual links in the diagram, but rather the collective probabilities belonging to the primitive cycles. There is also a tendency to disguise the contribution of the block determinants to the zeta function, but this simply means that we have a complementary approach to its interpretation.

10.5 Reduced evolution matrix

The zeta function approach is useful for determining the characteristic equation of a matrix, or even of a family of matrices with a similar structure. Its best application lies in contexts where the parameters associated with all cycles beyond a certain length are known to be negligible, or at least thought to be so. Information about eigenvalues and quantities derived from them is available, but eigenvectors are beyond the scope of the zeta function. Since reciprocal eigenvalues determine the poles of the zeta function, it would seem to be best suited for determining the first few of the largest eigenvalues of a family of matrices.

The reduced evolution matrix has a constant column sum, so its largest eigenvalue is a foregone conclusion. The equilibrium eigenvector is an object of interest, but cannot be encountered via the zeta function. The second largest eigenvalue determines the stability of equilibrium, making it a worthwhile quantity to determine.

The non-zero elements e_{xy} of the reduced evolution matrix are those for which there are cells i and j for which $\varphi(ixj) = y$. Diagonal elements must therefore satisfy

$$\varphi(ixj) = x$$

and therefore are given by the number of ways that x is surrounded by acceptable neighbors; but in any event are determined by the still lifes.

Diagonal elements of E^2 are those for which

$$\varphi(ixj) = y, \varphi(kyl) = x,$$

so that if there is a cycle of length two, there will be a corresponding diagonal element. Continuing the analysis, it is clear that there will be diagonal elements in the powers of E for all the periods whose cycle is equal to the block length. The fact that the least pole of the zeta function is known will exercise a certain restraint on the possible values which the cycle probabilities can assume.

It is in the nature of a reduced evolution matrix that the actual cycles will not just be counted, but overcounted to the extent that they can be broken open and embedded in a larger neighborhood whose central segement follows the same cycle; such is the influence of the dangling cells in the formulas above. Nevertheless this discrepancy is expected to diminish as longer blocks are used, or if the reduced evolution matrix for iterated evolution were used.

Let us work out some of the reduced evolution matrices for Wolfram's (2,1) Rule 22: Its symbolic de Bruijn matrix is

$$\begin{bmatrix} 0 & 1 & . & . \\ . & . & 1 & 0 \\ 1 & 0 & . & . \\ . & . & 0 & 0 \end{bmatrix}.$$

giving the two matrices

$$\alpha = \begin{bmatrix} 1 & . & . & . \\ . & . & . & 1 \\ . & 1 & . & . \\ . & . & 1 & 1 \end{bmatrix}, \beta = \begin{bmatrix} . & 1 & . & . \\ . & . & 1 & . \\ 1 & . & . & . \\ . & . & . & . \end{bmatrix}.$$

from whose products the number of counterimages of any given configuration can be determined. These products give directly the row sums of the reduced evolution matrix, but the individual matrix elements have to be worked out on their own account. Evolution matrices describe evolution differently, sorting counterimages by their central strings, rather than by the cells with which they begin and end.

The reduced evolution matrix for 1-blocks is

$$\begin{bmatrix} 2 & 3 \\ 2 & 1 \end{bmatrix}.$$

The zeta function is

$$\begin{aligned} \zeta(t) &= \frac{1}{(1-4t)(1+t)} \\ &= 1 + 3t + 17t^2 + \ldots \end{aligned}$$

For 2-blocks it is

$$\begin{bmatrix} 1 & 1 & 1 & 4 \\ 1 & 1 & 1 & . \\ 1 & 1 & 1 & . \\ 1 & 1 & 1 & . \end{bmatrix}$$

with the same zeta function, in which the presence of two zero eigenvalues is not visible

$$\begin{aligned}\zeta(t) &= \frac{1}{(1-4t)(1+t)} \\ &= 1+3t+17t^2+63t^3+257t^4+\ldots\,.\end{aligned}$$

For 3-blocks it is

$$\begin{bmatrix} 1 & . & . & 2 & 1 & 1 & 2 & 4 \\ 1 & . & . & . & . & . & 2 & . \\ . & 2 & 1 & . & . & . & . & . \\ . & 2 & 1 & . & 1 & 1 & . & . \\ 1 & . & . & 2 & . & 1 & . & . \\ 1 & . & . & . & . & 1 & . & . \\ . & . & 1 & . & 1 & . & . & . \\ . & . & 1 & . & 1 & . & . & . \end{bmatrix}.$$

Chapter 11

Cellular automata via de Bruijn diagrams

Graph theory plays several important roles in the theory of cellular automata, one of which consists in describing the evolution of the automaton, and another of which consists in relating local properties to global properties. Evolution is described by local rules mapping cell neighborhoods into its subsequent state; because successive neighborhoods overlap it is important to be able to take the overlap into account when relating the behavior of successive cells to one another. In illustration, the de Bruijn diagram and its subdiagrams are applied to the study of cellular automata in one dimension.

An automaton has *states* and *mappings* between its states. The concept arose from electrical switching theory, but thanks to the pioneering work of Claude Shannon [207], boolean algebra and probability theory were seen to have an important role in the theory of switching networks. Speculation concerning the relationship between logical processes and the physiology of the nervous system, modelled by networks of cells, had early origins in the work of Warren McCulloch and Walter Pitts [156]. Numerous subsequent investigators have contributed to the subject, in forms ranging all the way from abstract mathematical logic and formal language theory to concrete engineering and biological simulations and experiments.

A *cellular* automaton lends itself to rigorous analysis because of the symmetry inherent in a crystallographic lattice of individual automata—the cells—and the assumption that individual changes of state depend solely on the states of a limited number of neighbors. At first sight the extensive overlapping between neighborhoods of different cells seemed to confuse the issue, but then it was found that certain ideas used in shift register theory and various techniques from information theory took care of the overlapping in a very natural and elegant way.

Without going into the details of automata theory, our objective is to present graph theory from a point of view useful for automata theory, while also exhibiting some important theorems for this application.

11.1 Five representations

Graphs have varied representations, each of which has its advantages and its limitations.

11.1.1 Mathematical definition

From a purely abstract point of view, a graph is a pair (N, L) consisting of *nodes* N and *links* L. The nodes are just a set, often taken to be the positive integers from 0 to $n-1$; the links are a subset of $N \times N$

11.1.2 Simple drawing

The nodes are laid out as dots on a sheet of paper, frequently labelled by the members of the set N. The links are represented by arrows joining the nodes, the arrow runs *from* node i *to* node j if $(i, j) \in L$. Note that there is no requirement that both (i, j) and (j, i) belong to L, nor that (i, i) either belongs or does not belong.

It is a classical problem in graph theory, to decide whether a graph may be drawn in two dimensions so that its links do not cross; this point does not interest us except insofar as the aesthetics of laying out the graph is concerned.

11.1.3 Topological matrix

The topological matrix M_{ij} is defined by

$$M_{ij} = \begin{cases} 1 & (i,j) \in L \\ 0 & \text{otherwise} \end{cases} .$$

The most important application of the topological matrix consists in the fact that its powers describe paths between nodes containing multiple links, according to the power involved. Since the elements of such matrices are integers the powers are useful for counting total numbers of paths, but not to classifying the paths into different categories. The reason for this is apparent from examining the definition of matrix multiplication,

$$\begin{aligned} [M]^2_{ij} &= \sum_k M_{ik} M_{kj} \\ &= \sum_k \text{link}(ik)\text{link}(kj) \\ &= \sum_k \text{link}(ikj). \end{aligned}$$

Diagonal elements of this matrix count loops; the trace yields the total number of loops, counting each of them once for every node which they contain. Thus

$$\begin{aligned}\phi(t) &= \sum_{n=1}^{\infty}\frac{t^n}{n}\text{Trace}(M^n)\\ &= -\text{Trace}(\ln(I-tM)).\end{aligned}$$

is a generating function for the prime loop count. By prime loop is meant one for which portions of the loop are not retraced. By using a determinantal identity for the matrix exponential, it is possible to obtain a generating function involving the characteristic equation of M (and using a matrix version of ϕ).

$$\begin{aligned}\zeta(t) &= \text{Det}(e^{\phi(Mt)})\\ &= \text{Det}(I-tM)^{-1}\\ &= \chi(t)^{-1},\end{aligned}$$

where n is the dimension of M and χ is its characteristic polynomial. Actually this is a variant definition of the characteristic polynomial, often seen in integral equation theory; by associating t with M rather than I the limit of large n is better behaved.

11.1.4 Symbolic equations

To describe a graph symbolically, let the nodes be represented by variables, such as the capital letters X, Y, Z, and the links by constants, such as the lower case letters a, b, c. *Entrance equations* for the graph may be written in the form

$$X = Ya + Zb + \lambda$$

which say that one arrives at the node X by having previously arrived at the node Y and then following the link a, by having arrived at Z followed by b, or by simply remaining at the node X. The latter alternative is used when X is a terminal point, otherwise the term λ would be dropped from the equation.

Conversely *exit equations* take the form

$$X = aY + bZ + \lambda$$

with λ marking initial nodes. In either event, the entire diagram is described by a system of equations which can be written in the appropriate symbolic matricial form,

$$\begin{aligned}\mathbf{M} &= \mathbf{AM} + \mathbf{R}\\ \mathbf{M} &= \mathbf{MA} + \mathbf{R},\end{aligned}$$

whose symbolic matricial solutions according to Arden's lemma,

$$\begin{aligned}\mathbf{M} &= \mathbf{A}^*\mathbf{R}\\ \mathbf{M} &= \mathbf{RA}^*,\end{aligned}$$

have been described by John Conway [48], and in considerably greater detail by Backhouse and Carré [10]. The essence of the method consists in a recursive application of the identity

$$\begin{bmatrix} A & B \\ C & D \end{bmatrix}^* = \begin{bmatrix} \alpha & \beta \\ \gamma & \delta \end{bmatrix},$$

to a suitably partitioned **A**, where

$$\begin{aligned} \alpha &= (A^*BD^*C)^*A^* \\ \beta &= (A^*BD^*C)^*A^*BD^* \\ \gamma &= (D^*CA^*B)^*D^*CA^* \\ \delta &= (D^*CA^*B)^*D^*. \end{aligned}$$

Trying to split **A** in half is an efficient way to reduce the amount of calculation required; removing the first row and column yields very tractable iterative formulas. Alternatively, rewriting **A** as a sum via sundry regular expression identities yields analogues of the LR factorization of matrices, wherein the stars of triangular factors are much easier to compute than for full matrices.

11.1.5 Regular expression

The solutions of linear symbolic equations are regular expressions, which represent sets. In other words regular expressions furnish a concise description of the paths through a diagram rather than the diagram itself. Comparisons between regular expressions therefore translate into comparisons between the sets which they represent, and conversely. As a result, boolean operations such as intersection and complementation are possible on regular expressions; union already forms part of the algebra of regular sets.

Likewise, due to the distributive and associative laws, it is possible for diverse regular expressions to represent the same set, creating the problem of recognizing the equality of two different regular expressions.

11.2 Comparison of diagrams

One can discuss the ordering and the equivalence of diagrams. From the first, one might stipulate that one diagram is larger than another if it has all the nodes of the smaller; likewise all the links. If $D_1 = (N_1, L_1)$ is one diagram and $D_2 = (N_2, L_2)$ is another, then $D_1 \leq D_2$ would mean

$$N_1 \subset N_2 \text{ and } L_1 \subset L_2.$$

In short, subdiagrams would be smaller. However we are also interested in comparing paths through diagrams, and not necessarily the diagrams themselves. So mappings between diagrams could hold our interest because of wanting to embed a given path in an equivalent diagram which could be more favorable.

There are many other reasons for such an interest in mappings. Drawings can be laid out quite differently as the groupings of nodes are changed or the crossing of links is avoided; regular expressions can be rearranged according to the distributive and associative laws, and so on. At the very least, mappings between different formulations are desirable, to be able to compare them to one another, and to recognize equivalent formulations.

11.2.1 Order relation

Consequently a more general point of view might hold that the larger graph contains all the paths of the smaller, especially the loops. Actually, the image of a loop has to be a loop because any point of closure has the same image as the point of departure; but additional loops may exist in the image. To repeat, we want to say that a larger graph has more paths, even if they are degenerate; such a viewpoint is still consistent with embedment into a graph that the first sense considers larger.

Consider (writing ab for (a,b) whenever it is unambiguous to improve legibility)

$$\{a, b; ab, ba\}$$

and

$$\{a, b, c, d; ab, bc, cd, da\};$$

these graphs are diagrammed in Figure 11.1.

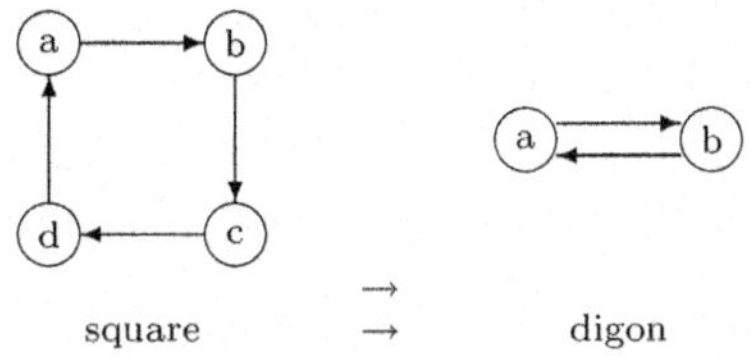

Figure 11.1: A digon is larger than a square, which has no loops of length 2.

The second, larger, is a digon with its vertices linked to one another; the first, smaller, is a square with successive vertices linked. Mapping a to a, b to b, c to a, and d to b, it is seen that cycles around the square runs twice around the digon, with similar relationships for other paths.

The interesting point is that the larger is not a superset of the larger, but rather it contains a functional image. One might say that $D_1 \leq D_2$ means that there exists a function such that

$$\begin{aligned} f(N_1) &\subseteq N_2 \\ \text{link}(a, b) &\Rightarrow \text{link}(f(a), f(b)). \end{aligned}$$

This definition reduces to the previous one when f is the injection map that inserts a subset into the set containing it, but allows a comparison which extends far beyond pairs of diagrams in which one is a subset of the other.

Strictly, it must be proved that this general comparison is an order relation. The only complicated point is proving antisymmetry, that when $f(X) \subseteq Y$ and

$g(Y) \subseteq X$ (and links are conserved), X and Y are at least equivalent if not identical. The mapping $f(x) = 2x$ applied in succession to the integers, even integers, and the multiples of four shows that this need not be true in general, although it works out well enough for finite sets.

11.2.2 Least upper bound, or union

Consider the graphs

$$\{a, b; abba\}$$

represented by a digon, and

$$\{a, b, c; ab, bc, ca\}$$

represented by a triangle. The loops of the first have length 2, those of the second, length 3. It would seem that loops belonging to one or the other would belong to their union, in which links would exist if and only if they had already existed in the components.

$$(N_1, L_1) \cup (N_2, L_2) = (N_1 \cup N_2, L_1 \cup L_2)$$

The sense of inclusion in the definition of the order relation was chosen so that one of the two connected components could be ignored while making comparisons. In general, it is profitable to consider a graph as a union of its disconnected parts and to observe that conversely, nontrivial unions are disconnected.

The topological matrix of the union of two diagrams is the direct sum of the topological matrices of the constituents.

11.2.3 Greatest lower bound, or intersection

Consider the same graphs,

$$\{a, b; ab, ba\}$$

representing a digon, and

$$\{a, b, c; ab, bc, ca\}$$

representing a triangle. The loops of the first have length 2, those of the second, length 3. It would seem that loops common to the two would have length 6; what is the diagram which would express this fact? The evident candidate is the cartesian product of the two graphs, linkage between its points occurring only if both coordinates were linked, for which we use the symbol $L_1 \sharp L_2$.

$$(N_1, L_1) \cap (N_2, L_2) = (N_1 \times N_2, L_1 \sharp L_2)$$

Effectively, linkage in the cartesian product is defined by

$$\text{link}((a, b), (c, d)) = \text{link}(a, c) \& \text{link}(b, d).$$

In this example, the intersection has nodes

$$\{(a, a), (a, b), (a, c), (b, a), (b, b), (b, c)\},$$

with links

$$\{((a,a),(b,b)),((b,b),(a,c)),((a,c),(b,a)),$$
$$((b,a),(a,b)),((a,b),(b,c)),((b,c),(a,a))\}.$$

Graphed by usual conventions, a folded hexagon results, manifesting the required cycle of length 6, as seen in Figure 11.2.

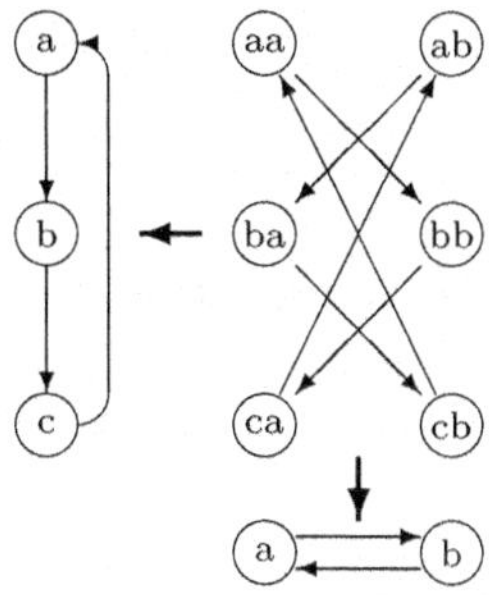

Figure 11.2: Greatest lower bound of a digon and a triangle.

The topological matrix of the intersection of two diagrams is the tensor product of the topological matrices of the constituents.

11.3 Mappings and equivalences

As an interesting point, consider the proposition that a graph is the intersection of two copies of itself, yet the cartesian square or the tensor square of a diagram has quite a few more nodes. However they can be grouped into components, one single image of the graph repeated once for each node. A similar anomaly appears with the union of a graph with itself, both situations showing that the ordering is incompletely defined unless the role of functional mappings and equivalences is clearly understood.

To promote such an understanding, consider the cartesian product of the digon with itself. The nodes

$$\{(a,a),(a,b),(b,a),(b,b)\}$$

are linked by

$$\{((a,a),(b,b)),((b,b),(a,a)),$$
$$((a,b),(b,a)),((b,a),(a,b))\}$$

which are seen to form two independent digons; this is shown in Figure 11.3. Either of them can be mapped onto the original digon by one or another of various mappings, or the pair can be mapped in two to one fashion by either of the two coordinate projections.

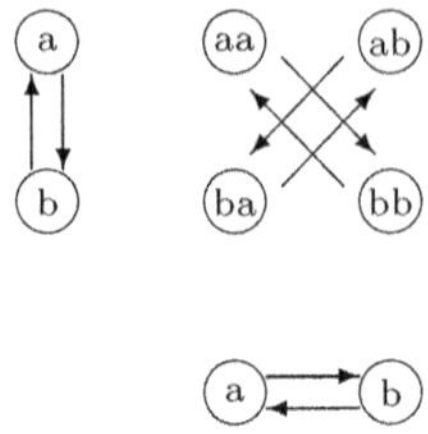

Figure 11.3: Greatest lower bound of a digon with itself.

Evidently graphs can have many representations, making it desirable to have ways of comparing them to one another; functional mappings among the nodes, which preserve linkage, provide a good mechanism.

11.3.1 Homomorphisms

A mapping from the nodes of one graph to those of another which preserves linkages is called a *homomorphism*, whose definition was already introduced while defining the ordering of graphs. In short, the larger graph is a homomorphic image of the smaller, even though the scarcity of its nodes will run counter to this intuition.

However, there is a subtle point which concerns the mappings of non-links — should they also be respected? As stated, the definition allows a single node linked to itself to be the maximum of all graphs; the requisite mapping is simply a constant mapping. All images are thereby linked, even when their counterimages are not. Another instance where the difference is felt concerns the difference between equivalence relations and congruence relations.

Equivalence classes are the constant value contours of functions; congruence classes are contours of homomorphisms. It might be interesting to impose a uniformity requirement, such as stipulating that

$$if \left\{ \begin{array}{ccc} a & \equiv & b \\ x & \equiv & y \end{array} \right. \quad \text{then} \quad \text{link}(a,x) \;\Rightarrow\; \text{link}(b,y).$$

In other words, links would always have to be uniform throughout congruence classes, but a moment's thought shows that this need not even be true for all cartesian products, where one would surely expect such a result to be valid. Uniformity would disallow too many reasonable non-links; if non-links as well as links had to be preserved, this congruence relation would characterize such a homomorphism.

A graph of the sort required, where every member of a set of p elements is linked to all the members of a set of q elements is traditionally named $K_{p,q}$; $K_{p,p}$ is simply K_p.

As homomorphisms have been defined, the only requirement which two congruence classes have to satisfy is that if their images are linked, at least one pair of representatives have to be linked; this is still enough of a requirement to distinguish between equivalences and congruences. Let P and Q be congruence classes; then

$$\text{link}(P, Q) \equiv (\exists\, p \in P, q \in Q) \ni \text{link}(p, q). \qquad (11.1)$$

The discarded version would have read:

$$\text{link}(P, Q) \equiv (\forall\, p \in P, q \in Q)\text{link}(p, q). \qquad (11.2)$$

A new diagram constructed from the congruence classes of another will have fewer nodes, still preserving the links of the original diagram. Such replacements are often sought, in the interest of economy. In the extreme case of a maximum congruence relation, the final graph will have a minimum number of nodes, a unique representation of the original diagram and all others homomprohic to it.

Consider again the intersection of two diagrams, while supposing that $(a, b) \equiv (c, d)$ means that $b = d$ and that $\text{link}((u, v), (w, x))$ means $\text{link}(v, x)$ with respect to equivalence classes. Then the projection $\pi_2(a, b) = b$ is a mapping from the intersection to its second factor whose equivalence classes correspond to the first factor; there is a similar projection π_1 for the first factor. Either is a congruence relation because the mapping is defined uniformly for the entire equivalence class. Figure 11.4 illustrates this relationship.

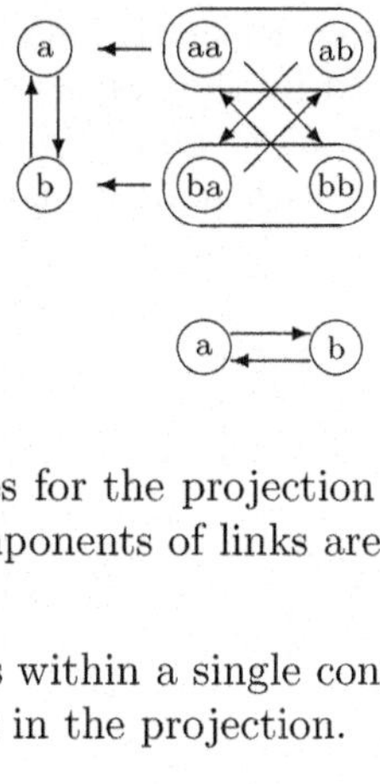

Figure 11.4: Congruence classes for the projection of a cartesian product onto its second coordinate; vertical components of links are respected.

Note that there are no links within a single congruence class, as a consequence of which there are no self-loops in the projection.

11.3.2 Matrix representation

A rectangular matrix can represent a mapping between the nodes of one diagram and another; each column belonging to a node will have nonzero elements in the rows corresponding to the nodes mapping into it. In fact, the rows will be either orthogonal or identical, according to whether images are different or equal. The following relation is the matrix version of the mapping of a square to a digon shown in Figure 11.1:

$$\begin{aligned}
\begin{bmatrix} \cdot & 1 & \cdot & \cdot \\ \cdot & \cdot & 1 & \cdot \\ \cdot & \cdot & \cdot & 1 \\ 1 & \cdot & \cdot & \cdot \end{bmatrix}
\begin{bmatrix} 1 & \cdot \\ \cdot & 1 \\ 1 & \cdot \\ \cdot & 1 \end{bmatrix}
&= \begin{bmatrix} \cdot & 1 \\ 1 & \cdot \\ \cdot & 1 \\ 1 & \cdot \end{bmatrix} \\
&= \begin{bmatrix} 1 & \cdot \\ \cdot & 1 \\ 1 & \cdot \\ \cdot & 1 \end{bmatrix}
\begin{bmatrix} \cdot & 1 \\ 1 & \cdot \end{bmatrix}.
\end{aligned}$$

Likewise the mapping of the cartesian product shown in Figure 11.4 takes the form:

$$\begin{aligned}
\begin{bmatrix} \cdot & \cdot & \cdot & 1 \\ \cdot & \cdot & 1 & \cdot \\ \cdot & 1 & \cdot & \cdot \\ 1 & \cdot & \cdot & \cdot \end{bmatrix}
\begin{bmatrix} 1 & \cdot \\ 1 & \cdot \\ \cdot & 1 \\ \cdot & 1 \end{bmatrix}
&= \begin{bmatrix} \cdot & 1 \\ \cdot & 1 \\ 1 & \cdot \\ 1 & \cdot \end{bmatrix} \\
&= \begin{bmatrix} 1 & \cdot \\ 1 & \cdot \\ \cdot & 1 \\ \cdot & 1 \end{bmatrix}
\begin{bmatrix} \cdot & 1 \\ 1 & \cdot \end{bmatrix}.
\end{aligned}$$

Link imaging is built into the matrix equation; the test of whether the mapping is a homomorphism is that the product matrix resemble the connectivity matrix of the image diagram, as determined by the respective arrangements of the nonzero elements. For example, if nodes a, b, and c of the square of Figure 11.1 were mapped into node a of the digon with node d mapping into node b, the equation above would become:

$$\begin{bmatrix} \cdot & 1 & \cdot & \cdot \\ \cdot & \cdot & 1 & \cdot \\ \cdot & \cdot & \cdot & 1 \\ 1 & \cdot & \cdot & \cdot \end{bmatrix}
\begin{bmatrix} 1 & \cdot \\ 1 & \cdot \\ 1 & \cdot \\ \cdot & 1 \end{bmatrix}
= \begin{bmatrix} 1 & \cdot \\ 1 & \cdot \\ \cdot & 1 \\ 1 & \cdot \end{bmatrix}$$

$$\begin{bmatrix} 1 & \cdot \\ 1 & \cdot \\ 1 & \cdot \\ \cdot & 1 \end{bmatrix}
\begin{bmatrix} \cdot & 1 \\ 1 & \cdot \end{bmatrix}
= \begin{bmatrix} \cdot & 1 \\ \cdot & 1 \\ \cdot & 1 \\ 1 & \cdot \end{bmatrix}.$$

The discrepancy in the $(1, 1)$ element of the products is due to a linking to b in the square, then mapping to a in the digon; alternatively it maps to a in the digon which can only link to the digon's b.

In mnemonic form, the matrix representation of a homomorphism is

$$\sum_l \text{link}(i,l)(f(l) = j) \leq \sum_l (f(i) = l)\text{link}(l,j),$$

normalized according to taste. Notice how the rows of the lefthand matrix scan the nodes of the source diagram to see if there are any links to be followed. The right hand side jumps directly to the destination to see whether there are any links that will complete a commutative diagram. The inequality supposes that additional links may form bridges. The combination definitely favors the $\exists$ version of the congruence relation.

In matrix form, the criterion for a homomorphism from the graph whose connectivity matrix is A to that whose matrix is B is for there to exist a conformable matrix X such that

$$AX \leq XB. \tag{11.3}$$

In either of these formulations, inequality is sufficient to ensure the existence of links in the image set, but when equality prevails it means that there are no additional links in the image set. In other words, with equality both links and nonlinks are preserved.

It should be noted that inequality for nonnegative (indeed, arbitrary real) matrices is defined by requiring all corresponding pairs of matrix elements to satsfy the stated relationship. Generally, the matrix X of Eq. 11.3, consisting of zeroes and ones, rectangular if the orders differ, describes the mapping between the nodes of the two graphs. Note that numerical inequality of the matrix elements follows the sense of the mapping: homomorphisms increase monotonically; the connectivity matrix of the larger graph (the one left multiplied by X) has larger matrix elements, even though it may have fewer nodes.

Although Eq.11.3 is extremely elegant, it is also capable of great mischief. Equality is the preferred condition, since it implies conservation of both links and non-links; however the definition of graph ordering requires the additional inequality.

Sometimes matrices X fulfill Eq.11.3 without representing functions. A proper X must be row stochastic, otherwise nodes could have multiple images. If X were nevertheless column stochastic, the equation ought to be rearranged to take advantage of the implied inverse function.

The multiplicity of links often interferes with the interpretion of Eq. 11.3; to define homomorphism it is the pattern of zeroes which is relevant; matrix operations should be performed with boolean algebra followed by a numerical comparison, or else the nonzero elements should be normalized to ones before comparison. One is fortunate when ordinary algebra suffices for an entire calculation, which is the only way that additional conclusions regarding eigenvalues and eigenvectors will be valid.

11.3.3 Regular expression

Since regular expressions describe only paths through a diagram but not the nodes themselves, a homomorphism between regular expressions is a bit harder to describe. Probably one would convert any given expression to a diagram, map the diagram, then construct a new regular expression. As it is, unions form a part of the definition of regular expressions, but the computation of intersections and complements typically reverts to a diagram or a machine.

11.4 Dual diagrams

Just as links join pairs of nodes in a diagram, it can be said that nodes join pairs of links. If links were drawn as nodes and conversely, a new diagram, called the dual, could be established. Sometimes a dual is more concise than the original; and sometimes it just reveals its information in a more desirable form.

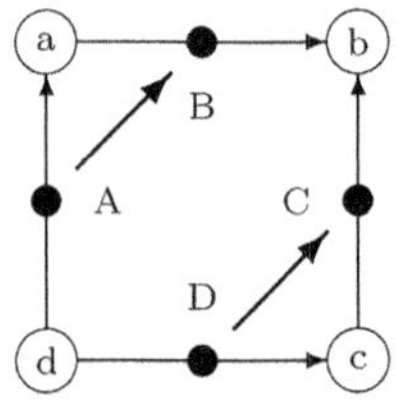

Figure 11.5: The dual of a diagram lets links be nodes.

11.4.1 The tail-head factorization

The matrix representation of a diagram can be factored into a product of matrices, generally rectangular, which relate the nodes and links to one another, rather than nodes to nodes or links to links. Using capital letters for links, small letters for nodes, the matrix elements of the factors are generally boolean propositions such as "node a lies at the tip of link B." Then the node linking matrix is the product $\mathcal{L} = \mathcal{T}\mathcal{H}^T$,

$$\text{link}(i, j) = \sum_K \text{tail}(i, K)\text{head}^T(K, j),$$

whilst the link joining matrix is $\mathcal{N} = \mathcal{H}^T\mathcal{T}$, or

$$\text{node}(I, J) = \sum_k \text{head}^T(I, k)\text{tail(k, J)}.$$

Figure 11.5 shows a diagram upon which its dual has been superimposed; the $\mathcal{T}\mathcal{H}^T$ factorization for $\mathcal{L}$ is:

$$\begin{bmatrix} \cdot & 1 & \cdot & \cdot \\ \cdot & \cdot & \cdot & \cdot \\ \cdot & \cdot & 1 & \cdot \\ 1 & \cdot & \cdot & 1 \end{bmatrix} \begin{bmatrix} 1 & \cdot & \cdot & \cdot \\ \cdot & 1 & \cdot & \cdot \\ \cdot & 1 & \cdot & \cdot \\ \cdot & \cdot & 1 & \cdot \end{bmatrix} = \begin{bmatrix} \cdot & 1 & \cdot & \cdot \\ \cdot & \cdot & \cdot & \cdot \\ \cdot & 1 & \cdot & \cdot \\ 1 & \cdot & 1 & \cdot \end{bmatrix}.$$

Reversing the order of the factors produces $\mathcal{N}$:

$$\begin{bmatrix} 1 & \cdot & \cdot & \cdot \\ \cdot & 1 & \cdot & \cdot \\ \cdot & 1 & \cdot & \cdot \\ \cdot & \cdot & 1 & \cdot \end{bmatrix} \begin{bmatrix} \cdot & 1 & \cdot & \cdot \\ \cdot & \cdot & \cdot & \cdot \\ \cdot & \cdot & 1 & \cdot \\ 1 & \cdot & \cdot & 1 \end{bmatrix} = \begin{bmatrix} \cdot & 1 & \cdot & \cdot \\ \cdot & \cdot & \cdot & \cdot \\ \cdot & \cdot & \cdot & \cdot \\ \cdot & \cdot & 1 & \cdot \end{bmatrix}.$$

There is a temptation to repeat the process to obtain a series of higher order duals; likewise one wonders about the reverse, whereby a given diagram could be exhibited as the dual of another.

11.4.2 Duals and homomorphism

The head matrix $\mathcal{H}$ is column stochastic, because only one node can be the head of a link; of course several links can terminate at the same node. Similarly the tail matrix $\mathcal{T}$ is row stochastic, there being only one tail to a given link. Strictly stochastic means that elements sum up to one, but the only possibility for integer elements in such a sum are zeroes and ones; just one single one may be present.

Dedicating the letters R and C to such matrices is more mnemonic than using $\mathcal{H}$ and $\mathcal{T}$; passing to the dual matrix M^D then amounts to finding a CR factorization

$$M = CR, \tag{11.4}$$

then writing

$$M^D = RC. \tag{11.5}$$

The second dual would result from writing

$$RC = C_1R_1 \tag{11.6}$$

with

$$M^{DD} = R_1C_1, \tag{11.7}$$

and so on.

Note that if Equation 11.5 is multiplied on the right by R, the result

$$M^DR = RCR = RM, \tag{11.8}$$

asserts that M^D is homomorphic to M. So also is the second dual (and all the rest, for that matter), inasmuch as

$$\begin{aligned} M^{DD}(R_1R) &= (R_1C_1)R_1R \\ &= R_1(C_1R_1)R \\ &= R_1(RC)R \\ &= (R_1R)M. \end{aligned}$$

The matrix R which generates the homomorphism from the dual to its graph associates each link with its terminal node.

11.4.3 The head-tail factorizarion

Going in the other direction, it would be nice to know when the CR factorization is possible, independently of whether RC was available for comparison or not.

The rules of matrix multiplication are such that when R is used as a left factor, it simply rearranges the rows of the left factor. Thus in any product $RA = N$, A would have the same rows as N. But if A were of type C, different rows would have to have their ones in different columns, a restriction which could be summarized by saying that the rows were orthogonal. Consequently the rows of N would have to be either identical or orthogonal; lacking this characteristic it could not have an RC factorization. Conversely, dual diagrams necessarily show the evidences of such a composition.

Given the necessary conditions, factors R and C can be constructed; a lack of uniqueness will be evident in some of the steps. To begin with, C must have at least the distinct rows of N (including a zero row if necessary) arranged in some order. But if N has one or more columns of zeroes, some rows containing ones must be provided lest C be nonstochastic.

The final number of rows chosen determines the number of columns of R, which is filled in at last by placing the rows of C in their proper place in N. Zero rows in N must be created by copying a row from C—not using zero coefficients in R—in order to keep R stochastic.

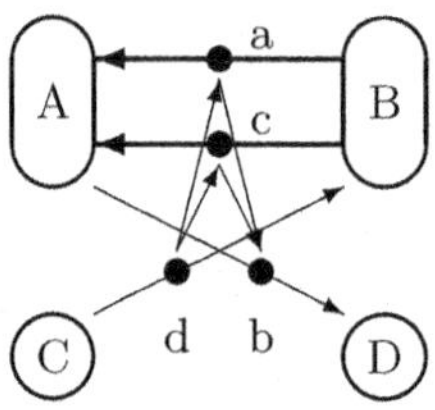

Figure 11.6: The antidual of a diagram lets nodes be links.

If the diagram of Figure 11.5 is given this treatment, we find the antidual shown in Figure 11.6, calculated as follows:

$$\begin{bmatrix} 1 & . & . & . \\ . & . & . & 1 \\ 1 & . & . & . \\ . & 1 & . & . \end{bmatrix} \begin{bmatrix} . & 1 & . & . \\ 1 & . & 1 & . \\ . & . & . & 1 \\ . & . & . & . \end{bmatrix} = \begin{bmatrix} . & 1 & . & . \\ . & . & . & . \\ . & 1 & . & . \\ . & . & . & . \end{bmatrix};$$

reversing the order yields the graph to which it is dual:

$$\begin{bmatrix} . & 1 & . & . \\ 1 & . & 1 & . \\ . & . & . & 1 \\ . & . & . & . \end{bmatrix} \begin{bmatrix} 1 & . & . & . \\ . & . & . & 1 \\ 1 & . & . & . \\ . & 1 & . & 1 \end{bmatrix} = \begin{bmatrix} . & . & . & 1 \\ 2 & . & . & . \\ . & 1 & . & . \\ . & . & . & . \end{bmatrix}.$$

Interestingly, this antidual contains a double link, whose importance is evident from Figure 11.6.

For a thorough presentation of duals, which they call "line graphs," Robert Hemminger and Lowell Beinecke's survey [115] should be consulted. Among other things they cite criteria which avoid the presence of self-loops and multiple links in the dual diagram. Essentially there is a list of configurations such as the example presented here, which have to be avoided. The dichotomy "identical or orthogonal" which figures in the construction of the antidual was formulated by Paul Richards [195] in 1967.

11.4.4 Counting paths

There is an interesting relation between the number of paths in a diagram and the number to be found in the dual diagram. Recall that the elements of powers of the connectivity matrix M represent the number of paths joining the row index node to the column index node; if the respective sums are taken, the total number n of paths results. In matrix form, taking u to be a vector whose components are all ones,

$$n(M) \quad = \quad u^T M u. \tag{11.9}$$

Given that u is an eigenvector (with unit eigenvalue) of a row stochastic matrix just as u^T is a similar eigenvector of a column stochastic matrix, it follows from the RC decomposition of the dual that

$$\begin{aligned} u^T M^2 u &= u^T RCRCu \\ &= u^T CRu \\ &= u^T M^D u. \end{aligned}$$

Hence $n(M^2) = n(M^D)$; entirely similar reasoning reveals that

$$n(M^k) \quad = \quad n((M^D)^{k-1}). \tag{11.10}$$

Although the matrices figuring in the equation are not likely to be identical to their duals, they both contain the same number of 1's; in particular the number of paths of length 2 in any diagram is the same as the number of links in its dual diagram. Note that this is a property of *directed* graphs, not necessarily of general graphs.

Another useful property of duality is the relation between Hamiltonian paths and Eulerian paths. A Hamiltonian path is one which visits each node of the diagram once (clearly, the diagram must be connected), while an Eulerian path is one which traverses every link once (in the proper direction) even though nodes are repeated during the voyage. A Hamiltonian path, one of which is usually hard to find, becomes an Eulerian path in the antidual diagram, criteria for which are generally easier to obtain. The result would be more useful if more graphs actually had antiduals.

11.4.5 Eigenvalues and eigenvectors

A further consequence of equality in Eq. 11.3 is that except for a possible discrepancy associated with zero, the eigenvalues and eigenvectors of the connectivity matrix associated with a homomorphic image are the same as for the diagram itself. Suppose that A and B are the matrices and that z is an eigenvector of B.

$$Bz \quad = \quad \lambda z.$$

Then

$$\begin{aligned} XBz &= AXz \\ &= \lambda Xz, \end{aligned}$$

making Xz an eigenvector of A with the same eigenvalue. A similar relation, valid in the opposite direction, holds for the row eigenvectors. Precaution is necessary when λ vanishes, as it must when A and B have different dimensions; the numbers of eigenvectors associated with zero will differ.

11.5 Reverses and complements

Besides the chain of duals, there are other related graphs which can be derived from an originally given graph.

11.5.1 Reversed graph

The definition of a graph which we have employed is not the usual one to be found in the literature; technically our graphs are called *digraphs* (a contraction of "directed graph"). The reasons are historical, inasmuch as the concern with a graph was the linkages between is nodes which were originally concieved as a mutual connection between the two nodes involved.

From the point of view of an algebraic treatment of the connectivity matrix, symmetrical linkages are desirable because they lead to symmetrical matrices, for which there are much more convenient estimates of eigenvalues and procedures for obtaining the eigenvectors and eigenvalues.

On the other hand, actual eigenvalues (other than the largest and possibly the second lagest) do not have much direct application to graphs, nor is the decomposition according to eigenvectors particularly relevant. Consequently there is little reason to insist upon symmetical linkages, particularly given the existence of a very complete theory of positive (asymmetrical) matrices which does include the most useful aspects of eigenvalues and eigenvectors.

Applications to automata envisage directed links, just as the development given so far supposes, and certainly encompass the symmetrical case — just include every link twice, once in each direction. A symmetrical graph is just a special case; every graph can be symmetrized by adjoining reverse links, and one can define the *reverse* of any graph (N, L): merely reverse every link.

$$\begin{aligned} (N, L)^R &= (N, L^T) \\ L^T &= \{(x, y) \ni (y, x) \in L\} \end{aligned} \tag{11.11}$$

The connectivity matrix of a reversed graph is the transposed matrix, described by the reversed regular expression, in which each sequence is written backwards.

11.5.2 Complementary graph

There is sufficient occasion to mention that certain links are *not* in a graph that it is convenient to construct a new graph composed entirely and exclusively of the missing links. The result is the *complement* of the given graph.

$$\begin{aligned} (N, L)^C &= (N, \bar{L}) \\ \bar{L} &= \{(x, y) \ni (y, x) \notin L\} \end{aligned} \tag{11.12}$$

If it is not desired that all the links running in the wrong direction be included in the complement, the graph should be symmetrized before complementation.

At least when a graph contains single links without multiplicity, the connectivity matrix of the reversed graph is $A-U$, supposing that A is the matrix of the original graph and that U is a conformable square matrix, all of whose components are ones.

In terms of Eq. 11.3, for a homomorphism to preserve non-links (that is, the complementary graph) requires that

$$(U_1 - A)X \quad \leq \quad X(U_2 - B). \tag{11.13}$$

However, $XU_2 = U_x$ is a consequence of the definition of a function, that every point have one and only one image. Likewise $U_1X \geq U_x$ holds when the function is *onto* and X lacks a zero column; together

$$XU_2 \quad \leq \quad U_1X. \tag{11.14}$$

The subscripts on U remind us that the dimension of each of the matrices may be different. Subtracting Eq. 11.13 from Eq. 11.14 and reversing the equality to compensate the sign produces

$$AX \quad \geq \quad XB. \tag{11.15}$$

Comparison now yields the derivation of the assertion that equality in Eq. 11.3:

$$AX \quad = \quad XB \tag{11.16}$$

corresponds to the preservation of both links and non-links; note that X must be *onto* and not just *into* to get the result.

The strong homomorphism expressed in Eq. 11.16 guarantees that eigenvalues and eigenvectors of the image matrix belong to the source matrix as well: consider

$$BZ \quad = \quad \lambda Z \tag{11.17}$$
$$XBZ \quad = \quad \lambda XZ \tag{11.18}$$
$$A(XZ) \quad = \quad \lambda(XZ). \tag{11.19}$$

Similar relations, in the opposite direction, hold for eigenrows.

11.6 The de Bruijn diagram

It is nearly impossible to do any work with linear cellular automata without encountering the fact that their neighborhoods overlap, so that it is preferable to begin with this circumstance as a fundamental fact. Shift register theory is a discipline based on the treatment of overlapping sequences, so it is not surprising that its fundamental tool, the de Bruijn diagram, would be useful for automata theory.

Although this diagram is explained in Solomon Golomb's fundamental book [95] on shift registers, it actually has a venerable history which has been traced by Anthony Ralston in an article in the *American Mathematical Monthly* [191].

11.6.1 Definition

The nodes of the de Bruijn diagram are sequences of symbols from some alphabet, just as regular expressions are. They can even be sequences of nodes from a specific graph. The links of the diagram describe how such sequences may overlap. Different degrees of overlap lead to different diagrams, the simplest of which overlap according to the gain or loss of a single initial or terminal symbol. Thus the binary sequence 0011 overlaps the sequence 0110 by losing 0 as its initial symbol, whereas the second sequence overlaps through the loss of 0 as a terminal symbol. The link can be labelled according to the displaced symbol, as well as by the composite sequence 00110, and a variety of other ways. The intended application generally governs the choice.

When the symbols are consecutive integers they can be treated as elements of a ring or perhaps a finite field; the ease of discussing their properties arithmetically or algebraically makes the choice eminently worthwhile. For example, the topological matrix of a de Bruijn diagram becomes simply

$$\begin{aligned} M_{ij} &= 1 \quad j = \begin{cases} ki \\ ki+1 \\ \cdots \\ ki+k-1 \end{cases} \qquad \pmod{k^{2r}} \\ &= 0 \quad \text{otherwise.} \end{aligned}$$

11.6.2 Subdiagrams

Through the intermediary of a de Bruijn diagram, graphs can represent configurations, or classes of configurations, in cellular automata. The links of a de Bruijn diagram are naturally associated with the neighborhoods of an automaton using the same symbols, which associates the links with a step of evolution in an equally natural fashion. If there is any reason to discriminate between neighborhoods, the same discrimination may define a subdiagram of the de Bruijn diagram. A good example of the process arises from the still lifes—configurations of cells which do not change during the course of evolution.

The generic de Bruijn diagram for a (2,1) automaton (one having two states per cell with a single neighbor on each side, according to Stephen Wolfram's notation [250]) has four nodes corresponding to the four two-cell partial neighborhoods, with eight links representing the full three-cell neigborhoods. Such a diagram is shown in Figure 11.7.

The still life diagram may be extracted from the generic diagram by removing all links which do not conserve their central cell during the automaton's evolution. Loops in the subdiagram composed of the remaining links define sequences, rather than individual cells, which are invariant to evolution. By reading off such sequences, all the still lifes for the automaton are automatically determined.

In general any boolean combination of the cells of the neighborhood and the evolved cell may determine the subdiagram, leading to chains of cells all of which fulfill the same property. As an example, one important binary automaton evolves according to the so-called Rule 22 [250]:

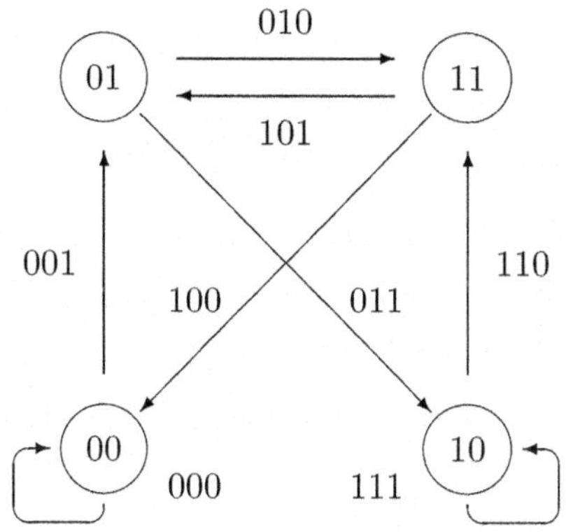

Figure 11.7: The generic (2,1) de Bruijn diagram.

neighborhood	000	001	010	011	100	101	110	111
image	0	1	1	0	0	1	0	0

The de Bruijn diagram has eight links connecting four nodes (namely, 00, 01, 10, and 11); keep or reject the links according to whether they conserve their central cell, with results summarized by the following table:

neighborhood	000	001	010	011	100	101	110	111
keep	yes	no	yes	no	no	yes	no	no

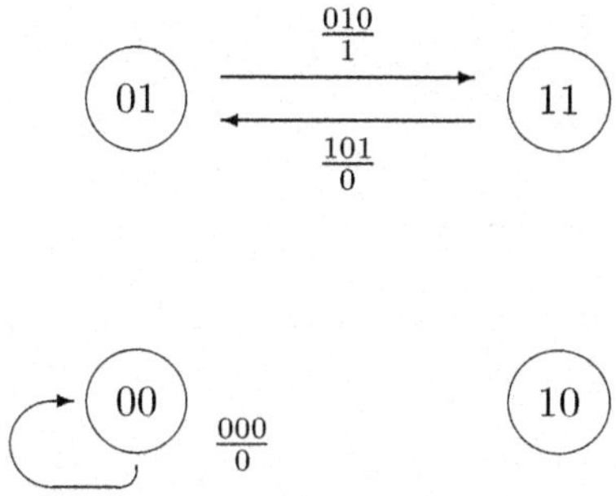

Figure 11.8: Subdiagram generated by Rule 22 still lifes.

Figure 11.8 shows the resulting subdiagram, which evidently consists of two disjoint loops; the first, of length 1, leads to an unending sequence of zeroes while the second, of length 2, leads to an alternating sequence of zeroes and ones. These are exactly the still life configurations for this rule.

There is an inverse application for Figure 11.8; suppose that it had beed decided beforehand that one needed a rule with quiescent zeroes and the combination $(01)^*$ as a still life. If so, the loops in the figure are forced, but the remaining five links can be assigned arbitrary values. All told there are thirty two rules with the required still lifes. To the extent that the required loops can be found, rules can be located which meet almost any desired specification.

11.6.3 A master diagram

Evidently the subdiagram consists of two disjoint loops; the first, of length 1, leads to an unending sequence of zeroes while the second, of length 2, leads to an alternating sequence of zeroes and ones. These are exactly the still life configurations for this rule.

Perhaps it is worth emphasizing the point that there are many diagrams which can be associated with a cellular automaton, a significantly large proportion of which are subdiagrams of a de Bruijn diagram. Many of the remainder are associated with extensions of de Bruijn diagrams; altogether de Bruijn diagrams constitute a central reference point in a context that might have seemed to be open to any arbitrary kind of diagram.

The literature of automata theory shows varying degrees of awareness of the diagram; much contains intuition that was never formalized, but as time passes references have become more centralized and specific. For example, there are treatments of automata in terms of formal language theory [251, 148] which refer to "regular expression diagrams," which are recognizable as subdiagrams of a de Bruijn diagram. Masakazu Nasu [169] refers to them by name and invokes their properties; Erica Jen [131] refers to them, and it is likely that a diligent search would turn up still further instances of their use.

By using longer neighborhoods, the results of more than one generation of evolution may be incorporated in the criterion defining the subdiagram. In that way arbitrary combinations of shifts and periodicities may be studied, not to mention even more elaborate combinations. The following list mentions some of the additional possibilities.

1. evolution to a constancy after n generations.

2. evolution of the cells of binary automaton into their complements; permutation of the values of the cells of an arbitrary automaton.

3. a quiescent state is one which does not change when its entire neighborhood is quiescent; a nilpotent rule (of order n) is one for which all cells become quiescent after n generations.

4. a configuration is idempotent (of order n) if, for the nth generation and thereafter, the values of its cells do not change. A rule is idepotent if all configurations are idempotent.

5. a configuration is superluminal if it shifts a distance greater than the radius of the neighborhoods in each generation.

11.7 The de Bruijn matrix

The topological matrix of a de Bruijn diagram is quite regular, although it is asymmetrical and differs from circulant matrices, tridiagonal matrices, or other specialized forms which have been extensively analyzed in the literature. De Bruijn matrices $D_{k,s}$ or D_s for short, are characterized by k, the number of symbols from which sequences may be formed, and s—the number of stages—which is the length of the sequence. In these terms, there will be k links entering each node, as well as k links leaving each node. In turn there will be k^s nodes, with altogether $k^{(s+1)}$ links to join them, corresponding to sequences of length $s+1$.

The uniform distribution of the numbers of incoming and outgoing links means that the topological matrix has constant row sums as well as constant column sums; therefore divided by k would be a doubly stochastic matrix with maximum eigenvalue 1 having a uniform probability distribution for its eigenvector.

It is worth presenting one or two examples and then the general case.

11.7.1 One stage binary matrix

	0	1
0	1	1
1	1	1

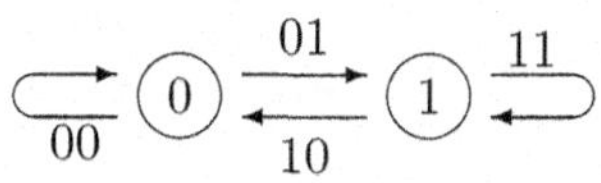

Figure 11.9: Single stage binary de Bruijn diagram.

A one stage de Bruijn diagram connects every node with every other node, making a full diagram with k nodes and k^2 links. Its topological matrix is filled with 1's; a binary example is shown in Figure 11.9. The minimal equation satisfied by the binary matrix, $D_{2,1}^2 = 2U_{2,1}$, is a special case of the general equation $D_{k,1}^2 = kD_{k,1}$. Here and subsequently U is a matrix all of whose elements are ones, just as I will stand for the unit matrix.

11.7.2 Two stage binary matrix

Not every node of a two stage diagram or beyond links to all the other nodes; their topological matrices present the form of a series of staircases, corresponding to the diagram shown in Figure 11.7.

$$D_2 = \begin{bmatrix} 1 & 1 & \cdot & \cdot \\ \cdot & \cdot & 1 & 1 \\ 1 & 1 & \cdot & \cdot \\ \cdot & \cdot & 1 & 1 \end{bmatrix}$$

Simple calculation reveals that the minimal equation is $D^3 = 2D^2$ with eigenvalues 2, 0, 0, and principal vectors

$$R_2 = \begin{bmatrix} 1 & 1 & 1 & 1 \\ 1 & -1 & -1 & 1 \\ 1 & 1 & -1 & -1 \\ 1 & -1 & 1 & -1 \end{bmatrix}$$

given that one of the zero eigenvalues has a 2×2 Jordan block.

11.7.3 Three stage binary matrix

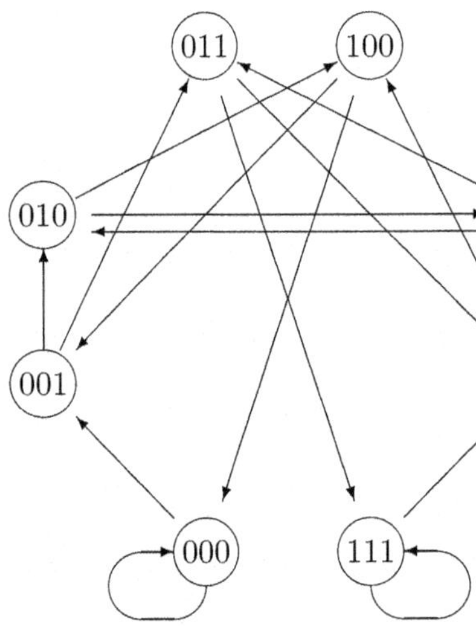

Figure 11.10: Generic (2,3/2) de Bruijn diagram.

By the third stage the staircase is fully evident; each step contains k 1's, likewise there are k staircases:

$$D_3 = \begin{bmatrix} 1 & 1 & . & . & . & . & . & . \\ . & . & 1 & 1 & . & . & . & . \\ . & . & . & . & 1 & 1 & . & . \\ . & . & . & . & . & . & 1 & 1 \\ 1 & 1 & . & . & . & . & . & . \\ . & . & 1 & 1 & . & . & . & . \\ . & . & . & . & 1 & 1 & . & . \\ . & . & . & . & . & . & 1 & 1 \end{bmatrix}.$$

The minimal equation is readily determined to be $U^4 = 2U^3$, and as before the Jordan normal form prevails.

There are two interesting factorizations, both of which are quite general: $D_3 = P_3U_3$ according to

$$D_3 = \begin{bmatrix} 1 & \cdot & \cdot & \cdot & \cdot & \cdot & \cdot & \cdot \\ \cdot & \cdot & 1 & \cdot & \cdot & \cdot & \cdot & \cdot \\ \cdot & \cdot & \cdot & \cdot & 1 & \cdot & \cdot & \cdot \\ \cdot & \cdot & \cdot & \cdot & \cdot & \cdot & 1 & \cdot \\ \cdot & 1 & \cdot & \cdot & \cdot & \cdot & \cdot & \cdot \\ \cdot & \cdot & \cdot & 1 & \cdot & \cdot & \cdot & \cdot \\ \cdot & \cdot & \cdot & \cdot & \cdot & 1 & \cdot & \cdot \\ \cdot & \cdot & \cdot & \cdot & \cdot & \cdot & \cdot & 1 \end{bmatrix} \times \begin{bmatrix} 1 & 1 & \cdot & \cdot & \cdot & \cdot & \cdot & \cdot \\ 1 & 1 & \cdot & \cdot & \cdot & \cdot & \cdot & \cdot \\ \cdot & \cdot & 1 & 1 & \cdot & \cdot & \cdot & \cdot \\ \cdot & \cdot & 1 & 1 & \cdot & \cdot & \cdot & \cdot \\ \cdot & \cdot & \cdot & \cdot & 1 & 1 & \cdot & \cdot \\ \cdot & \cdot & \cdot & \cdot & 1 & 1 & \cdot & \cdot \\ \cdot & \cdot & \cdot & \cdot & \cdot & \cdot & 1 & 1 \\ \cdot & \cdot & \cdot & \cdot & \cdot & \cdot & 1 & 1 \end{bmatrix}$$

and $D_3 = Q_3 U_3$ according to

$$D_3 = \begin{bmatrix} \cdot & 1 & \cdot & \cdot & \cdot & \cdot & \cdot & \cdot \\ \cdot & \cdot & \cdot & 1 & \cdot & \cdot & \cdot & \cdot \\ \cdot & \cdot & \cdot & \cdot & \cdot & 1 & \cdot & \cdot \\ \cdot & \cdot & \cdot & \cdot & \cdot & \cdot & \cdot & 1 \\ 1 & \cdot & \cdot & \cdot & \cdot & \cdot & \cdot & \cdot \\ \cdot & \cdot & 1 & \cdot & \cdot & \cdot & \cdot & \cdot \\ \cdot & \cdot & \cdot & \cdot & 1 & \cdot & \cdot & \cdot \\ \cdot & \cdot & \cdot & \cdot & \cdot & \cdot & 1 & \cdot \end{bmatrix} \times \begin{bmatrix} 1 & 1 & \cdot & \cdot & \cdot & \cdot & \cdot & \cdot \\ 1 & 1 & \cdot & \cdot & \cdot & \cdot & \cdot & \cdot \\ \cdot & \cdot & 1 & 1 & \cdot & \cdot & \cdot & \cdot \\ \cdot & \cdot & 1 & 1 & \cdot & \cdot & \cdot & \cdot \\ \cdot & \cdot & \cdot & \cdot & 1 & 1 & \cdot & \cdot \\ \cdot & \cdot & \cdot & \cdot & 1 & 1 & \cdot & \cdot \\ \cdot & \cdot & \cdot & \cdot & \cdot & \cdot & 1 & 1 \\ \cdot & \cdot & \cdot & \cdot & \cdot & \cdot & 1 & 1 \end{bmatrix}.$$

Several relationships among these matrices are evident; for example:

$$\begin{aligned} D_n &= P_n(I_{n-1} \otimes D_1) \\ &= (D_1 \otimes I_{n-1})P_n \\ &= Q_n(I_{n-1} \otimes D_1) \\ &= (D_1 \otimes I_{n-1})Q_n \end{aligned}$$

wherein I is the unit matrix. The factors satisfy several identities, of which the most useful are:

$$U_n^2 = 2U_n,$$

$$P_n + Q_n = D_n.$$

P and Q are noncommuting permutation matrices; nevertheless the noncommutative binomial theorem can still be used to evaluate powers of D. Whatever scheme is used, the essential observation is that powers of D exhibit broader steps and more numerous ladders until the matrix is solidly filled with 1's; from then on successive powers merely differ by the factor k.

11.7.4 General de Bruijn matrix

Thus there are both factorizations and sum decompositions for the de Bruijn matrix, all readily obtained in a way that indicates many more representations are possible, none quite as symmetrical as the ones shown. For $k > 2$ there are additional permutation matrices which participate in the decompositions, and the blocks in the matrix U are correspondingly larger. The factored forms show that the determinant of any de Bruijn matrix is zero, but it is not difficult to obtain the entire minimal equation, which is

$$D_{k,n}^{n+1} = kD_{k,n}^{n}, \tag{11.20}$$

nor is it difficult to verify that

$$Tr(D_{k,n}^{n}) = k^{n} \tag{11.21}$$

for all values of n. This is the count of loops of length n, wherein each loop is weighted by the number of nodes it contains. It must be borne in mind that *all* loops are counted, including the degenerate cases where the same cycle of nodes is traversed several times.

11.7.5 RC and CR factorizations

Bearing in mind the relation between a diagram and its dual established in Section 11.4.2, it is not surprising to see that all the de Bruijn diagrams for k states form a chain of duals. In fact, D_0, the null string occupying its single node (see Figure 11.11), could begin the chain.

	λ
λ	2

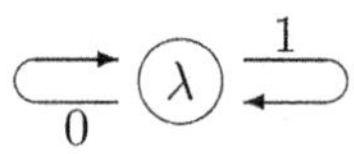

Figure 11.11: Zero stage binary de Bruijn diagram.

In any event we can start the chain with

$$\begin{aligned}
2 &= D_0 \\
D_0 &= \begin{bmatrix} 1 & 1 \end{bmatrix} \begin{bmatrix} 1 \\ 1 \end{bmatrix} \\
\begin{bmatrix} 1 \\ 1 \end{bmatrix} \begin{bmatrix} 1 & 1 \end{bmatrix} &= \begin{bmatrix} 1 & 1 \\ 1 & 1 \end{bmatrix} (= D_1) \\
D_1 &= \begin{bmatrix} 1 & 1 & . & . \\ . & . & 1 & 1 \end{bmatrix} \begin{bmatrix} 1 & . \\ . & 1 \\ 1 & . \\ . & 1 \end{bmatrix} \\
\begin{bmatrix} 1 & . \\ . & 1 \\ 1 & . \\ . & 1 \end{bmatrix} \begin{bmatrix} 1 & 1 & . & . \\ . & . & 1 & 1 \end{bmatrix} &= \begin{bmatrix} 1 & 1 & . & . \\ . & . & 1 & 1 \\ 1 & 1 & . & . \\ . & . & 1 & 1 \end{bmatrix} (= D2)
\end{aligned}$$

$$D_2 = \begin{bmatrix} 1 & 1 & . & . & . & . & . & . \\ . & . & 1 & 1 & . & . & . & . \\ . & . & . & . & 1 & 1 & . & . \\ . & . & . & . & . & . & 1 & 1 \end{bmatrix} \begin{bmatrix} 1 & . & . & . \\ . & 1 & . & . \\ . & . & 1 & . \\ . & . & . & 1 \\ 1 & . & . & . \\ . & 1 & . & . \\ . & . & 1 & . \\ . & . & . & 1 \end{bmatrix}$$

wherein the pattern of the factorization is fairly evident. Suppose that I_n is a $k^n \times k^n$ unit matrix and that u is a k-dimensional vector with unit components. Then evidently

$$\begin{aligned}(u \otimes I_n)(I_n \otimes u^T) &= D_{n+1} \\ D_n &= (I_n \otimes u^T)(u \otimes I_n).\end{aligned}$$

From this it is immediately clear that the whole chain of binary de Bruijn matrices has the single nonzero eigenvalue k belonging to D_0. Note that the factors in the Kronecker product cannot be multiplied together because they are not conformable; the formulas stand as written.

11.7.6 Hamiltonian and Eulerian paths

Hamiltonian paths in a diagram are those which visit each vertex once and only once; Hamiltonian loops are those for which the tour returns to the starting point. In contrast Eulerian paths are those which use every edge in the diagram once and only once, loops likewise completing a circuit by returning to the start.

An Eulerian problem, to decide whether such paths exist or not, is readily soluble; for loops, each vertex must have as many incoming as outgoing links, and the diagram as a whole must be connected (which is certainly true of a de Bruijn diagram). For paths, the terminal nodes could be separated, allowing one extra outgoing link at the initial node plus a surplus incoming link at the terminal node. Euler's classical result refers to undirected links; I. J. Good [96] proved the analogous directed variant in 1946.

A Hamiltonian problem is usually intractable, but for de Bruijn diagrams, the dual of a diagram of n stages is the diagram of $n+1$ stages. Consequently, running through all the links in any diagram is equivalent to running through all the nodes of its successor, allowing the Eulerian result to solve the equivalent Hamiltonian problem for an additional stage.

Demonstrating existence is much easier than counting, in turn simpler than exhibiting, paths. In 1946 N. J. de Bruijn [60] obtained a recursion relation to count the Hamiltonian paths in a binary diagram exploiting duality, which he called *doubling*; a sufficiently significant result to associate his name with the graph. Additional discussion can be found in Marshall Hall, Jr.,'s *Combinatorial Theory* [109].

When diagrams are both Hamiltonian and Eulerian, every Hamiltonian circuit can be promoted to an Eulerian circuit and the interesting question is to determine in how many ways. Shift register theory esteems the actual sequences highly; a good part of Golomb's book [95] concentrates on obtaining them. Automata theory is less concerned with this particular detail, but it is still useful to know some procedures.

An upper bound on the number of Hamiltonian cycles is always that number of cycles whose length matches the number of nodes. Since the diagram of order n has k^n nodes, by Eq. 11.21 this number is k^{k^n}. For binary diagrams, de Bruijn's value is $2^{2^{n-1}-n}$. Aside from number theoretic deformations at low values of n, comparison of the results would lead one to believe that $(1/k)^{th}$ of all candidates are Hamiltonian.

Similarly, the number of loops whose length equals the total number of links bounds the number of Eulerian cycles; in both cases, degenerate loops must be discounted to get the true count.

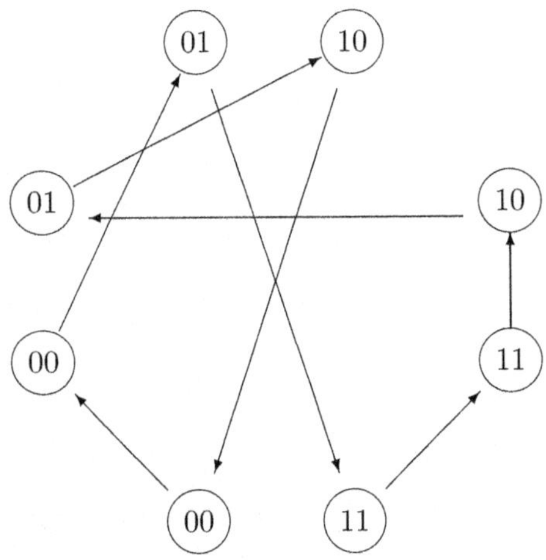

Figure 11.12: A Hamiltonian path in the (2,3/2) de Bruijn diagram.

Figure 11.12 illustrates several of the points which have been discussed; it shows one of the two Hamiltonian paths associated with the matrix D_3, but the nodes have been labelled according to D_2. Homomorphism is apparent—the numerical node indices have been reduced modulo 4, while the residues modulo 2 (linkages) have not changed. If duplicate nodes were superimposed and reference made to Figure 11.7, one of the Eulerian loops of the latter would manifest itself; a Hamiltonian path wrapped around the image twice.

There is an embedment in the opposite direction too—the $(2, 1)$ diagram fits into just half of the $(2, 3/2)$ diagram. All the arrows are there; homomorphism fails when they don't all close up within the image of $(2, 1)$ itself, the trick is to match up two copies of the image to get the next diagram. In fact, when this arrangement presents itself one should always try to arrange an inverse homomorphism.

11.8 Subset diagrams

So far no great importance has been attached to labelling the links in a graph, although it is clearly possible to do so. But often labelled links are supposed to belong to different classes, implying that the links should be distinguished accordingly. In turn the classes would have individual connectivity matrices. Instead of just tracing paths through the diagram, it could be required that the labels follow a certain sequence, with the attendant question of whether the required path actually exists; a question that could be resolved by multiplying the corresponding connectivity matrices in the prescribed order.

Frequently the question is more general — does the required path exist anywhere at all in the diagram? One approach would be to list all the paths, so as to search the list for the presence of the candidate. Something of the nature can hardly be avoided, so it is a matter of organizing the technique, which is usually known as Moore's subset construction [163], which was introduced in 1956 by Edward F. Moore for a different reason.

11.8.1 Scalar subset diagram

The nodes are grouped into subsets, note being taken of the subsets to which one can arrive through systematic departures from all the nodes in any given subset. The result is a new graph, with subsets for nodes and links summarizing all the places that one can get to from all the different combinations of starting points. Sometimes, but far from always, the possible destinations narrow down as one goes along; in any event one has all the possibilities cataloged.

One point to be observed is that if one thinks that there should be a link at a certain node and there is not, the link should be drawn to the empty set instead; a convention which assures every label of having a representation at every node in the subset diagram.

Let a and b be nodes in a given diagram, S a subset, and $|S|$ the cardinality of S; then the formal definition of its subset diagram is

$$\Sigma_i(S) \quad = \quad \begin{cases} \phi & S = \phi \\ \{b \mid \mathrm{link}_i(a,b)\} & S = \{a\} \\ \bigcup_{a \in S} \Sigma_i(a) & |S| > 1 \end{cases} . \tag{11.22}$$

There is another important reason for working with subsets. Labelled links resemble functions, by associating things with one another. But if two links with the same label emerge from a single vertex, they can hardly represent a function. Forging the subset of all destinations, leaves one single link between subsets, bringing functionality to the subset diagram even though it did not exist originally. Including the null set ensures that every point has an image, avoiding partially defined functions.

Once the subset diagram has been formed, if a path leads from the universal set to the empty set, that is conclusive evidence that such a path exists nowhere in the original diagram. Another application—the one originally envisioned by Moore—is to determine whether there are paths leading to the unit classes. Such a paths, if they existed, could be used to force an automaton into a predetermined state, no matter what its original condition.

11.8.2 Vector subset diagram

Moore's traditional discussion gives no internal structure to the nodes of the subset diagram, but we will find it convenient to regard each node as a vector whose components are indexed by the nodes of the original graph. Then links in the subset graph are represented by projections applied to the topological matrix of the original graph; precisely, projection onto the subset belonging to the label in question. Each row of the graph will have k nonzero submatrices (if that is the number of labels) except that if there is no link corresponding to some particular label, its submatrix will be zero too.

The column containing the submatrix for label k will correspond to the subset of nodes which are destinations for that label. Should the destination subsets coincide for two or more labels, neither common nor impossible, the multiplicity should be entered into the submatrix. Alternatively, one could separate the subset matrices for each link and later sum them as needed.

If there were n nodes in the original graph, the topological matrix of the subset diagram will have dimension $2^n \times 2^n$. For a g-stage de Bruijn diagram, $n = 2^g$. As usual, powers of the topological matrix describe chains with multiple links, but the most interesting matrix elements will be the ones in the top row, so to speak. They connect the full set with other subsets, including the empty set. If it is immaterial where a path begins or ends, the submatrix connecting the full set with itself suffices. Restrictions placed upon the choices of initial and terminal nodes will lead to other submatrices.

Were it not for the projections, the vector connection matrix V would be the tensor product of the de Bruijn matrix with the scalar connection matrix. But the projected elements are missing; so introduce row and column indices of the form (σ, m), where σ, τ are subsets whose own linkage is defined by

$$\text{Link}(\sigma,\tau) = \begin{cases} 1 & \tau = \text{link}^{-1}(\sigma) \\ 0 & \text{otherwise} \end{cases} . \tag{11.23}$$

Then,

$$V_{\sigma m,\tau n} = \text{Link}(\sigma,\tau)\text{link}(m,n). \tag{11.24}$$

11.8.3 Rule 126

To illustrate the use of the subset diagram, consider the automaton for $(2,1)$ Rule 126:

neighborhood	000	001	010	011	100	101	110	111
image	0	1	1	1	1	1	1	0

In other words, all neighborhoods evolve into ones with the exception of 111 and the quiescent neighborhood 000, which evolve into zeroes. The de Bruijn diagram for this rule is $D_{2,2}$, which has the symbolic variant (dots represent the null word ϕ):

$$\begin{bmatrix} 0 & 1 & \cdot & \cdot \\ \cdot & \cdot & 1 & 1 \\ 1 & 1 & \cdot & \cdot \\ \cdot & \cdot & 1 & 0 \end{bmatrix} = \begin{bmatrix} 0 & \cdot & \cdot & \cdot \\ \cdot & \cdot & \cdot & \cdot \\ \cdot & \cdot & \cdot & \cdot \\ \cdot & \cdot & \cdot & 0 \end{bmatrix} + \begin{bmatrix} \cdot & 1 & \cdot & \cdot \\ \cdot & \cdot & 1 & 1 \\ 1 & 1 & \cdot & \cdot \\ \cdot & \cdot & 1 & \cdot \end{bmatrix}.$$

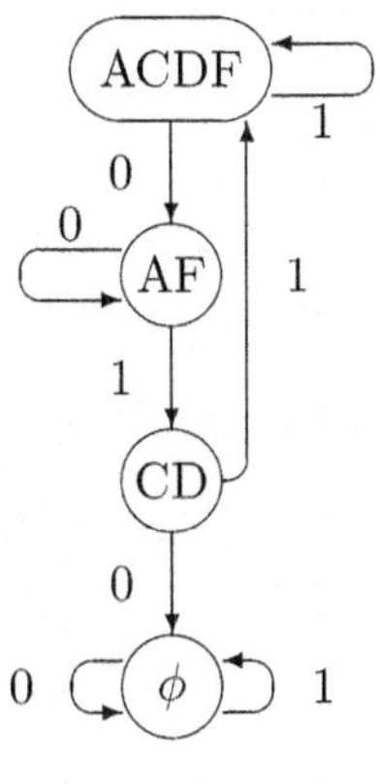

Figure 11.13: Subset diagram for Rule 126.

Since the nodes $A = 00$, $C = 01$, $D = 10$, and $F = 11$ are linked by the symbols 0 and 1, the subset diagram can be constructed. It actually has 2^4 nodes for the sixteen subsets, but to save space we show only the part connected to the full subset (ϕ is the empty set). First, note the destination of each arrow leaving individual nodes:

node	0 leads to	1 leads to
A	A	C
C	ϕ	D, F
D	ϕ	A, C
F	F	D

Then, form vectors whose elements are the tail nodes and whose components are indexed by the head nodes; an element is null if there is no incoming link from any node.

subset	0 leads to	1 leads to
$\{ACDF\}$	(A, ϕ, ϕ, F)	$(D, A + D, C + F, C)$
$\{AF\}$	(A, ϕ, ϕ, F)	(ϕ, A, F, ϕ)
$\{CD\}$	(ϕ, ϕ, ϕ, ϕ)	(D, D, C, C)
$\{\}$	(ϕ, ϕ, ϕ, ϕ)	(ϕ, ϕ, ϕ, ϕ)

This structure is actually the vector subset diagram; the scalar subset diagram arises by enumerating the non-null indices (we don't care how many incoming arrows there are nor where they originate, as long as there is at least one):

subset	0 leads to	1 leads to
$\{ACDF\}$	$\{AF\}$	$\{ACDF\}$
$\{AF\}$	$\{AF\}$	$\{CD\}$
$\{CD\}$	$\{\}$	$\{ACDF\}$
$\{\}$	$\{\}$	$\{\}$

its topological matrix, having indexed the subsets in the order shown, is $S = a + b$ according to

$$\begin{bmatrix} 1 & 1 & \cdot & \cdot \\ \cdot & 1 & 1 & \cdot \\ 1 & \cdot & \cdot & 1 \\ \cdot & \cdot & \cdot & 2 \end{bmatrix} = \begin{bmatrix} \cdot & 1 & \cdot & \cdot \\ \cdot & 1 & \cdot & \cdot \\ \cdot & \cdot & \cdot & 1 \\ \cdot & \cdot & \cdot & 1 \end{bmatrix} + \begin{bmatrix} 1 & \cdot & \cdot & \cdot \\ \cdot & \cdot & 1 & \cdot \\ 1 & \cdot & \cdot & \cdot \\ \cdot & \cdot & \cdot & 1 \end{bmatrix}.$$

The matrix is upper triangular, because the null set is linked only to itself. Only the 3×3 submatrix in the upper left hand corner is really necessary. The 2 in the lower right corner is not strictly required, attesting only to the fact that *no* path from the null set has anywhere else to go, thereby conserving the uniform row sum.

Finally, the topological matrix of the vector subset diagram is

$$V = \left[\begin{array}{cccc|cccc|cccc} \cdot & 1 & \cdot & \cdot & 1 & \cdot & \cdot & \cdot & \cdot & \cdot & \cdot & \cdot \\ \cdot & \cdot & 1 & 1 & \cdot & \cdot & \cdot & \cdot & \cdot & \cdot & \cdot & \cdot \\ 1 & 1 & \cdot & \cdot & \cdot & \cdot & \cdot & \cdot & \cdot & \cdot & \cdot & \cdot \\ \cdot & \cdot & 1 & \cdot & \cdot & \cdot & \cdot & 1 & \cdot & \cdot & \cdot & \cdot \\ \hline \cdot & \cdot & \cdot & \cdot & 1 & \cdot & \cdot & \cdot & \cdot & 1 & \cdot & \cdot \\ \cdot & \cdot & \cdot & \cdot & \cdot & \cdot & \cdot & \cdot & \cdot & \cdot & \cdot & \cdot \\ \cdot & \cdot & \cdot & \cdot & \cdot & \cdot & \cdot & \cdot & \cdot & \cdot & \cdot & \cdot \\ \cdot & \cdot & \cdot & \cdot & \cdot & \cdot & \cdot & 1 & \cdot & \cdot & 1 & \cdot \\ \hline \cdot & \cdot & \cdot & \cdot & \cdot & \cdot & \cdot & \cdot & \cdot & \cdot & \cdot & \cdot \\ \cdot & \cdot & 1 & 1 & \cdot & \cdot & \cdot & \cdot & \cdot & \cdot & \cdot & \cdot \\ 1 & 1 & \cdot & \cdot & \cdot & \cdot & \cdot & \cdot & \cdot & \cdot & \cdot & \cdot \\ \cdot & \cdot & \cdot & \cdot & \cdot & \cdot & \cdot & \cdot & \cdot & \cdot & \cdot & \cdot \end{array}\right].$$

11.8.4 Structural observations

The topological matrix of the vector subset diagram manifests some interesting structural details. It is not quite a tensor product, but the nonzero submatrices fall exactly on the locations of the nonzero elements of the scalar matrix. The submatrices are of two types, corresponding to the decomposition of the symbolic de Bruijn matrix, being derived from them through multiplication on the right by a projection onto the subset defining the row. In the matrix shown, the projector for the first row is the identity matrix, for the second it preserves components D and F, while the third goes with A and C.

Moreover a supermatrix whose diagonal submatrices are the appropriate projectors commutes with the vector subset matrix, just as it commutes with each of the two summands resulting from the symbolic decomposition of the de Bruijn matrix. Incorporation of this projector as a factor allows writing the vector matrix as a sum of two tensor products, which in turn facilitates the evaluation of powers of the vector matrix and products of its summands.

Suppose that S is the scalar subset matrix, Π the projector, $V = \Pi\mathcal{V}$ the vector subset matrix, and that the de Bruijn matrix D is decomposed into the sum $A + B$ according to a criterion such as the value of the evolved cell. Then we would have

$$V = \mathcal{A} + \mathcal{B},$$

$$\mathcal{A} = \Pi(A \otimes a),$$

$$\mathcal{B} = \Pi(B \otimes b);$$

for example the factorization below defines $\mathcal{B}$:

$$\mathcal{B} = \left[\begin{array}{cccc|cccc|cccc}
1 & \cdot & \cdot & \cdot & \cdot & \cdot & \cdot & \cdot & \cdot & \cdot & \cdot & \cdot \\
\cdot & 1 & \cdot & \cdot & \cdot & \cdot & \cdot & \cdot & \cdot & \cdot & \cdot & \cdot \\
\cdot & \cdot & 1 & \cdot & \cdot & \cdot & \cdot & \cdot & \cdot & \cdot & \cdot & \cdot \\
\cdot & \cdot & \cdot & 1 & \cdot & \cdot & \cdot & \cdot & \cdot & \cdot & \cdot & \cdot \\
\hline
\cdot & \cdot & \cdot & \cdot & 1 & \cdot & \cdot & \cdot & \cdot & \cdot & \cdot & \cdot \\
\cdot & \cdot & \cdot & \cdot & \cdot & \cdot & \cdot & \cdot & \cdot & \cdot & \cdot & \cdot \\
\cdot & \cdot & \cdot & \cdot & \cdot & \cdot & \cdot & \cdot & \cdot & \cdot & \cdot & \cdot \\
\cdot & \cdot & \cdot & \cdot & \cdot & \cdot & \cdot & 1 & \cdot & \cdot & \cdot & \cdot \\
\hline
\cdot & \cdot & \cdot & \cdot & \cdot & \cdot & \cdot & \cdot & \cdot & \cdot & \cdot & \cdot \\
\cdot & \cdot & \cdot & \cdot & \cdot & \cdot & \cdot & \cdot & \cdot & 1 & \cdot & \cdot \\
\cdot & \cdot & \cdot & \cdot & \cdot & \cdot & \cdot & \cdot & \cdot & \cdot & 1 & \cdot \\
\cdot & \cdot & \cdot & \cdot & \cdot & \cdot & \cdot & \cdot & \cdot & \cdot & \cdot & \cdot
\end{array}\right]
\left[\begin{array}{cccc|cccc|cccc}
\cdot & 1 & \cdot & \cdot & \cdot & \cdot & \cdot & \cdot & \cdot & \cdot & \cdot & \cdot \\
\cdot & \cdot & 1 & 1 & \cdot & \cdot & \cdot & \cdot & \cdot & \cdot & \cdot & \cdot \\
1 & 1 & \cdot & \cdot & \cdot & \cdot & \cdot & \cdot & \cdot & \cdot & \cdot & \cdot \\
\cdot & \cdot & 1 & \cdot & \cdot & \cdot & \cdot & \cdot & \cdot & \cdot & \cdot & \cdot \\
\hline
\cdot & \cdot & \cdot & \cdot & \cdot & \cdot & \cdot & \cdot & \cdot & 1 & \cdot & \cdot \\
\cdot & \cdot & \cdot & \cdot & \cdot & \cdot & \cdot & \cdot & \cdot & \cdot & 1 & 1 \\
\cdot & \cdot & \cdot & \cdot & \cdot & \cdot & \cdot & \cdot & 1 & 1 & \cdot & \cdot \\
\cdot & \cdot & \cdot & \cdot & \cdot & \cdot & \cdot & \cdot & \cdot & \cdot & 1 & \cdot \\
\hline
\cdot & 1 & \cdot & \cdot & \cdot & \cdot & \cdot & \cdot & \cdot & \cdot & \cdot & \cdot \\
\cdot & \cdot & 1 & 1 & \cdot & \cdot & \cdot & \cdot & \cdot & \cdot & \cdot & \cdot \\
1 & 1 & \cdot & \cdot & \cdot & \cdot & \cdot & \cdot & \cdot & \cdot & \cdot & \cdot \\
\cdot & \cdot & 1 & \cdot & \cdot & \cdot & \cdot & \cdot & \cdot & \cdot & \cdot & \cdot
\end{array}\right].$$

An immediate consequence of this representation is the fact that in any product of supermatrices, all the submatrices are formed from products of the two terms into which the de Bruijn matrix is decomposed, even though they may eventually have to be projected to obtain a final result. This establishes a formal relation between the methods of this section and the procedures of Erica Jen [131] or Backhouse and Carré [10].

11.9 Use of the subset diagrams

An excellent application of the subset construction to linear cellular automata is to the search for the Garden of Eden—configurations which cannot evolve from any predecessor. The problem has two settings—for finite or periodically closed automata in which it is trivial due to finiteness, and for infinite automata. Later, the problem can be generalized to counting the counterimages, calculating the frequency distribution of the multiplicity of counterimages, and establishing the relationship between singlevaluedness and invertibility.

11.9.1 Garden of Eden

To apply the subset diagram to the construction of counterimages, begin with the de Bruijn diagram whose links are the neighborhoods of the automaton. The nodes would be parital neighborhoods whose overlap is attested by the links, and the links can be labelled by the value of the cell which evolves from them. This is a new labelling, different from that obtained from the cell by which the neighborhood is extended. For the latter, all the outgoing nodes have distinct labels, but not for the former unless that is the nature of the rule. Possible ambiguities will be resolved in the subset diagram. Starting from the universal node, following the path dictated by the new configuration, it is possible to obtain all the ancestral configurations.

For example, using Rule 126 and the diagram of the last section, the chain 010 leads form $\{ACDF\}$ to $\{AF\}$ to $\{CD\}$ to $\{\}$, the null set. Therefore there is no way that 010 can be part of any evolved configuration. On the other hand a sequence of 0's leads repeatedly to the subset $\{AF\}$, so there are always counterimages of 0^*, the quiescent configuration.

Since the universal subset is the starting point, the counterimage might conceivably begin with any pair of states, but since C and D are missing from the repeating subset, there is no ancestor which terminates with either 01 or 10, a mixed pair of states. Such restrictions can prejudice the evolution of finite or cyclical chains of states, yielding additional ancestorless configurations that would not occur for infinite automata.

Moreover, if the infinite automaton is to be "quiescent at infinity," by having only a finite number of excited states, it may turn out that a similar ancestor is impossible.

Altogether, the principal value of the scalar subset diagram is to establish such things as

1. the shortest excluded words, the occurrence of any one of which creates a Garden of Eden configuration,
2. a maximum length for a minimal excluded word, which is the number of nodes in the portion of the subset diagram connected to the full subset,
3. whether exclusion occurs in stages, as key segments are built up.
4. a regular expression describing excluded words.

11.9.2 Counting counterimages

A vector subset diagram whose links have been labelled according to the evolution of an automaton allows the gathering of fairly subtle information. Its powers classify the counterimages they are counting *exactly* by initial and terminal partial neighborhoods, in addition to lumping them into subsets as the scalar matrix does.

Often this is a more elaborate classification than is needed; what really count are the de Bruijn fragments comprising the submatrices, classifying ancestors according to their initial and final partial neighborhoods. Generally further sorting by subsets of partial neighborhoods only serves to identify and count the Garden of Eden configurations quickly.

For example, continuing to use the example from Rule 126, the de Bruijn product for the sequence 0^* is

$$\begin{bmatrix} 1 & \cdot & \cdot & \cdot \\ \cdot & \cdot & \cdot & \cdot \\ \cdot & \cdot & \cdot & \cdot \\ \cdot & \cdot & \cdot & 1 \end{bmatrix}^n = \begin{bmatrix} 1 & \cdot & \cdot & \cdot \\ \cdot & \cdot & \cdot & \cdot \\ \cdot & \cdot & \cdot & \cdot \\ \cdot & \cdot & \cdot & 1 \end{bmatrix},$$

which makes it clear that an infinite quiescent configuration has exactly two ancestors, itself and 1^*; in any event, 01 and 10 are excluded both as initial sequences and as final sequences from finite configurations such as on a ring. Of course, this information can also be elicited from the scalar subset matrix by looking at all the paths originating from unit classes.

Counterimages can be counted three ways: quiescent at infinity, cyclic, or completely general. Let Π be a typical product of de Bruijn fragments generating the counterimages of a configuration; then

$$t = [\Pi]_{q,q} = Tr(\pi_q \Pi), \tag{11.25}$$

where q is the completely quiescent partial neighborhood, yields the number of counterimages quiescent at infinity. The single nonzero element of the projection operator π_q occupies the qth position on the diagonal; the trace formulation provides a uniformity of treatment for all three cases.

For a cyclic automaton, it is appropriate to recognize that a cycle must begin and end on the same partial neighborhood by calculating (for unit matrix I)

$$t = Tr(\Pi) = Tr(I\Pi). \tag{11.26}$$

Finally, the appropriate unrestricted formula would be

$$t = u^T \Pi u = Tr(U\Pi), \tag{11.27}$$

where u is a vector, U a matrix, solidly filled with 1's. In fact, all three counts can be described by either a trace or an inner product, both of which are linear functionals for matrices.

Taking into account all possible products representing ancestors of a configuration of length n requires evaluating

$$\alpha_1 = k^{-n} Tr(U(a+b)^n), \tag{11.28}$$

but $a + b = D$, the de Bruijn matrix which satisfies $DU = kU$ and $D^{k+r} = k^r D^k$. Altogether, then,

$$\alpha_1 = k^2. \tag{11.29}$$

The unsurprising average of ancestors per configuration being k^2, whenever some configuration has fewer, another must have more. When the imbalance is sufficiently severe, Garden of Eden configurations are required to compensate the configurations with multiple ancestors—a trivial result for finite automata but with interesting consequences for infinite automata.

The second and higher moments of the distribution of counterimages can be obtained from some additional matrix algebra. In the following sequence of substitutions, let P be one of the k^n (noncommuting) factors in the expansion of $D^n = (a+b)^n$.

$$\begin{aligned}
\alpha_2 &= k^{-n} \sum (Tr(UP))^2 \\
&= k^{-n} \sum Tr(UP \otimes UP) \\
&= k^{-n} \sum Tr((U \otimes U)(P \otimes P)) \\
&= k^{-n} Tr((U \otimes U) \sum P \otimes P) \\
&= k^{-n} Tr((U \otimes U)(a \otimes a + b \otimes b)^n).
\end{aligned}$$

If ρ is the dominant eigenvalue of $a \otimes a + b \otimes b$, a term proportional to ρ^n will result for large n.

Higher moments will depend on higher tensor powers of a and b, while the same formula persists with a suitably subscripted ρ and α; for example

$$\begin{aligned}
\alpha_3 = {} & k^{-n} Tr((U \otimes U \otimes U) \times \\
& (a \otimes a \otimes a + b \otimes b \otimes b).
\end{aligned} \tag{11.30}$$

In another direction, if the connectivity matrix had been decomposed into three classes of links a, b, and c, the second moment would have taken the form

$$\begin{aligned} \alpha_2 &= k^{-n}Tr((U \otimes U) \times \\ &\quad (a \otimes a + b \otimes b + c \otimes c)^n). \end{aligned} \tag{11.31}$$

11.10 Pair diagram

The cartesian product of graphs has many applications. As the upper bound of two graphs, it compares paths between two different diagrams (or two paths in the same diagram). It also appears in second moment (and therefore variance) calculation of path frequencies. As an example, Figure 11.14 shows the pair diagram (cartesian product of a graph with itself) for the (2, 1) de Bruijn diagram of Figure 11.7.

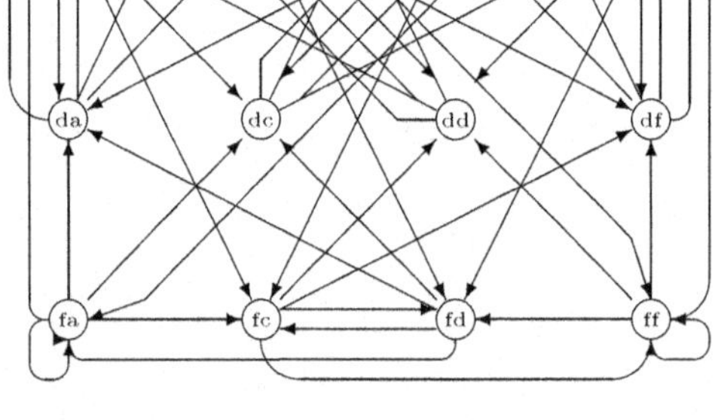

Figure 11.14: Sixteen node pair diagram for a (2, 1) de Bruijn diagram, an image of which is visible along the diagonal as well as in horizontal and vertical section.

Its connectivity matrix is

$$P = \left[\begin{array}{cccc|cccc|cccc|cccc}
1 & 1 & \cdot & \cdot & 1 & 1 & \cdot & \cdot & \cdot & \cdot & \cdot & \cdot & \cdot & \cdot & \cdot & \cdot \\
\cdot & \cdot & 1 & 1 & \cdot & \cdot & 1 & 1 & \cdot & \cdot & \cdot & \cdot & \cdot & \cdot & \cdot & \cdot \\
1 & 1 & \cdot & \cdot & 1 & 1 & \cdot & \cdot & \cdot & \cdot & \cdot & \cdot & \cdot & \cdot & \cdot & \cdot \\
\cdot & \cdot & 1 & 1 & \cdot & \cdot & 1 & 1 & \cdot & \cdot & \cdot & \cdot & \cdot & \cdot & \cdot & \cdot \\
\hline
\cdot & \cdot & \cdot & \cdot & \cdot & \cdot & \cdot & \cdot & 1 & 1 & \cdot & \cdot & 1 & 1 & \cdot & \cdot \\
\cdot & \cdot & \cdot & \cdot & \cdot & \cdot & \cdot & \cdot & \cdot & \cdot & 1 & 1 & \cdot & \cdot & 1 & 1 \\
\cdot & \cdot & \cdot & \cdot & \cdot & \cdot & \cdot & \cdot & 1 & 1 & \cdot & \cdot & 1 & 1 & \cdot & \cdot \\
\cdot & \cdot & \cdot & \cdot & \cdot & \cdot & \cdot & \cdot & \cdot & \cdot & 1 & 1 & \cdot & \cdot & 1 & 1 \\
\hline
1 & 1 & \cdot & \cdot & 1 & 1 & \cdot & \cdot & \cdot & \cdot & \cdot & \cdot & \cdot & \cdot & \cdot & \cdot \\
\cdot & \cdot & 1 & 1 & \cdot & \cdot & 1 & 1 & \cdot & \cdot & \cdot & \cdot & \cdot & \cdot & \cdot & \cdot \\
1 & 1 & \cdot & \cdot & 1 & 1 & \cdot & \cdot & \cdot & \cdot & \cdot & \cdot & \cdot & \cdot & \cdot & \cdot \\
\cdot & \cdot & 1 & 1 & \cdot & \cdot & 1 & 1 & \cdot & \cdot & \cdot & \cdot & \cdot & \cdot & \cdot & \cdot \\
\hline
\cdot & \cdot & \cdot & \cdot & \cdot & \cdot & \cdot & \cdot & 1 & 1 & \cdot & \cdot & 1 & 1 & \cdot & \cdot \\
\cdot & \cdot & \cdot & \cdot & \cdot & \cdot & \cdot & \cdot & \cdot & \cdot & 1 & 1 & \cdot & \cdot & 1 & 1 \\
\cdot & \cdot & \cdot & \cdot & \cdot & \cdot & \cdot & \cdot & 1 & 1 & \cdot & \cdot & 1 & 1 & \cdot & \cdot \\
\cdot & \cdot & \cdot & \cdot & \cdot & \cdot & \cdot & \cdot & \cdot & \cdot & 1 & 1 & \cdot & \cdot & 1 & 1
\end{array}\right],$$

whose tensor product form is clearly visible.

It is not always necessary to distinguish between the members of a pair; the arrows in a graph require links to be defined as ordered pairs of nodes, but that does not necessarily require that a pair of links be taken in any particular order, for example.

Figure 11.15 shows the unordered pair graph of the two-stage de Bruijn diagram for $(2, 1)$ automata, whose connectivity matrix is the symmetrized Kronecker product of the individual connectivity matrices. It has fewer nodes, yet the full Kronecker product is homomorphic to it, giving it computational and visual advantages.

Both versions satisfy Richard's criterion [195] for duality, presumably participating in the same kind of dual chains as the de Bruijn diagrams themselves.

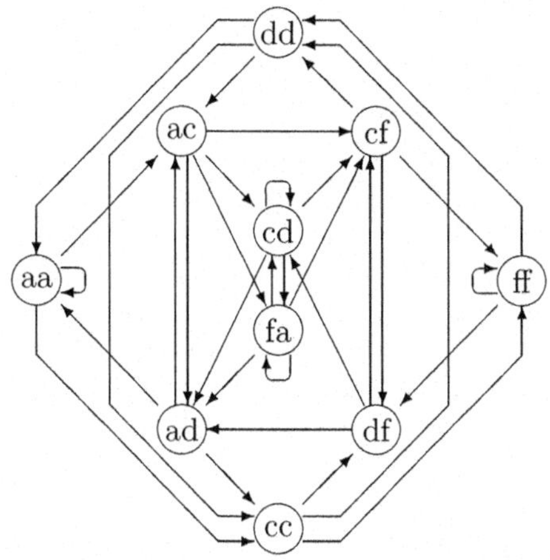

Figure 11.15: Ten node unordered pair diagram for a $(2, 1)$ de Bruijn diagram, an image of which (the diagonal) is visible around the circumference.

The connectivity matrix for unordered pairs is

$$p = \begin{bmatrix}
1 & 1 & . & . & 1 & . & . & . & . & . \\
. & . & 1 & 1 & . & 1 & 1 & . & . & . \\
1 & 1 & . & . & 1 & . & . & . & . & . \\
. & . & 1 & 1 & . & 1 & 1 & . & . & . \\
. & . & . & . & . & . & . & 1 & 1 & 1 \\
. & . & 1 & 1 & . & 1 & 1 & . & . & . \\
. & . & . & . & . & . & . & 1 & 1 & 1 \\
1 & 1 & . & . & 1 & . & . & . & . & . \\
. & . & 1 & 1 & . & 1 & 1 & . & . & . \\
. & . & . & . & . & . & . & 1 & 1 & 1
\end{bmatrix},$$

while the homomorphism matrix between the two versions is

$$X = \begin{bmatrix}
1 & . & . & . & . & . & . & . & . & . \\
. & 1 & . & . & . & . & . & . & . & . \\
. & . & 1 & . & . & . & . & . & . & . \\
. & . & . & 1 & . & . & . & . & . & . \\
. & 1 & . & . & . & . & . & . & . & . \\
. & . & . & . & 1 & . & . & . & . & . \\
. & . & . & . & . & 1 & . & . & . & . \\
. & . & . & . & . & . & 1 & . & . & . \\
. & . & 1 & . & . & . & . & . & . & . \\
. & . & . & . & . & 1 & . & . & . & . \\
. & . & . & . & . & . & . & 1 & . & . \\
. & . & . & . & . & . & . & . & 1 & . \\
. & . & . & 1 & . & . & . & . & . & . \\
. & . & . & . & . & . & 1 & . & . & . \\
. & . & . & . & . & . & . & . & 1 & . \\
. & . & . & . & . & . & . & . & . & 1
\end{bmatrix}.$$

Multiplicities spoil comparing PX to Xp:

$$\begin{bmatrix}
1 & 2 & . & . & 1 & . & . & . & . & . \\
. & . & 1 & 1 & . & 1 & 1 & . & . & . \\
1 & 2 & . & . & 1 & . & . & . & . & . \\
. & . & 1 & 1 & . & 1 & 1 & . & . & . \\
. & . & 1 & 1 & . & 1 & 1 & . & . & . \\
. & . & . & . & . & . & . & 1 & 2 & 1 \\
. & . & 1 & 1 & . & 1 & 1 & . & . & . \\
. & . & . & . & . & . & . & 1 & 2 & 1 \\
1 & 2 & . & . & 1 & . & . & . & . & . \\
. & . & 1 & 1 & . & 1 & 1 & . & . & . \\
1 & 2 & . & . & 1 & . & . & . & . & . \\
. & . & 1 & 1 & . & 1 & 1 & . & . & . \\
. & . & 1 & 1 & . & 1 & 1 & . & . & . \\
. & . & . & . & . & . & . & 1 & 2 & 1 \\
. & . & 1 & 1 & . & 1 & 1 & . & . & . \\
. & . & . & . & . & . & . & 1 & 2 & 1
\end{bmatrix} \neq \begin{bmatrix}
1 & 1 & . & . & 1 & . & . & . & . & . \\
. & . & 1 & 1 & . & 1 & 1 & . & . & . \\
1 & 1 & . & . & 1 & . & . & . & . & . \\
. & . & 1 & 1 & . & 1 & 1 & . & . & . \\
. & . & 1 & 1 & . & 1 & 1 & . & . & . \\
. & . & . & . & . & . & . & 1 & 1 & 1 \\
. & . & 1 & 1 & . & 1 & 1 & . & . & . \\
. & . & . & . & . & . & . & 1 & 1 & 1 \\
1 & 1 & . & . & 1 & . & . & . & . & . \\
. & . & 1 & 1 & . & 1 & 1 & . & . & . \\
1 & 1 & . & . & 1 & . & . & . & . & . \\
. & . & 1 & 1 & . & 1 & 1 & . & . & . \\
. & . & 1 & 1 & . & 1 & 1 & . & . & . \\
. & . & . & . & . & . & . & 1 & 1 & 1 \\
. & . & 1 & 1 & . & 1 & 1 & . & . & . \\
. & . & . & . & . & . & . & 1 & 1 & 1
\end{bmatrix}.$$

But the discussion surrounding Equations 11.3 and 11.16, gives a choice: 1) reduce PX to a boolean matrix whose nonzero entries are ones, or 2) weight the links in p with the number of links per pair to get a numerically valid comparison.

11.11 Probabilistic de Bruijn matrix

The topological matrix of a de Bruijn diagram defines the connectivity of a diagram by indicating the different branches that arise when symbols are added and deleted from a sequence, but it takes no account of the likelihood that one or the other of the symbols will be chosen. If probabilities are incorporated in the matrix elements, it is just as easy to calculate the probability that a long string will occur as to establish the possibility of its presence in the first place.

11.11.1 Positive matrices

There is an extensive lore regarding the properties of positive matrices, dating from the work of Frobenius and Perron at the beginning of the century, and ably summarized in several contemporary textbooks, such as *Matrix Iterative Analysis* by Varga [221], *Non-Negative Matrices* by Seneta [206], and *Nonnegative Matrices in the Mathematical Sciences* by Berman and Plemmons [21]. A recent addition to this series is Minc's *Nonnegative Matrices.*

Applied to topological matrices, the theory yields estimates for such things as the increase in the numbers of paths through a network as the path length increases. Applied to probabilistic matrices it allows the determination of equilibrium configurations and the rapidity of approach to equilibrium from arbitrary initial configurations.

It is not surprising that the theory of positive matrices has strong applications to Markov chains and similar probabilistic concepts, or even that much of the motivation for their study came from this direction.

11.11.2 Largest eigenvalue

Gerschgorin estimates are somewhat easier to work with for positive matrices. Either of the two forms may be used

1. all eigenvalues lie within disks centered at the origin, whose radii are either the different row sums, or the different column sums. Whichever of the two yields the more restrictive results may be chosen.

2. all the eigenvalues lie within disks centered at the diagonal matrix elements, whose radii are the sums of the remaining elements in the respective row or column. Again, the more restrictive result may be used.

For positive matrices the additional information is available that the maximum eigenvalue must surpass the minimum row sum as well as being unable to reach the maximum row sum. Indeed the actual value is bounded away from these limits unless they coincide, there being results of various degrees of complexity for the actual bounds. The situation is described in Varga's book [221], along with references to original papers.

When working with a stochastic matrix, it is already a foregone conclusion that the maximum eigenvalue is 1.0, but the estimates are still valuable for non-stochastic matrices. The equilibrium eigenvector as well as the maximum eiganvalue can often be found by the expedient of successively squaring the matrix.

11.11.3 Second eigenvalue and convergence

For strictly positive matrices the maximum eigenvalue is unique and bounded away from the remaining eigenvalues. Nevertheless the size of the second largest eigenvalue is influential in determining such matters as the rate of convergence of the technique of successive squaring, and in any event governs the rate of approach to the equilibrium eigenvector when the matrix is applied iteratively to a trial vector.

There are two familiar formulations for bounds on the modulus of the second eigenvalue (which neither has to be real nor unique). One is essentially equivalent to using the spectral theorem to subtract out the projection operator belonging to the largest eigenvalue in order to apply the Gerschgorin limits anew [21, p. 51]. The second interprets the matrix as an operator on a projective space, for which the largest eigenvector becomes a fixed point and the remaining eigenvalues refer to movement within the projective space.

The second approach is quite reasonable considering that eigenvectors in linear spaces correspond to fixed points in projective spaces. The stability of the fixed point is determined to a certain extent by the Jacobian — corresponding to the derivative — of the mapping at the fixed point, but in more detail by considering the remaining eigenvalues individually. In other words, the Jacobian matrix rather than just the Jacobian determinant should be consulted.

Birkhoff [24] has emphasized the projective point of view, but the most accessible account is probably the painstaking derivation from first principles by Bauer [12].

In any event, suppose that **A** is a strictly positive matrix for which

$$K = \max_{ijkl} \sqrt{\frac{a_{ij}a_{kl}}{a_{il}a_{kj}}}$$

$$M = \max_{ij} a_{ij}$$

$$m = \min_{ij} a_{ij}$$

and that $\rho(A)$ is the Frobenius-Perron, or dominant, eigenvalue of **A**. Then any other eigenvalue λ must satisfy

$$|\lambda| \leq \frac{K-1}{K+1}\rho(A) \leq \frac{M-m}{M+m}\rho(A).$$

Hopf [121] obtained an alternative version,

$$|\lambda| \leq \frac{K-1}{K+1}\rho(A) \leq \frac{M^2-m^2}{M^2+m^2}\rho(A).$$

The bounds for both estimates are actually attained for certain matrices, making them optimal in the absence of further information concerning **A**. When the eigenvector corresponding to $\rho(A)$ is known, some further improvement is possible [13].

These results would not apply directly to a de Bruijn matrix because of its sparsity. Yet it has a power filled solidly with 1's whose oscillation vanishes, forcing the second largest eigenvalue of the power to be identically zero, in agreement with the characteristic equation. Less extreme conclusions govern probabilistic de Bruijn matrices, but uniformity would still surely speed any approach to equilibrium.

11.11.4 Equilibrium eigenvector

Crude estimates of the equilibrium eigenvector may be obtained by successive squaring, but better estimates are usually needed. A fairly recent article [50] utilizes an approximation to the matrix resolvent — itself both an eigenvector matrix and a positive matrix for a suitable range of its parameter — to characterize polyhedra confining the equilibrium eigenvector. Namely, if **L** is a positive matrix of lower bounds to the matrix **A**, and λ is an upper bound to $\rho(A)$, the rows (or columns) of the resolvent

$$R(\lambda) = (\lambda I - L)^{-1}$$

can be used to form the cages.

11.12 Determinant and inverse

A doubly stochastic one-stage de Bruijn diagram would have a probability matrix

$$\mathcal{P}_1 = \begin{bmatrix} p & q \\ q & p \end{bmatrix}$$

with $q = 1 - p$, whose eigenvalues would be $\lambda = 1,\ p - q$; being symmetric, the matrix of both left and right eigenvectors would be

$$\frac{1}{2}\begin{bmatrix} 1 & 1 \\ 1 & -1 \end{bmatrix}.$$

The probability matrix of a two stage diagram is a little more complicated.

$$\mathcal{P}_2 = \begin{bmatrix} p_1 & q_2 & \cdot & \cdot \\ \cdot & \cdot & p_3 & q_4 \\ q_1 & p_2 & \cdot & \cdot \\ \cdot & \cdot & q_3 & p_4 \end{bmatrix}$$

with $q_i = 1 - p_i$, and the characteristic polynomial

$$\lambda^4 - \lambda^3(p_1 + p_4) - \lambda^2(p_1p_4 - p_2p_3) +$$

$$\lambda(p_2d_{34} + p_3d_{12}) - d_{12}d_{34}$$

with two determinants defined by

$$\begin{aligned} d_{12} &= (p_1p_2 - q_1q_2) \\ d_{34} &= (p_3p_4 - q_3q_4) \end{aligned}$$

The determinant of the de Bruijn matrix is the product of two smaller determinants corresponding to the evident blocks in the de Bruijn matrix; in this we have a special case of a quite general result. The p's and q's were defined as they are with the thought of equating the subscript pairs (1,2) and (3,4) to get a doubly stochastic matrix, but the best interpretation of the vanishing of the small determinants is that the biases of the probabilities in their submatrices are equal.

The constant term in the characteristic equation will be zero, giving a single zero root, if either determinant is zero. The coefficient of λ will also be zero if both are zero, producing a double zero root. In order to get three equal roots and thus open up the possibility of the Jordan normal form, requires some similarity between the two submatrices in addition.

The representations of the de Bruijn matrix as sums and products need very little change to apply to the probabilistic versions as well. The factorizations give immediate formulas for determinants; when they do not vanish, the block diagonal form of of one of the factors, together with the fact that the other is a permutation matrix quickly reveals a form for the inverse matrix. Taking as an example the matrix of the last section, we have

$$\mathcal{P}_2^{-1} = \begin{bmatrix} \frac{p_2}{d_{12}} & \cdot & \frac{-q_2}{d_{12}} & \cdot \\ \frac{-q_1}{d_{12}} & \cdot & \frac{p_1}{d_{12}} & \cdot \\ \cdot & \frac{p_4}{d_{34}} & \cdot & \frac{-q_4}{d_{34}} \\ \cdot & \frac{-q_3}{d_{34}} & \cdot & \frac{p_3}{d_{34}} \end{bmatrix}.$$

The Gerschgorin limit of this matrix requires its maximum eigenvalue to be less than the greatest reciprocal determinant $1/d_{12}$ or $1/d_{34}$ (all with respect to absolute values). Generalizing, this means that *every* eigenvalue of $\mathcal{P}_n$ must be greater than the lesser of these determinants. Of course, if one of them is zero, we know that the lower bound is reached and $\mathcal{P}_n$ is singular.

In the other direction, unless some p_i or q_i is zero, every matrix element of $\mathcal{P}_n^n$ will be strictly positive, although in the probabilistic case no two of them have to be equal. This ensures that the maximum eigenvalue, $\lambda = 1$, will be unique. Conversely, if some probability or coprobability vanishes, the possibility of degeneracy exists (but is not obligatory).

11.13 Chacteristic equation

Unfortunately information about the other eigenvalues, particularly the second largest of them, is not so readily available—either directly or from the factored forms. However, formulas for the coefficients of the characteristic equation can be derived from the symbolic form of the de Bruijn matrix elements; their expression is even more elegant if a new rule of composition is introduced for regular expresions, particularly for sequences of symbols.

11.13.1 Merged product

Let us suppose a and b are letters, and that x is a sequence, possibly null and possibly empty. We define the merged product of the sequences s and t, denoted $s \vee t$, by

$$s \vee t = \begin{cases} axb & s = ax \text{ and } t = xb \\ \phi & \text{otherwise} \end{cases}$$

Likewise, define the overlap, $s \wedge t$, by

$$s \wedge t = \begin{cases} x & s = ax \text{ and } t = xb \\ \phi & \text{otherwise} \end{cases}$$

These are just the combinations of indices required to work with the de Bruijn matrices. For example, the $2^n \times 2^n$ column-stochastic matrix is defined in terms of n- and $(n+1)$-block probabilities by the formula

$$[\mathcal{C}_n]_{ax,yb} = \begin{cases} \frac{p(ax \vee yb)}{p(ax)} & p(ax) \neq 0 \\ 0 & \text{otherwise} \end{cases}$$

Correspondingly the row-stochastic matrix is defined by

$$[\mathcal{R}_n]_{ax,yb} = \begin{cases} \frac{p(ax \vee yb)}{p(xb)} & p(xb) \neq 0 \\ 0 & \text{otherwise} \end{cases}$$

Unless stated otherwise, we will assume that a probabilistic de Bruijn matrix is column-stochastic. Since it is not always convenient to show the matrix elements as a quotient, but it is essential to know which of them intrinsically vanish, let us write

$$[\mathcal{C}_n]_{ax,yb} = m(ax, yb)$$

and note that it vanishes unless $x = y$, or alternatively when $ax \vee yb = \phi$.

11.13.2 Trace

The trace of $\mathcal{C}_n$, which is the coefficient of λ^{n-1} in the characteristic equation, satisfies

$$\begin{aligned} Tr(\mathcal{C}) &= \sum_{ax} m(ax, ax) \\ &= \sum_{i|ax=xi} m(ax, xi) \\ &= \sum_{i} m(i^*, i^*). \end{aligned}$$

The essential element in all of these calculations is to factor the column index in such a way that it can match a row index and yet produce a non-zero merged product.

11.13.3 Second coefficient

$$\begin{aligned} a_2 &= \sum_{ax \le by} (m(ax,ax)m(by,by) - \\ &\quad m(ax,by)m(by,ax)) \\ &= \sum_{i|ax=xi} \sum_{j|by=yj} m(ax,xi)m(by,yj) - \\ &\quad \sum_{i|by=xi} \sum_{j|ax=yj} m(ax,xi)m(by,yj) \\ &= \sum_{i \le j} m(i^*,i^*)m(j^*,j^*) - \\ &\quad \sum_{i \le j} m((ij)^*)m((ji)^*) \\ &= \sum_{i \le j} \begin{vmatrix} m(i^*,i^*) & m((ij)^*) \\ m((ij)^*) & m(j^*,j^*) \end{vmatrix}. \end{aligned}$$

In essence, this sum runs over all cycles of length 2 in the de Bruijn diagram. The solution to the requirements $by = xi$ and $ax = yj$ is $y = (ab)^*$; however this notation must be understood to designate only as much of the sequence as is needed. Thus $(01)^*$ could represent 0, 01, 010, 0101, and so on, but not 101, for example.

11.13.4 Principal minors

Striking out rows and columns intersecting on a diagonal will always leave some surviving elementary determinants, until so many have been stricken out that it is no longer possible to form a complete determinant. The coefficient of the elementary determinants will depend on how much of the remaining elementary determinant can be salvaged.

In general, the determinant of the de Bruijn $|m(ix,yj)|$ factorizes, so that

$$|m(ix,yj)| = \prod_{x,x} |m(ix,xj)|,$$

where the factors are $k \times k$ determinants obtained by varying the indices i and j with fixed x. To get the minor of the hz, hz diagonal element, we need to consider two cases. If this pivot does not belong to any subdeterminant, it can be ignored. If it does, we drop the z, z term from the product, replacing it by the internal minor from within the omitted factor. Thus

$$a_1 = \sum_{hz=wh} cofactor(m(hz,hz)) \prod_{x \ne z} |m(ix,xj)|.$$

11.13.5 Determinant

Although we already have the determinant of the probabilistic de Bruijn matrices from their factored forms, it is worth observing that it can be calculated directly by following the symbolic procedure.

$$|m(ix, yj)| = \prod_{x,x} |m(ix, xj)|.$$

11.14 Summary

In conclusion, there are two aspects of graph theory which are important for automata theory. One, typified by the use of classical de Bruijn diagrams and their subdiagrams, portrays the relationships between components of the automaton, such as the individual neighborhoods or partial neighborhoods, and the evolution of the automaton. This application emphasizes several aspects of graph theory, motivating the introduction of operations on graphs which adequately describe properties of the automaton.

Among these are the ordering of graphs, homomorphisms between graphs, and such operations as the formation of cartesian products and dual graphs. The characterization of homomorphism in terms of matrices embodied in Equations 11.3 and 11.16 has long been known to mathematicians with varying degrees of formality, but deserves to be emphasized anew. The definition of duals via the RC factorization is more original, probably because it relates better to "digraphs" than to the structure traditionally called a graph.

Consequently, it is a step away from the classical treatments of graph theory, such as Ore's *American Mathematical Society* monograph [179]; treatments which were more content with simply describing and classifying graphs.

Not only that, but the requirements of the application to automata theory oblige us to consider several alternative approaches to graphs not ordinarily considered part of graph theory, such as the description of graphs by symbolic equations, and their representation by regular expressions. Likewise the topological matrix, or connectivity matrix, is intimately related to properties considered essential to the understanding of automata, requiring a working knowledge of the class of positive matrices.

The second aspect of graph theory concerns its relation to probability theory, via the topological matrix, probabalistic evolution, and probabilistic de Bruijn diagrams. Strictly, interest lies in the application of detailed properties of positive matrices and the Frobenius-Perron theory to all the structures which have been constructed with the aid of graph theory in the earlies sections of the article.

Probabilistic de Bruijn matrices describe correlations between overlapping cell blocks in an automaton. Yet another class of diagram, evolution diagrams, has its probabalistic version; thus either type of diagram has a probabilistic version which could be of concern. Some interesting recently published papers have dealt with establishing and then solving equations for self-consistent probabilities for clusters of cells [107, 242].

So, although it is not part of graph theory, automata theory brings about an interesting application of the classical Perron-Frobenius theory to certain classes of graphs — the de Bruijn diagrams and evolutionary diagrams — which might otherwise have escaped such attention; and in the process some interesting new characteristics of graphs are revealed.

Chapter 12

Reversible cellular automata

A reversible cellular automaton is one whose evolution, and therefore the entire past history of any configuration, can be uniquely deciphered. There are degrees of reversibility, depending upon whether the configurations considered are arbitrary, periodic, or quiescent at infinity; which are subsidiary to more general questions of injectivity and surjectivity, within a general perspective of the ancestry of configurations. Reversibility is examined within this general context, expanded to include the frequency distribution of ancestors and its moments. It is argued that the coupling of zero variance (judged from the maximum eigenvalue of the second moment matrix) with zero frequency for zero ancestors (surjectivity) is the fundamental concept. An ideal theoretic property inherent in decompositions of the de Bruijn matrix suffices to prove the coupling for automata. Surjectivity in different contexts, injectivity, and degrees of multiple valuedness all follow from this central result. Although the article is intended as a review, it is far from a complete historical survey; the presentation is uniformized through the use of graphs, de Bruijn diagrams and matrices wherever possible.

When John von Neumann [177] worked out the theory of universal constructors within the framework of cellular automata in the mid fifties, philosophical questions soon arose concerning ancestorless patterns, which surely could not be "constructed" through the intermediary of other configurations occupying the same system. That such predicaments could arise under certain circumstances was first appreciated by Edward F. Moore [164], whose article of 1962 was reprinted in a collection of influential papers edited by Arthur W. Burks [38] in 1970. Even though the only requirement which von Neumann was interested in satisfying was that a device with the computational power of a Turing machine should be capable of producing copies of itself — not necessarily that it had to fill space with arbitrary designs — the latter concept is an interesting one which has received further study.

In short order the question turned into the relationship between the *local* mapping in an automaton, which assigns successors to neighborhoods, and the *global* mapping, which produces the succession of configurations from one generation to the next. In contemporary mathematical terms, the interesting questions are whether the global mapping is *injective*, formerly called *one-to-one*, and also whether it is *surjective*, formerly designated *onto*. In the latter case every configuration has at least one ancestor, and there is no Garden of Eden. For finite sets the equivalence of the two characteristics follows from counting arguments which no longer apply to infinite sets.

To pose the problem properly for infinite automata, some thought has to be given to the meaning of infinity, often with respect to boundary conditions which are relaxed or assumed to lose their influence as the automaton grows in size. Three choices usually considered are: 1) cyclic boundary conditions, 2) quiescence at infinity, or 3) a quite general, unspecified arrangement (sometimes with a topology which minimizes the importance of exceedingly remote cells).

In a sense, these are external considerations, subordinate to other, internal, aspects of the problem. Fundamental to the nature of cellular automata is the difference in size between cells and the neighborhoods which determine their evolution, resulting in a permanent discrepancy between the area of a region and that from which it has evolved. Paying attention to the external aspects may try to avoid the difficulty by folding or smoothing the margins, but it is probably better to resolve the internal aspects first, adjusting them to the external requirements at the end.

Although it is quite easy to work out ancestral neighborhoods for one or even a few cells, problems of consistency are bound to arise because the neighborhoods always overlap. Fortunately, especially in one dimension, there are matricial techniques based on the de Bruijn diagram which organize the relationships involved, and the calculation of ancestors can procede. Most commonly inconsistencies arise which cannot be resolved, leading to the Garden of Eden, and the conclusion that most rules of evolution are not surjective. For such rules, most configurations have multiple ancestors whilst others have none.

Among the exceptions, the freedom of choice which exists in the margin usually leads not just to a single ancestor for each configuration, but to several distinct ancestors. But for a still smaller subset of rules, the freedom is to greater or lesser degree illusory, even to the extent that the state of one of the cells in the ancestral neighborhood — the central cell, for instance — can be deduced once the state of the evolved cell and some of its neighbors is known. The rule of association can be used to run the automaton backward, for which such an automaton would be called reversible.

Historically, various attempts were made to define the problem, and to encounter examples of various types of behavior; some of the conflicts inherent in which were discussed by Sven Skyum [211] in a 1975 article aptly titled "Confusion in the Garden of Eden."

Interestingly, similar studies had grown out of the topic of symbolic dynamics and ergodic theory, originating with Henri Poincaré in the nineteenth century, continued in the first half of the twentieth by George Birkhoff [25], formalized by Walter Gottschalk and Gustav A. Hedlund [97] in a 1955 Colloquium Publication of the American Mathematical Society. In 1969 Hedlund [114] published a thorough summary of the properties of automorphisms and endomorphisms of the shift dynamical system, albeit written in a very terse mathematical language with many topological features.

Indeed, continuous mappings of the shift *are* cellular automata, providing one of the few known instances of an unequivoval application of the theory of cellular automata to another field; ironically the applications to symbolic dynamics and ergodic theory were worked out in exhaustive detail much before the subject of cellular automata became particularly organized, and quite unbeknownst to the majority of its practitioners.

In 1978, about a decade after Hedlund's fundamental paper, Masakazu Nasu [169] undertook to describe these results with the aid of graph theory, without relying on topology, and incorporating some ideas from automata theory. During this same period a variety of other articles appeared; but in the meantime, a more direct approach to the explicit construction of reversible automata was invented, then applied to some physical problems, by Edward Fredkin, Tommaso Toffoli, and others.

Recent interest in automata has been greatly stimulated by Stephen Wolfram's computer experiments, first reported in *Reviews of Modern Physics* [249] in 1983; familiarity with his reprint collection [252] of 1986 is assumed, as well as his use of (k, r) whenever k-state cells form a neighborhood of radius r (possibly half-integral) is followed, likewise his k-ary enumeration of evolutionary rules.

This paper also relies heavily on the properties of nonnegative matrices, an exposition of whose theory can be found in half a dozen or more contemporary textbooks, such as Gantmacher [89], Bellman [18], Varga [221], Seneta [206], Berman and Plemmons [21], or Minc [162].

12.1 Fredkin's model

Edward Fredkin discovered that the parity, or exclusive or, operation could lead to an interesting replication of patterns in cellular automata [91, 112, 64]; he also devised a model for reversible automata [219]. Reputedly differential equations were his model; first order equations typically have real exponential solutions which are not time reversible, whereas second order equations often have solutions composed of readily reversible pairs of complex conjugates.

12.1.1 Two-generation rule

The proposed rule of evolution involved two generations of cells: let x_i^t represent the state of the ith cell in generation t. Then if

$$x_i^{t+1} = \varphi(x_{i-1}^t, x_i^t, x_{i+1}^t)$$

were the rule of evolution of any ordinary $(2, 1)$ automaton, the new rule

$$\begin{aligned} x_i^{t+1} &= \varphi(x_{i-1}^t, x_i^t, x_{i+1}^t) \oplus x_i^{t-1} \\ &= \Phi(x_i^{t-1}, x_{i-1}^t, x_i^t, x_{i+1}^t) \end{aligned} \tag{12.1}$$

spans two generations, but can be inverted by writing

$$\begin{aligned} x_i^{t-1} &= \varphi(x_{i-1}^t, x_i^t, x_{i+1}^t) \oplus x_i^{t+1} \\ &= \Phi^{-1}(x_i^{t+1}, x_{i-1}^t, x_i^t, x_{i+1}^t), \end{aligned} \tag{12.2}$$

on account of the algebraic properties of the exclusive or. Any other invertible function of two variables could have been used in place of $\oplus$; for boolean variables the only other candidate is exclusive nor (equivalence).

12.1.2 Cartesian product rule

Toffoli's own scheme [218] increases the dimensionality of the automaton rather than the number of generations spanned by its evolutionary rule. However, the simplest solution to the problem of exhibiting explicitly reversible rules of evolution may well lie in increasing the number of states of the automaton, by forming their cartesian products, whose rule of evolution needs to span but a single generation:

$$\begin{aligned}(x_i, y_i)^{t+1} & \\ &= (y_i, \varphi(y_{i-1}, y_i, y_{i+1}) \oplus x_i) \\ &= \Phi((x_{i-1}, y_{i-1})^t, (x_i, y_i)^t, (x_{i+1}, y_{i+1})^t).\end{aligned} \tag{12.3}$$

Thus a single cell of a $(4, 1)$ automaton may hold two successive generations of a $(2, 1)$ automaton by forming pairs whose components are to be interpreted as

$$(x_i^{t-1}, x_i^t);$$

substitution of these values into Eq. 12.3 reproduces Eq. 12.1, in its turn reversible by the rule

$$\begin{aligned}(x_i, y_i)^{t+1} & \\ &= (\varphi(x_{i-1}^t, x_i^t, x_{i+1}^t) \oplus y_i^t, y_i^t) \\ &= \Phi^{-1}((x_{i-1}, y_{i-1})^t, (x_i, y_i)^t, (x_{i+1}, y_{i+1})^t).\end{aligned} \tag{12.4}$$

Similar formulas would allow the formation of reversible (k^2, r) rules from arbitrary (k, r) rules, the more so by taking advantage of additional alternatives for $\oplus$. Nor is it difficult to invent schemes whereby the equivalent of three generations of evolution could be incorporated into (k^3, r) automata, and so on and on for even longer histories.

To understand possible complications which can arise, consider Fredkin's idea applied to a $(2, 1/2)$ automaton via a rule in which sequential letters replace subscript indices for improved legibility:

$$\Phi((a, b), (c, d)) = (d, \varphi(b, d) \oplus a).$$

Evolution loses the value of c, preserves d intact, yet permits the recovery of a if b were known explicitly. Supposing a third cell (e, f) provides two cells in the second generation,

$$(d, \varphi(b, d) \oplus a) \quad \text{and} \quad (f, \varphi(d, f) \oplus c),$$

from which c can now be recovered via $c = \varphi(d, f) \oplus (\varphi(d, f) \oplus c)$ with the aid of the explicitly known d and f. Since d is already known, the cell (c, d) has been reconstructed, as could be done for all the rest of the cells of a configuration.

In each case evolution "remembers" the second component of the *right* cell, but the *left* ancestor is the one recovered by the reversed rule. Typical of neighborhoods with half integral radius, this asymmetry seems to have no further significance. It is often seen in rules with shifting.

12.1.3 Rules spanning three generations

The next rule depends upon three generations of evolution, following the same convention as before:

$$\Phi((a,b,c),(d,e,f)) \quad = \quad (e,c,\varphi(c,f)\oplus a). \tag{12.5}$$

Time reversal cannot be achieved as before by a rule with a two cell neighborhood, because the new state is independent of both b and d. Again supposing a cell (g,h,i) to the right of the first two, all three generating the evolved pair

$$(e,c,\varphi(c,f)\oplus a) \quad \text{and} \quad (h,f,\varphi(f,i)\oplus d),$$

the lack of b prevents recovery of (a,b,c), while the lack of i prevents releasing d from the combination in which it is bound (admittedly, a degenerate φ would help) in order to recover (d,e,f). Not even (g,h,i) can be obtained, for the lack of g; even less information is available going further afield.

However, a third cell in the second generation, involving yet a fourth cell from the first generation, would have the form (h,i,x) wherein the value of x is immaterial, but i combines with f to release d from $\varphi(f,i)\oplus d$ revealing (d,e,f). Altogether, we have a reversible $(8,1/2)$ rule which requires an $(8,1)$ (three cell neighborhood) rule for its reversion. Altogether, the inverse rule appears to be

$$\begin{aligned}&\Phi^{-1}((a,b,c),(d,e,f),(g,h,i))\\ &\quad = \quad (\varphi(e,h)\oplus f,a,e).\end{aligned} \tag{12.6}$$

Evidently considerable ingenuity can be exercised in devising reversible automata, but it would be desirable to have criteria of necessity or sufficiency for the process. One significant aspect of the formulas shown so far is their uniformity in the respect that every state has the same number of counterimages as any other. Additionally, the rules are entirely local; even though the size of the neighborhood of an inverse rule may be different from that of the original rule, it is fixed; moreover distant boundary conditions do not affect it.

12.2 Some relevant diagrams

One approach to encountering reversible automata is to scan all the automata of a given order, to find those for which there are unique ancestors, and to work out their rules of inversion. The task is formidable, considering that there are $k^{k^{2r+1}}$ different rules for (k,r) automata; S. Amoroso and Y. N. Patt [6] reported a survey of $(2,3/2)$ automata in 1972, which uncovered eight instances of nontrivial reversible automata, whose principle of operation differed from the later ideas of Fredkin.

Previously they had known that $(2, 1/2)$ or $(2, 1)$ automata lacked nontrivial rules (shift, complement, identity).

Two different algorithms formed the content of their article; one determined the existence of the Garden of Eden (or else concluded that every configuration has ancestors), the other could verify that a configuration had only one ancestor. Formalizing the auxiliary diagrams and tables which were incidental to their presentation would enhance our understanding of the procedures and could facilitate the derivation of additional results.

12.2.1 De Bruijn diagrams

By a de Bruijn diagram [191] one understands a graph whose nodes are sequences of symbols from some alphabet, for example the set of states of an automaton. The sequences all have the same length, often called the number of stages because of an application to shift register theory [95]; for automata they are partial neighborhoods. In one dimension, a single cell could be dropped from one end or the other of a given neighborhood, although other decompositions are conceivable (dropping or adding *two* cells, say).

Many authors have used de Bruijn diagrams to study cellular automata, among them Nasu [169], Erica Jen [130], Wentian Li [148], and Wolfram [251]. There does not seem to be any evidence of an attempt to formulate two-dimensional de Bruijn diagrams.

The nodes of a de Bruijn diagram are linked according to their overlap, resulting in a map which reveals the possible succession of contents in a moving window as it advances across an even longer sequence. For automata, the most convenient choice of window size — the number of stages — is the length of a full neighborhood, that being the basic unit participating in the evolutionary process.

A de Bruijn diagram is also represented by its connectivity matrix; for this purpose it is convenient to represent the states of a (k, r) automaton by the digits $0, 1, \ldots k-1$, its partial neighborhoods by $2r$-digit numbers radix k. Multiplication by k shifts cells, arithmetic modulo k^{2r} drops the leftmost cell, adding a single digit number adjoins a new rightmost cell; altogether

$$D_{ij} = \begin{cases} 1 & j = \begin{cases} ki \\ ki+1 \\ \cdots \\ ki+(k-1) \end{cases} \\ 0 & \text{otherwise} \end{cases} \qquad (12.7)$$

As an example, $(2, 1)$ automata have four different partial neighborhoods 00, 01, 10, and 10, overlapping each other to form eight full neighborhoods of three cells each. Regarding these binary strings as numerical indices produces

$$D = \begin{bmatrix} 1 & 1 & . & . \\ . & . & 1 & 1 \\ 1 & 1 & . & . \\ . & . & 1 & 1 \end{bmatrix},$$

wherein dots replace zeroes for clarity.

12.2.2 The subset diagram

By its very definition, a de Bruijn diagram serves as a guide to the sequencing of strings of symbols, from which any other set of objects can be given the same sequencing. A trivial example would be to recover the original symbols from the central element of each string, while the purpose of using neighborhoods from an automaton is to obtain the sequence of evolved cells. In the same way, a de Bruijn matrix serves as a skeleton in which the rules of matrix multiplication (for noncommutative elements) ensures that the elements of a product respect the sequence of the corresponding de Bruijn diagram.

In this spirit, alternative labels can be affixed to the links in a de Bruijn diagram; reading one label instead of the other while following along some path associates the labels with one another. In particular, relating the central cell of a neighborhood to the value of the evolved cell associates configurations with their ancestors; whether or not there was any ancestor at all turns into the question of discovering whether the corresponding path exists somewhere in the diagram.

Moore's subset construction

The subset construction [163] was devised to solve this precise problem, so that the transcription of Amoroso and Patt's first algorithm [6] into graphical language is readily accomplished.

The subset diagram of a $2r$ stage k-ary de Bruijn diagram has $2^{k^{2r}}$ nodes; if all configurations of this length possess ancestors, then all configurations of whatsoever length will also do so. If some do not, they can be described; in fact there is a regular expression describing every configuration in the Garden of Eden, which is nothing more than the description of the path leading from the full set to the null set in the subset diagram. Frequently required information can be found without having to use the entire subset diagram; the connected component of the full set is often all that is needed.

As an example, consider the connectivity matrix of the subset diagram belonging to the $(2,1)$ Rule 22. With four partial neighborhoods, the subset diagram has sixteeen nodes. Ranking the subsets in order of size places the full subset as number 1 and the empty set as number 16; it is likewise convenient to group intermediate subsets of similar size.

With such an arrangement links between different groups can be discerned, according to variations in the total number of links connecting nodes in the original diagram, which is represented in the subset diagram by the unit classes. Rather than joining to each other, the unit classes will link to the empty set or to larger classes to reflect inhomogenieties in the original linkage.

$$\Sigma = \left[\begin{array}{c|cccc|cccccc|cccc|c}
1 & 1 & . & . & . & . & . & . & . & . & . & . & . & . & . & . \\ \hline
. & 1 & 1 & . & . & . & . & . & . & . & . & . & . & . & . & . \\
. & . & . & 1 & . & . & . & . & 1 & . & . & . & . & . & . & . \\
1 & . & . & . & . & 1 & . & . & . & . & . & . & . & . & . & . \\
. & . & . & . & 1 & . & 1 & . & . & . & . & . & . & . & . & . \\ \hline
. & . & . & . & . & . & . & 1 & 1 & . & . & . & . & . & . & . \\
. & . & . & . & . & 2 & . & . & . & . & . & . & . & . & . & . \\
. & . & . & 1 & . & . & . & . & . & . & . & . & 1 & . & . & . \\
. & . & . & . & . & . & 1 & . & . & 1 & . & . & . & . & . & . \\
. & . & . & . & . & . & . & . & . & . & 1 & . & . & 1 & . & . \\
. & . & . & . & 1 & . & . & . & . & . & . & 1 & . & . & . & . \\ \hline
. & . & . & . & . & . & . & . & . & . & . & 1 & 1 & . & . & . \\
. & . & . & . & . & . & . & . & . & . & . & . & . & 1 & 1 & . \\
. & . & . & . & . & . & . & . & . & . & . & 1 & 1 & . & . & . \\
. & . & . & . & . & . & . & . & . & . & 1 & . & . & . & . & 1 \\ \hline
. & . & . & . & . & . & . & . & . & . & . & . & . & . & . & 2
\end{array}\right] .$$

Σ^8 is the lowest power of Σ with an element linking the full subset to the empty set; one of the Garden of Eden configurations thus revealed is the sequence 10101001.

Note that Σ is partially tridiagonal, in that there are nonzero submatrices only along its principal diagonal, superdiagonal, and subdiagonal. This is not a necessary structure inasmuch as the imbalance between the number of links at the nodes of the de Bruijn diagram could have been more severe than it is, nor is it especially helpful in analyzing this particular matrix.

Nevertheless, the rows corresponding to the unit classes are significant, inasmuch as they suffice to determine all the remaining rows.

Formal definition

The indexing shown is one of convenience for visual display, but a more mathematical arrangement would be to use n-bit binary numbers as indices. The ith bit of the index would be 0 or 1 according to whether the ith node is included in the subset or not. With such symbolism the remaining rows could be explicitly constructed from those associated with the unit classes.

Even without this element of sophistication, the rule of formation of the subset matrix is easily described. Begin with the observation that the fundamental requirement for the definition of any function is that each argument has just one image. Links in a graph define a function when there is just one link of a given class leading onward from a given node; the subset diagram is created when this condition is violated, by combining multiple links into a single link between subsets.

Even though the linkage of a given graph does not define a function, the subset linkage is always functional and, thanks to the inclusion of the null set as a destination, defined for the entire subset graph. Each class of links defines one function, say Σ_0 or Σ_1. The subset diagram describes the union $\Sigma_0 \cup \Sigma_1$, which itself is not functional.

Let a and b be nodes, S a subset, and $|S|$ the cardinality of S; then

$$\Sigma_i(S) \quad = \quad \left\{ \begin{array}{ll} \phi & S = \phi \\ \{b \mid \mathrm{link}_i(a,b)\} & S = \{a\} \\ \bigcup_{a \in S} \Sigma_i(a) & |S| > 1 \end{array} \right. .$$

General characteristics

From this definition three properties of the subset diagram are evident. First, if there is a chain leading from any subset to the empty set, a similar chain will lead from any smaller (included) subset to the empty set. Conversely, if the unit classes lack links to the empty set, no other subset will have such a link either, and there will be no Garden of Eden.

Such will be the case when every node of the de Bruijn diagram has a link associated with each different kind of cell; that is, when it is row stochastic or itself a function. Similarly, the full subset will link with no others, depriving the rule of a Garden of Eden, if every node has an incoming link for each kind of cell; such rules have column stochastic de Bruijn matrices.

Rules for which the Garden of Eden fails to exist for more subtle reasons are much rarer, correspondingly more interesting, and require much more careful scrutiny of the subset diagram for their discovery.

Second, there is a certain residue of the connectivity of the de Bruijn diagram, in the sense that given any source and any destination, there will always be a subset containing the destination accessible from any subset containing the source; but the destination can hold additional nodes.

Third, the subset diagram may not be connected; even if it is, it is interesting to know the largest subsets accessible from a given subset, as well as the smallest. Consider the subsets of largest cardinality accessible from a given unit class $\{a\}$, and another unit class $\{b\}$.

Since there is a chain running from $\{a\}$ to a subset containing b and thence onward to a subset even larger than the largest subset accessible from b directly, the resulting contradictions can only be resolved if the cardinalities of all the maximal subsets are the same.

Unfortunately this information alone does not ensure that the subsets encountered along a path are monotonic, not even after a maximal subset has finally been reached.

12.2.3 The reversed subset diagram

The subset diagram which has just been discussed pertains to extending a partial neighborhood on the right, but an extension to the left could just as easily be made. Rule 22 is symmetric by reflection, but the partial neighborhoods are not, leaving the lefthanded subset matrix similar to the righthanded version by a permutation rearranging the subsets. For asymmetrical ryles, such a similarity transformation would not exist.

$$\Sigma' = \left[\begin{array}{c|cccc|cccccc|cccc|c}
1 & 1 & . & . & . & . & . & . & . & . & . & . & . & . & . & . \\
\hline
. & 1 & . & 1 & . & . & . & . & . & . & . & . & . & . & . & . \\
1 & . & . & . & . & . & 1 & . & . & . & . & . & . & . & . & . \\
. & . & 1 & . & . & . & . & . & 1 & . & . & . & . & . & . & . \\
. & . & . & . & 1 & 1 & . & . & . & . & . & . & . & . & . & . \\
\hline
. & . & . & . & . & . & 1 & . & 1 & . & . & . & . & . & . & . \\
. & . & . & . & . & . & . & 1 & 1 & . & . & . & . & . & . & . \\
. & . & 1 & . & . & . & . & . & . & . & . & . & . & 1 & . & . \\
. & . & . & . & . & 1 & . & . & . & . & 1 & . & . & . & . & . \\
. & . & . & . & 1 & . & . & . & . & . & . & 1 & . & . & . & . \\
. & . & . & . & . & . & . & . & . & . & 1 & . & 1 & . & . & . \\
\hline
. & . & . & . & . & . & . & . & . & . & . & 1 & . & 1 & . & . \\
. & . & . & . & . & . & . & . & . & . & . & 1 & . & 1 & . & . \\
. & . & . & . & . & . & . & . & . & . & . & . & 1 & . & 1 & . \\
. & . & . & . & . & . & . & . & . & 1 & . & . & . & . & . & 1 \\
\hline
. & . & . & . & . & . & . & . & . & . & . & . & . & . & . & 2
\end{array}\right].$$

Since subsets of partial neighborhoods defining the rows are linked to those of the columns by *incoming* links, it would be more natural to use the transpose of the matrix shown to follow incoming links back through a configuration.

12.2.4 The pair diagram

Whether a configuration has multiple ancestors can be resolved with the subset diagram, but an alternative, the pair diagram, is easier to interpret. The nodes of a pair diagram are pairs of nodes from the original diagram; labelled links join pairs whenever both members of the pair are joined by links with that same label. Paths in the pair diagram correspond to pairs of paths in the original diagram; they do not even have to originate at the same node. However, the same path taken twice can always be found amongst any others, so that the original diagram is always embedded in the pair diagram.

The pertinent question is whether there are any other, and how many, paths besides. The pair diagram is the appropriate arena for the second of Amoroso and Patt's constructions [6]. One might also refer to Tom Head's recent proposal [113].

The pair diagram for the $(2, 1)$ Rule 22 automaton is subordinate to the maximal matrix which would correspond to Rule 0 or Rule 255, the location of the remainder of whose elements is shown by $*$'s in the following matrix:

$$\Pi = \left[\begin{array}{cccc|cccc|cccc|cccc}
1 & * & . & . & * & 1 & . & . & . & . & . & . & . & . & . & . \\
. & . & * & 1 & . & . & 1 & * & . & . & . & . & . & . & . & . \\
* & 1 & . & . & 1 & * & . & . & . & . & . & . & . & . & . & . \\
. & . & 1 & 1 & . & . & * & * & . & . & . & . & . & . & . & . \\
\hline
. & . & . & . & . & . & . & . & * & 1 & . & . & 1 & * & . & . \\
. & . & . & . & . & . & . & . & . & . & 1 & * & . & . & * & 1 \\
. & . & . & . & . & . & . & . & 1 & * & . & . & * & 1 & . & . \\
. & . & . & . & . & . & . & . & . & . & * & * & . & . & 1 & 1 \\
\hline
* & 1 & . & . & 1 & * & . & . & . & . & . & . & . & . & . & . \\
. & . & 1 & * & . & . & * & 1 & . & . & . & . & . & . & . & . \\
1 & * & . & . & * & 1 & . & . & . & . & . & . & . & . & . & . \\
. & . & * & * & . & . & 1 & 1 & . & . & . & . & . & . & . & . \\
\hline
. & . & . & . & . & . & . & . & 1 & * & . & . & 1 & * & . & . \\
. & . & . & . & . & . & . & . & . & . & * & 1 & . & . & * & 1 \\
. & . & . & . & . & . & . & . & * & 1 & . & . & * & 1 & . & . \\
. & . & . & . & . & . & . & . & . & . & 1 & 1 & . & . & 1 & 1
\end{array}\right].$$

The diagram just described pertains to *ordered* pairs; a variant which is frequently encountered is the *unordered* diagram in which the distinction between

(a, b) and (b, a) is considered insignificant. A smaller diagram results; it can be gotten by coalescing the nodes of the larger diagram.

12.3 Vector subset diagram

The subset diagram is concerned with the existence of paths in the original diagram, whatever their point of origin, which it discovers by trying them all. However, it records them in terms of "at least one path," displayed according to the collection of nodes which could serve as an origin (row index) versus the collection of possible destinations (column index). The principal information revealed by the subset diagram is: what amount of choice in selecting a terminal right neighborhood remains, having prescribed both the leftmost partial neighborhood and that a configuration has evolved in a certain way? To the degree that none remains, the configuration cannot have had an ancestor.

Paths leading to the empty set reveal this information; other paths or loops indicate variations in the possible partial neighborhoods to be found at any stage in constructing ancestors. Indeed, Moore's original intention was locating paths leading to the unit classes — paths capable of isolating one single node from the original diagram, thus casting an automaton into a known state.

In a more accurate reckoning, when counting the paths becomes of interest, specific endpoints are required, and the dichotomy between "some" or "none" is no longer appropriate. One way to retain this additional information is to enlarge the nodes in the subset diagram, making each a "vector" whose components are the original nodes; matrices linking vector nodes preserve the original links. However, we should also consider splitting the original de Bruijn matrix into k parts, just because we are interested in the differences in what can be linked with one symbol in contrast to what can be linked with another; each part is assigned to its own subset.

For a formal definition of the entire vector subset diagram, define instead a connectivity matrix. Suppose that $\pi(S)$ is a $k^{2r} \times k^{2r}$ projection matrix whose elements are zero save for those diagonal elements for which the ith partial neighborhood belongs to S. In particular, the projector for the full set is the unit matrix, for the empty set, the zero matrix.

Then, using the matrix D of Eq. 12.7, the element of the vector subset matrix indexed by subsets S and T is just

$$V_{ST} = \pi(S)D\pi(T).$$

All these references to vectors and matrices can be avoided by establishing a suitable linkage within the cartesian product of the subset diagram with the original diagram, although they reappear when the connectivity matrix of the composite diagram is formed. For Rule 22 (or any other $(2,1)$ rule) the result would be a 64×64 matrix with 4×4 submatrices.

In any event, multiplying vector subset matrices induces a product of submatrices, from which most of the information which one desires can already be extracted. They may as well be used directly, discarding the intermediary of the vector subset matrix.

12.4 Symbolic de Bruijn matrices

Symbolic de Bruijn matrices and matrix fragments can be defined in several ways, each with its individual advantages. A good introduction to the techniques involved can be found in the article of R. C. Backhouse and B. A. Carré [10].

12.4.1 Neighborhood matrix

To form complex symbolic expressions, the language of regular expressions is convenient, whose operations are union, concatination, and iteration. Another operation which is sometimes useful is the overlapping (whose symbol could be $\diamond$) of two strings; if the words ax and yb each consist of letters a and b, joined to the shorter words x and y, then:

$$ax \diamond yb \quad = \quad \begin{cases} axb & x = y \\ \phi & \text{otherwise} \end{cases}$$

(ϕ is the empty word).

With such notation a symbolic de Bruijn matrix

$$D_{ij} \quad = \quad i \diamond j \tag{12.8}$$

could be defined; the symbolic form of the matrix shown above would be

$$D = \begin{bmatrix} \mathtt{000} & \mathtt{001} & \cdot & \cdot \\ \cdot & \cdot & \mathtt{010} & \mathtt{011} \\ \mathtt{100} & \mathtt{101} & \cdot & \cdot \\ \cdot & \cdot & \mathtt{110} & \mathtt{111} \end{bmatrix};$$

dots now represent the empty word. Such matrices would be multiplied by using $\diamond$ to multiply individual matrix elements. Subsequently the advantage of substituting images — such as the evolved cell — for the symbolic neighborhoods will be seen. The full neighborhood matrix is rarely used, but it serves to illustrate very explicitly the involvement of the row indices and the column indices with the individual matrix elements.

12.4.2 Evolution matrices

The multiplication of ordinary symbolic matrices follows the algebra of regular expressions; that is, multiplication concatinates symbols, union (addition) offers alternatives, the empty set acts like zero.

Moreover the nth power of a matrix M contains all the possible words of length n, distributed throughout the matrix according to the overlap of the indices of constituent letters as chains are built up to length n. All this overlap in M^n is no longer explicit, yet its matrix elements are still indexed by their initial and final partial neighborhoods.

To track the evolution of a configuration consider a variant of Eq. 12.8, which describes the evolution of individual neighborhoods:

$$\mathbf{S}_{ij} \quad = \quad \varphi(i \diamond j) \tag{12.9}$$

which can be decomposed into the sum

$$S = \sum_{i=0}^{k-1} S(k)$$

by defining

$$S(k)_{ij} = \begin{cases} k & \varphi(i \diamond j) = k \\ \phi & \text{otherwise} \end{cases} . \tag{12.10}$$

As an example, for Rule 22

$$S = \begin{bmatrix} 0 & 1 & . & . \\ . & . & 1 & 0 \\ 1 & 0 & . & . \\ . & . & 0 & 0 \end{bmatrix},$$

decomposable into the two matrices

$$S(0) = \begin{bmatrix} 0 & . & . & . \\ . & . & . & 0 \\ . & 0 & . & . \\ . & . & 0 & 0 \end{bmatrix} \text{ and } S(1) = \begin{bmatrix} . & 1 & . & . \\ . & . & 1 & . \\ 1 & . & . & . \\ . & . & . & . \end{bmatrix}.$$

Altogether, we see that 0 has five counterimages, 1 only three. From the product $S(0)S(0) = S(00)$,

$$\begin{aligned} S(0)^2 &= \begin{bmatrix} 0 & . & . & . \\ . & . & . & 0 \\ . & 0 & . & . \\ . & . & 0 & 0 \end{bmatrix} \begin{bmatrix} 0 & . & . & . \\ . & . & . & 0 \\ . & 0 & . & . \\ . & . & 0 & 0 \end{bmatrix} \\ &= \begin{bmatrix} 00 & . & . & . \\ . & . & 00 & 00 \\ . & . & . & 00 \\ . & 00 & 00 & 00 \end{bmatrix}. \end{aligned}$$

The string **00** is seen to have seven counterimages, and so on for all the other configurations.

12.4.3 Counterimage matrices

Since the evolved cell has been inserted into the symbolic matrix, it is hardly surprising that strings of evolved cells appear as elements of the product matrices. But it is decidedly inconvenient to have to look up all their indices to recover the original string, especially when the same information could have been encoded in the symbolic matrix from the beginning.

Therefore let a subscript such as w_t denote the tth letter of the word w, beginning with the first; $(i \diamond j)_{r+1}$ would be the state of the central cell of its neighborhood for integral r.

This time vary Eq. 12.8 by defining

$$\mathbf{s}_{ij} = (i \diamond j)_{r+1} \tag{12.11}$$

which is likewise decomposable into a sum

$$s = \sum_{i=0}^{k-1} s(k)$$

by defining

$$s(k)_{ij} = \begin{cases} (i \diamond j)_{r+1} & \varphi(i \diamond j) = k \\ \phi & \text{otherwise} \end{cases} . \quad (12.12)$$

Continuing to use Rule 22 as an example,

$$\begin{aligned} s(0)^2 &= \begin{bmatrix} 0 & . & . & . \\ . & . & . & 1 \\ . & 0 & . & . \\ . & . & 1 & 1 \end{bmatrix} \begin{bmatrix} 0 & . & . & . \\ . & . & . & 1 \\ . & 0 & . & . \\ . & . & 1 & 1 \end{bmatrix} \\ &= \begin{bmatrix} 00 & . & . & . \\ . & . & 11 & 11 \\ . & . & . & 01 \\ . & 10 & 11 & 11 \end{bmatrix} . \end{aligned}$$

There are still seven nonempty matrix elements. Expanded by adjoining the implicit boundary cells to obtain all seven precursors of the sequence 00, the evolution shown below results.

$$\begin{array}{ccccccc} 0000 & 0110 & 0111 & 1011 & 1101 & 1110 & 1111 \\ 00 & 00 & 00 & 00 & 00 & 00 & 00 \end{array} .$$

There is no reason to insist that the central cell be used in defining the matrices $s(i)$; frequently an ancestor at one side is chosen, particularly if the radius of the neighborhood is half integral.

12.4.4 Counting counterimages

Yet a third variation on the theme of symbolic de Bruijn matrices is actually a numerical matrix, whose purpose is to merely count counterimages, not to display them in some form. This final variant on Eq. 12.8,

$$D_{ij} = \begin{cases} 1 & \varphi(i \diamond j) \neq \phi \\ 0 & \text{otherwise} \end{cases} \quad (12.13)$$

gives an alternative definition of the de Bruijn matrix which in turn is decomposed into a sum

$$D = \sum_{i=0}^{k-1} N(k)$$

by defining

$$N(k)_{ij} = \begin{cases} 1 & \varphi(i \diamond j) = k \\ 0 & \text{otherwise} \end{cases} . \quad (12.14)$$

Still using Rule 22 as an example,

$$\begin{aligned} N(0)^2 &= \begin{bmatrix} 1 & . & . & . \\ . & . & . & 1 \\ . & 1 & . & . \\ . & . & 1 & 1 \end{bmatrix} \begin{bmatrix} 1 & . & . & . \\ . & . & . & 1 \\ . & 1 & . & . \\ . & . & 1 & 1 \end{bmatrix} \\ &= \begin{bmatrix} 1 & . & . & . \\ . & . & 1 & 1 \\ . & . & . & 1 \\ . & 1 & 1 & 1 \end{bmatrix} . \end{aligned}$$

Summing (rather than counting) the matrix elements again reveals seven counterimages.

12.5 The statistics of counterimages

Taking the numerical version $N(i)$ of the symbolic matrices $s(i)$ allows the number of counterimages to be counted, namely by summing the matrix elements of the product to be evaluated. By invoking the appropriate matrix algebra, various quantities of statistical interest can be obtained; the most important among these are averages and variances.

12.5.1 Averages

In fact there are three ways the sum yielding the number of counterimages can be carried out. If no importance is attached to the boundary cells of the sequence, all the elements should be summed. If the sequence is supposed to be cyclical, then the partial neighborhood with which the sequence starts should be the one with which it ends, and only the diagonal elements should be summed. Finally, if some specific boundaries are required, such as a quiescent partial neighborhood, then only that particular matrix element should be taken into consideration.

All of the summation techniques can be taken into account by choosing appropriate formulas from matrix algebra. Let $\langle u|$ be a row vector all of whose components are ones, $|u\rangle$ its transpose, and $U = |u\rangle\langle u|$ the matrix, all of whose elements are ones. Finally $|i\rangle$ could be a unit vector whose i^{th} component is the only one which is non-zero. Then the sum $s_g(M)$ of all elements of the matrix M is realized by the formulas

$$s_g(M) = \langle u|M|u\rangle = Tr(UM),$$

the diagonal sum $s_c(M)$ corresponds to

$$s_c(M) = \sum_i \langle i|M|i\rangle = Tr(IM)$$

and the quiescent sum, assuming that q is the quiescent state, is given by

$$s_q(M) = \langle q|M|q\rangle = Tr(|q\rangle\langle q|M).$$

The trace formulas are especially convenient.

The average number of ancestors of a configuration of length n can be obtained from summing all the possible products of matrices representing sequences of cells forming the configuration. Such a collection is simply the result of multiplying out the ordered (because of matrix noncommutativity) product

$$\begin{aligned} M_n &= k^{-n} \sum_i \prod_{j=1}^{n} N(i_j) \\ &= k^{-n} \left\{ \sum_i N(i) \right\}^n \end{aligned} \qquad (12.15)$$

Given that the de Bruijn matrix D_{kl} for k states and l stages ($l = 2r$ by our conventions) satisfies the minimal equation

$$D_{kl}^{l+1} = kD_{kl}^{l}$$

and additionally that $D_{kl}^{l} = U$, from which

$$D_{kl}^{l+m} = k^{m}U$$

we have the eminently credible results that independently of n,

$$\begin{aligned} s_g(M_n) &= (k^{2r})^2 && \text{(general)} \\ s_c(M_n) &= k^{2r} && \text{(cyclic)} \\ s_q(M_n) &= 1 && \text{(quiescent).} \end{aligned}$$

12.5.2 Second moment

The formulas for higher moments are more interesting. If R is a certain product of matrices, G the template for admissable boundary conditions, then the square of the number of ancestors represented by R is

$$(Tr(GR))^2.$$

However the relation

$$Tr(A)Tr(B) = Tr(A \otimes B),$$

($\otimes$ designates the tensor product [18]), yields

$$Tr(GR \otimes GR).$$

Identities for the tensor product allow transforming this expression into

$$Tr((G \otimes G)(R \otimes R)),$$

from which the first term can often be extracted as a constant factor from a sum of traces. If the sum in Eq. 12.15 is now carried out for all values of R, it is possible to recognize the expansion of

$$N = \sum_{i=0}^{k-1} N(i) \otimes N(i). \qquad (12.16)$$

This sum has to be evaluated in each particular case; the simple closed form of the general average was due to an algebraic identity which does not apply to the higher moments. Nevertheless, even when it has to be analyzed numerically, it constitutes one single matrix encompassing the behavior of all the ancestors. Particularly interesting is the fact that it is just the connectivity matrix of the pair diagram introduced in Sec 12.2.4.

12.5.3 Higher moments

It is not hard to establish the general formula for the third moment,

$$m_3 = \sum_{i=0}^{k-1} N(i) \otimes N(i) \otimes N(i) \qquad (12.17)$$

nor generally the analogous n^{th} moment in terms of the nth tensor power sum. However, the first two moments are the most important in view of the role of the average and the variance in statistical analyses. Specifically, configurations of zero variance in their number of ancestors are candidates for reversible rules.

12.6 Moments and frequencies

A distribution of data is characterized by its collection of moments as well as by the individual frequencies with which the data occur; these are really two complementary aspects of the same data set. Here, frequencies $n(i)$ are associated with the nonnegative integers i, namely the number of configurations with a given number of ancestors. The kth moment of this data is defined by

$$m_k = \sum_{i=0}^{r} n(i) i^k, \qquad (12.18)$$

where it is understood that $0^0 = 1$.

The second volume, second edition of William Feller's popular book on the theory of probability [80] contains some material on moments; an advanced theory can be found in the American Mathematical Society *Survey* written by J. A. Shohat and J. D. Tamarkin [209].

Regarding the right hand side of Eq. 12.18 as the inner product of two vectors, then displaying the equations for individual moments, reveals a matrix equation

$$\begin{bmatrix} m_0 \\ m_1 \\ m_2 \\ \cdot \\ m_r \end{bmatrix} = \begin{bmatrix} 1 & 1 & 1 & \dots & 1 \\ 0 & 1 & 2 & \dots & r \\ 0 & 1 & 4 & \dots & r^2 \\ \cdot & \cdot & \cdot & \dots & \cdot \\ 0 & 1 & 2^r & \dots & r^r \end{bmatrix} \begin{bmatrix} n(0) \\ n(1) \\ n(2) \\ \cdot \\ n(r) \end{bmatrix}$$

expressing the moments in terms of the frequencies. One desires a reversed relation, expressing frequencies using moments, the more so supposing that the moments could be calculated independently. Naturally this requires the determination of as many moments as frequencies, in order to produce a square matrix for inversion.

The matrix shown is a particular case of a type which is very well known in mathematics, the Vandermonde matrix. Its general form is

$$V = \begin{bmatrix} 1 & 1 & 1 & \dots & 1 \\ x_0 & x_1 & x_2 & \dots & x_r \\ x_0^2 & x_1^2 & x_2^2 & \dots & x_r^2 \\ \cdot & \cdot & \cdot & \dots & \cdot \\ x_0^r & x_1^r & x_2^r & \dots & x_r^r \end{bmatrix},$$

readily invertible through the use of the Lagrange interpolation polynomials $\pi_i(x)$, defined by

$$\begin{aligned} \pi(x) &= \prod_{j=0}^{r}(x - x_j) \\ \pi_i(x) &= \frac{\pi(x)}{(x - x_i)\pi'(x_i)} \\ &= \sum_{j=0}^{r} c_{ij}x^j. \end{aligned} \tag{12.19}$$

The inverse matrix transforms moments into frequencies:

$$\begin{bmatrix} n(0) \\ n(1) \\ n(2) \\ \cdot \\ n(r) \end{bmatrix} = \begin{bmatrix} c_{00} & c_{01} & c_{02} & \cdots & c_{0r} \\ c_{10} & c_{11} & c_{12} & \cdots & c_{1r} \\ c_{20} & c_{21} & c_{22} & \cdots & c_{2r} \\ \cdot & \cdot & \cdot & \cdots & \cdot \\ c_{r0} & c_{r1} & c_{r2} & \cdots & c_{rr} \end{bmatrix} \begin{bmatrix} m_0 \\ m_1 \\ m_2 \\ \cdot \\ m_r \end{bmatrix}. \tag{12.20}$$

It is important to appreciate the mechanism by which the product $V^{-1}V$ yields the identity matrix. There are $r+1$ interpolation polynomials to evaluate at $r+1$ points; each polynomial vanishes at all the "wrong" points, while assuming unit value at the "right" point, namely the one for which the subscripts match. The matrix V^{-1} contains rows of coefficients for such polynomials; in the matrix product they form inner products with the columns of V, composed of all the powers needed to evaluate the polynomial at some particular point. The resultant interplay of right and wrong creates the unit matrix.

Because the ancestral data is all evenly spaced, the interpolation polynomials specialize to the Newton polynomials $\nu_i(x)$,

$$\begin{aligned} \nu(x) &= \prod_{j=0}^{r}(x - j) \\ &= \sum_{j=0}^{r+1} S_{r+1}^{(j)} x^j \qquad (12.21) \\ \nu_i(x) &= \frac{(-1)^{r-i}\nu(x)}{i!(r-i)!(x-i)}, \end{aligned}$$

whose characteristics have been nicely described, for example in Berezin and Zhidkov's numerical analysis text [19]. The primary polynomial $\nu(x)$ is symmetric or antisymmetric with respect to $r/2$ according to its parity, oscillating with diminishing amplitude from the origin to the midpoint, and growing rapidly outside the interval $(0, r)$.

The coefficient $S_{r+1}^{(j)}$ in Eq. 12.21 is a Stirling number of the first kind, as described in Abramowitz and Stegun's mathematical handbook [1]. The explicit formula cited therein is fairly long; asymptotic formulas, some special values and tables are also presented.

Since we have already seen how to express the moments of the counterimage distribution in terms of tensor powers of the de Bruijn fragments, we have an alternative viewpoint from which to examine the frequency of counterimages.

An important special case is that of zero variance, in which all data equal x, say; in that case one would have $m_k = (r+1)x^k$, for which the inversion formula Eq. 12.20 would yield $n(i) = (r+1)\pi_i(x)$. Supposedly x would be an integer, so that x's actual frequency $r+1$ would be recovered, all other frequencies vanishing. Nevertheless the formula is valid for any sequence of moments which are powers of x, such as might arise from an eigenvalue of one of the tensor power sums.

In practice, the high moments of any distribution are dominated by the largest datum; accordingly we might expect the inversion of a moment distribution to involve the maximal frequency plus a tail composed of the lower moments weighted by coefficients from the upper left hand corner of the matrix V^{-1}. These coefficients approach limiting values for large values of r, which would be the case of principal interest; moreover they diminish factorially while the moments increase as powers, ensuring convergence. For example, combining Equations 12.19 and 12.21 yields the frequency $n(0)$ corresponding to the Garden of Eden,

$$n(0) = \frac{(-1)^{r-1}}{(r-1)!}\sum_{j=0}^{r} S_{r+1}^{(j+1)} m_j. \qquad (12.22)$$

We have just seen that $n(0)$ vanishes (no Garden of Eden) for distributions with zero variance; the converse would be that it *only* vanishes for zero variance, a conclusion whose likelihood is greatly enhanced by the existence and appearance of Eq. 12.22.

12.7 Some illustrative examples

Having succeeded in finding one single matrix for each automaton which summarizes important properties of the distribution of its ancestors, it is worthwhile examining some typical cases; because the 16×16 matrices of $(2,1)$ automata are unwieldy, it is convenient to make the same points with $(2,1/2)$ automata. We first consider an automaton with a Garden of Eden and configurations with arbitrarily large numbers of ancestors (or), then one which has no Garden of Eden but which is not reversible (exclusive or), and finally one which is reversible (right shift). In due course we shall encounter nontrivial reversible automata.

12.7.1 Rule 14

Matrices

Consider Rule 14, boolean or, amongst the $(2,1/2)$ automata. The matrices $N(i)$ become

$$N(0) = \begin{bmatrix} 1 & \cdot \\ \cdot & \cdot \end{bmatrix} \text{ and } N(1) = \begin{bmatrix} \cdot & 1 \\ 1 & 1 \end{bmatrix}.$$

Their tensor squares are

$$\otimes^2 N(0) = \begin{bmatrix} 1 & . & . & . \\ . & . & . & . \\ . & . & . & . \\ . & . & . & . \end{bmatrix} \text{ and } \otimes^2 N(1) = \begin{bmatrix} . & . & . & 1 \\ . & . & 1 & 1 \\ . & 1 & . & 1 \\ 1 & 1 & 1 & 1 \end{bmatrix};$$

consequently

$$N = \begin{bmatrix} 1 & . & . & 1 \\ . & . & 1 & 1 \\ . & 1 & . & 1 \\ 1 & 1 & 1 & 1 \end{bmatrix}.$$

Eigenvalues

Evidently this rule gives 0 one ancestor, 1 three. Their squares are 1 and 9, visible respectively as the number of nonzero elements in the individual tensor products. The element sum of N is 10; its eigenvalues are ± 1, $1 \pm \sqrt{3}$, contrasted to eigenvalues 0, 1 for $N(0)$ and $(1 \pm \sqrt{5})/2$ for $N(1)$.

The 2^n configurations of length n will have 2^{n+2} ancestors for an average of 4 each, while the sum of the squares of the number of ancestors will eventually grow according to 2.732^n. The growth could be as small as a factor of 2 per cell or as large as a factor of 4, according to whether all configurations have an equal number of ancestors, or all ancestors map into a single configuration.

The quiescent state will have asymtotically 1.57^n ancestors (whose square is 2.46^n), an increasingly negligible proportion.

Because the mean is always constant and small, the variance, $m_0^2\sigma^2 = m_0 m_2 - m_1^2$, will grow asymptotically at half (square root) the rate of the second moment, or by $(1.18)^n$, or 18% per additional cell.

Frequencies

An average splits any set of data; half is less than average, half greater. The frequency distribution for ancestors is uniformly positive with a mean which is always quite small in relation to the range of the data, leaving the frequency distribution highly skewed. The following tabulation of the 512 ancestors of the 256 configurations of length 8 illustrates the point:

number of ancestors i	instances $n(i)$	weight $n(i)(i - \text{av})$	comment
0	142	−284	
1	39	−39	
2	28	0	average
3	16	16	
4	4	8	
5	7	21	
6	2	8	
8	4	24	
9	1	7	
10	2	16	
13	4	44	
15	2	26	
16	2	28	
21	2	38	
89	1	87	
	256	0	total

At one extreme, there is a large number of ancestorless configurations, for example all those containing the string 010 somewhere amongst the eight cells; the other end boasts an outlier – 1* – with 89 ancestors. Indeed, some 5% of the configurations account for half the ancestors.

The lesson to be learned is that even a few points on the high end must be compensated by numerous points on the low end, the very lowest of which come from the Garden of Eden.

Moments

To avoid the large matrices of Sec. 12.6, consider just the sixteen ancestors of the eight configurations of three cells, for which in principle a 17×17 matrix is required. There are only four nonzero frequencies, shown in the following variant of the previous table:

i	$n(i)$
0	1
1	4
2	2
8	1

,

which gives rise to the following moments (the residuals remain when the powers of the dominant frequency are subtracted; the weights, to be used in Eq. 12.22, were approximated from Abramowitz and Stegun [1]):

moment	value	residual	c_{0j}	$c_{0j}m_j$
m_0	8	7	1.000	7
m_1	16	8	−3.381	−27
m_2	76	12	4.923	59
m_3	532	20	−4.159	−83
m_4	...	36	2.311	83 .
m_5	...	68	−0.902	−61
m_6	...	132	0.256	34
m_7	...	260	−0.054	−14
m_8	...	516	0.008	4
m_9	...	1028	−0.001	−1

Note how the residuals are eventually dominated by the second largest datum; of course the phenomonon would repeat if an attempt were made to separate more and more of the largest data from the rest. Even in this restricted example, at least ten moments are required to approximate $n(0)$ to the dubious accuracy of $n(0) = 1$, so the procedure may not be very practical. Nevertheless it is quite possible in theory, serving as a criterion whether $n(0)$ vanishes or not, and suggesting that such vanishing might be unlikely unless the variance were exactly zero.

12.7.2 Rule 6

Two further examples are illustrative. First consider Rule 6, evolving via exclusive or.

Matrices

The de Bruijn fragments for Rule 6 are doubly stochastic, which endows them with nice combinatorial properties:

$$N(0) = \begin{bmatrix} 1 & . \\ . & 1 \end{bmatrix} \text{ and } N(1) = \begin{bmatrix} . & 1 \\ 1 & . \end{bmatrix}.$$

The tensor squares are now

$$\otimes^2 N(0) = \begin{bmatrix} 1 & . & . & . \\ . & 1 & . & . \\ . & . & 1 & . \\ . & . & . & 1 \end{bmatrix} \text{ and } \otimes^2 N(1) = \begin{bmatrix} . & . & . & 1 \\ . & . & 1 & . \\ . & 1 & . & . \\ 1 & . & . & . \end{bmatrix};$$

with

$$N = \begin{bmatrix} 1 & . & . & 1 \\ . & 1 & 1 & . \\ . & 1 & 1 & . \\ 1 & . & . & 1 \end{bmatrix}.$$

Eigenvalues

Each state has two ancestors, the square of this number is 4; N has the minimum sum possible which is 8. Eigenvalues are readily calculated; $N(0)$ is a unit matrix, as is its tensor square which commutes with $N(1) \otimes N(1)$. $N(1)$ has eigenvalues ± 1; its tensor square their products in various combinations, so that altogether

N has eigenvalues 0 and 2; indeed it is essentially (projectively) idempotent since $N^n = 2^{n-1}N$. This is the least rate of increase possible for any $(2, 1/2)$ rule, conforming to the fact that every configuration of Rule 6 has exactly two ancestors.

Moments

Since every configuration has two ancestors, the moment distribution is concentrated on that value, and all the higher moments are powers of the first moment. In particular, the distribution has zero variance.

12.7.3 Rule 12

Finally consider Rule 12, right shift, which intuitively should be a reversible rule.

Matrices

The de Bruijn fragments for Rule 12 are stochastic, but not doubly stochastic.

$$N(0) = \begin{bmatrix} 1 & 1 \\ \cdot & \cdot \end{bmatrix} \text{ and } N(1) = \begin{bmatrix} \cdot & \cdot \\ 1 & 1 \end{bmatrix}.$$

Their tensor squares are

$$\otimes^2 N(0) = \begin{bmatrix} 1 & 1 & 1 & 1 \\ \cdot & \cdot & \cdot & \cdot \\ \cdot & \cdot & \cdot & \cdot \\ \cdot & \cdot & \cdot & \cdot \end{bmatrix} \text{ and } \otimes^2 N(1) = \begin{bmatrix} \cdot & \cdot & \cdot & \cdot \\ \cdot & \cdot & \cdot & \cdot \\ \cdot & \cdot & \cdot & \cdot \\ 1 & 1 & 1 & 1 \end{bmatrix};$$

consequently

$$N = \begin{bmatrix} 1 & 1 & 1 & 1 \\ \cdot & \cdot & \cdot & \cdot \\ \cdot & \cdot & \cdot & \cdot \\ 1 & 1 & 1 & 1 \end{bmatrix}.$$

Eigenvalues

Again there are two ancestors per state, giving squares of 4 and the minimal element count of 8 for N. This time $N(0)$ and $N(1)$ are idempotents with eigenvalues 0 and 1 but they do not commute. N is nevertheless "idempotent" according to $N^2 = 2N$; its element sum is 8 with powers growing at the minimal rate.

The difference is that Rule 12 is reversible, via a left shift, whereas Rule 6 is not reversible at all, each different value of a boundary cell resulting in a completely different ancestor (or no ancestor at all if the left and right boundaries conflict). Rule 12 is indifferent to its right boundary, but the ancestor is uniquely determined by the left boundary. Yet both rules are surjective.

Both these rules have a pair matrix with eight 1's, the minimal number possible; uniformly distributed throughout the rows and columns for Rule 6, but not for Rule 12. The difference is crucial, and is related to the observation that rules of the Fredkin type are all constructed from groups of variables, some of which are superfluous. Right shift does not depend upon the right cell, so the matrices $N(i)$ (and their tensor squares) have constant rows, bunching the 1's in a way that does not occur for Rule 6.

Moments

The moment distribution is the same as for Rule 6, also with zero variance.

12.8 A statistical Gerschgorin theorem

With the knowledge that the moments, especially the average and variance, of the distribution of ancestors of a cellular automaton can be gotten from the powers of readily obtained matrices, attention naturally turns to the matrices themselves. Powers, not to mention other properties, are most readily obtained from the eigenvalues and eigenvectors of any matrix, especially from the eigenvalue of largest absolute value.

12.8.1 Gerschgorin's theorem

Gerschgorin's theorem [221], one of the most serviceable estimates for the eigenvalues of a matrix, asserts that the largest row sum of absolute values of the matrix elements forms an upper bound for the absolute values of the eigenvalues. It results from applying the triangle inequality to the definition,

$$MX = \lambda X$$

once it has been written out in terms of the matrix elements,

$$\lambda x_i = \sum_j M_{ij} x_j;$$

namely

$$\lambda \leq \sum_j |M_{ij}| |\frac{x_j}{x_i}|.$$

Since not all the components of an eigenvector are zero and at least one of them is larger than (or at most equal to) the others, any inequality in which the largest component appears in the denominator can be further improved by increasing all the ratios to 1, leaving

$$\lambda \leq \sum_j |M_{ij}|.$$

Inasmuch as the identity of the largest component is usually unknown, the worst case has to suffice, giving the result which is usually quoted. Sometimes the diagonal element is associated with the eigenvalue, leading to a collection of disks centered on the diagonal elements whose radii are gotten from summing the off diagonal elements in the row.

12.8.2 Positive matrices

Further information is available if all the elements of M are positive, or if none is negative with an assurance of purely positive components for the eigenvector; then the minimum component can be used to reverse the inequality and reveal the least row sum as a lower bound to the eigenvalue. So it will be for the largest eigenvector of an irreducible nonnegative matrix, according to the classical theory of Perron and Frobenius [206].

Statistical information concerning the eigenvalue can be obtained from the equation just as easily as the traditional bounds; for example, suppose that the equations for each component are summed:

$$\lambda \sum_i x_i = \sum_{ij} M_{ij} x_j.$$

With the definitions

$$\xi_j = \frac{x_j}{\sum_i x_i}$$

and

$$C_j = \sum_i M_{ij}$$

one obtains

$$\lambda = \sum_i C_i \xi_i,$$

provided that the sum of the eigenvector components is not zero; again it is a foregone conclusion that the eigenvector of the largest eigenvalue of a nonnegative matrix can be so chosen.

Inasmuch as the right hand side has the form of an inner product, as long as the quantities involved are real, there is an angle θ and vectors Γ and Ξ for which

$$\lambda = |\Gamma||\Xi| \cos\theta.$$

Even if complex quantities were involved, there would only be a lost phase, which could be restored. In such an equation λ is circumscribed by the variances of the set of column sums, and of the eigenvector components.

12.8.3 The statistical theorem

Supposing that the required means and variances have been calculated, we could write

$$C_j = \gamma + c_j \tag{12.23}$$

$$\xi_j = \frac{1}{n} + u_j \tag{12.24}$$

$$\lambda = \gamma + n\sigma_c\sigma_\xi \cos\vartheta. \tag{12.25}$$

From our point of view, the interesting result is that

$$\gamma = \frac{1}{n}\sum_{i,j=1}^{n} M_{ij}, \tag{12.26}$$

and that deviations of λ from this value are expressible in terms of the variances, both of the column sums and the eigenvector components.

The factor n in Eq. 12.25 may be disconcertingly large unless it is compensated by considerable uniformity in the column sums and/or in the components of the eigenvector. When either, and especially both, are small, the eigenvalues are estimated by the element sum of the matrix. Since similar estimates apply to *all* eigenvalues (with suitable modifications for complex eigenvectors whose components do not sum to zero), it can be seen that the orientation of the eigenvector relative to the vector of column sums distinguishes between the different eigenvalues.

12.8.4 Examples

In practice it seems that σ_c is a large fraction of γ and that σ_ξ is on the order of the average $\bar{\xi} = 1/n$; thus γ approximates λ to $\log n$ decimals; the sign of the correction seems to be positive when the relative number of ancestors is fairly uniform, but negative when there is a large perponderance of ancestors for some of the states.

(2,1/2) automata

As an example, consider the following table of results for $(2, 1/2)$ automata, where a and b $(a + b = 4)$ are the respective numbers of neighborhoods evolving into 0 or 1 $(\gamma = (a^2 + b^2)/4)$; the first three columns show empirical results.

ab	min	max	ave	γ
04	4.000	4.000	4.000	4.000
13	2.533	2.737	2.635	2.500
22	2.000	2.000	2.000	2.000

(2,1) automata

For (2,1) automata, $a + b = 8$, while $(k^{2r})^2 = 16$ and $\gamma = (a^2 + b^2)/16$.

ab	min	max	ave	γ
08	4.000	4.000	4.000	4.000
17	2.938	3.381	3.102	3.125
26	2.400	2.800	2.580	2.500
35	2.132	2.695	2.253	2.125
44	2.000	2.480	2.155	2.000

(3,1/2) automata

As a further example, $(3, 1/2)$ automata, yield the following data, where a, b, and c (for which $a + b + c = 9$) are the respective numbers of neighborhoods evolving into 0, 1, or 2; $(k^{2r})^2 = 9$ with $\gamma = (a^2 + b^2 + c^2)/9$.

abc	min	max	ave	γ
009	9.000	9.000	9.000	9.000
018	7.047	7.519	7.204	7.222
027	5.747	6.110	5.949	5.889
036	5.000	5.531	5.107	5.000
045	4.562	5.266	4.676	4.556
117	5.095	6.002	5.704	5.667
126	4.362	5.283	4.719	4.556
135	3.813	4.615	4.121	3.889
144	3.707	4.854	3.920	3.667
225	3.707	4.854	3.970	3.667
234	3.259	5.000	3.602	3.222
333	3.000	4.002	3.430	3.000

12.8.5 Commentary

The trend seems to be favorable; generally the average value approximates the nominal value, but there are distortions at both ends of the range which can be attributed to the influence of the bounds to which the tensor squares are subject, but which would not affect a more random collection of matrices.

In order to have $\lambda = \gamma$ in Eq. 12.25 it is necessary that $\sigma_c = 0$, $\sigma_\xi = 0$, or $\cos\vartheta = 0$. In the first case all the column sums will be equal, equal to the eigenvalue, and the matrix will be stochastic. In the second case the vector $|u\rangle$ will be an eigenvector, the row sums will all be constant, and again the matrix will be stochastic. The final alternative is more interesting, requiring the residuals in the column sums to be orthogonal to the residuals in the eigenvector, for a matrix which is explicitly not stochastic.

12.9 General considerations

Both general properties of positive matrices and general statistical considerations favor sharper results than the ancestral distributions taken from de Bruijn matrices seem to justify. The reasons lie in the difference between positive matrices for which strict conclusions are possible, and nonnegative matrices, for which they *may* hold; and also in the strongly asymmetrical frequency distribution.

So, in spite of the encouragement given by matrix theory and Eq. 12.22, it is necessary to look elsewhere for actual proofs. An ideal theoretic result, that is, one which depends on collective properties and not on individual matrices themselves, has been published by Michel Dubois-Violette and Alain Rouet [68]; similar reasoning had already been used by Hedlund [114], was recounted by Nasu [169], and was probably the basis of Amoroso and Patt's assertion [6] regarding the balancing of counterimages (they cited an unpublished laboratory report).

Nasu has shown [172] that the argumemts, ostensibly valid for de Bruijn diagrams, actually apply to any graphs for which the numbers of incoming and outgoing nodes are themselves balanced. In all cases, the important connection lies between the variance and $n(0)$; otherwise understood as the fact that when every configuration has at least one counterimage, the number of counterimages is uniform.

12.9.1 The de Bruijn fragments

Once there are matrices from which the properties of counterimages may be deduced, the deduction usually takes the form of ascertaining eigenvalues, and occasionally, eigenvectors. For example, according to whether the largest eigenvalue λ_i of $N(i)$ is greater than, equal, or less than 1, the number of counterimages will inevitably increase, remain stable, or decrease. As it lies in the range from 1 to k, the rate of increase will range from modest to drastic.

The statistical Gerschgorin theorem, Eq. 12.26, estimates λ_i as $|\varphi^{-1}(i)|/k^{2r}$ (— — denotes cardinality), which would yield a value of 1 only for balanced counterimages. Although this estimate is subject to considerable variation, it gives a good indication of the degree to which a configuration profits from having cells with numerous ancestral neighborhoods.

Setting aside considerations of eigenvalues for the moment, consider the sum s_g; its value for each matrix or product of matrices is a nonnegative integer. Its least value m surely satisfies

$$m \leq \min_{i=0}^{k-1} \langle u|N(i)|u\rangle$$

and could well be as small as zero — the Garden of Eden case.

Consider further a configuration whose cells were in states $\{i_1, i_2, \ldots, i_s\}$, the number of whose ancestors would therefore be $m = \langle u|P|u\rangle$ for

$$P = N(i_1)N(i_2)\ldots N(i_s).$$

The extended configuration produced by adding one additional cell at the right in state i will have $m = \langle u|PN(i)|u\rangle$ ancestors; then the expected number of ancestors averaged over all such extensions is

$$\begin{aligned}\frac{1}{k}\sum_{i=0}^{k-1}\langle u|PN(i)|u\rangle &= \frac{1}{k}\langle u|P\left(\sum_{i=0}^{k-1} N(i)\right)|u\rangle, \\ &= \frac{1}{k}\langle u|PD|u\rangle \\ &= m\end{aligned}$$

because the $N(i)$ sum up to the de Bruijn matrix, whose eigenvector u has the eigenvalue k.

At this point, note that averages lie between the maximum and minimum of their data, reaching either extreme only if all the data are equal. This is the first part of Dubois-Violette and Rouet's [68] result. Such being the present situation, any single (and by induction, finite) extension of a minimal ancestor configuration also has a minimal number of ancestors.

In the worst case, an extended Garden of Eden is still a Garden of Eden, an observation already clear from the subset diagram, not to mention the consequence of multiplying by a zero matrix. The main point of this new result is that it is not possible to gain ancestors by taking longer configurations, then suddenly lose them; at least not for minimal configurations within a de Bruijn diagram. For others, counterexamples exist.

Although the proof shown is right sided, the left sided version is entirely similar. The proof is also completely symmetrical between maxima and minima, with the difference that a minimum is assured whereas a maximum is not.

The foregoing analysis then establishes the set $A_{\min}$ consisting of all P for which $s_g(P)$ takes the minimum value as a two sided ideal. By this we simply mean that

$$P \in A_{\min} \quad \Rightarrow \quad s_q(PQ) = s_g(QP) = m.$$

If $A_{\max}$ exists, it too must be a two sided ideal; from

$$P \in A_{\min}, \qquad Q \in A_{\max}$$

and taking n as the maximum number of counterimages follows

$$m \quad = \quad g_{\max}(PQ) = n.$$

The existence of ideals makes it hard to find matrices which are not absorbed into the ideal because those lying outside the ideal can only be formed from certain products of $N(i)$; of course when $m = 0$ the subset diagram provides a map showing just which matrices to choose and which to avoid. Furthermore, multiplication by zero is a drastic action whose role in defining ideals can be readily understood, but one might conjecture that those are the only kinds of ideals that there are. In other words, ideals depending on $m > 0$ would have to be null or else encompass the whole space.

12.9.2 The second moment

To obtain an ancestor census directly, it necessary to deal with k different matrices $N(i)$ together with all their possible products; their compliance with the general properties of nonnegative matrices is frequently complicated by their reduciblity. Fortunately the single matrix N characterizing the second moment subsumes the properties of all these individual matrices and their products; discovering its general properties avoids most of the more detailed calculations, although its own reducibility is still a possibility.

The basic conclusion that the eigenvalues of N lie in the range from k^{2r} to k^{2r^2} is quite easily established.

The lower limit follows from the fact that the diagram of identical pairs, $\{(a,a)\}$ is a subdiagram of the diagram of arbitrary pairs, $\{(a,b)\}$. The de Bruijn matrix of the former, whose maximum eigenvalue is k^{2r}, is a principal submatrix of the pair matrix, whose maximal eigenvalue must therefore be equal to or greater than k^{2r}. Since the lower limit corresponds to zero variance, it can only be reached when each of the matrices $N(i)$ contains exactly k^{2r} links (and thus N contains exactly $k(k^{2r})^2$ links); a necessary but not sufficient condition.

The upper limit follows from Gerschgorin's theorem, taking into account the fact that there are at most k^2 incoming (or outgoing) links from any node in the pair diagram. However, if each node has that many links, all neighborhoods must evolve into the same state, so that the upper limit is associated with automata evolving to a constant field within one generation.

The tables of Sec. 12.8 confirm these limits, as well as showing a general distribution of eigenvalues reflecting the number of ways that the total number of states of any given cell can be partitioned amongst the individual states.

12.9.3 Zero variance

We have seen that if the number of ancestors of whatever configuration is bounded, all states have the same number of ancestors. Otherwise the number of ancestors would increase as the length of the configuration increased, in agreement with the rate implied by the largest eigenvalue of N. The possibility is still open that there is no upper limit but that the lower limit might not be so small as zero, implying that every configuration has at least one ancestor. This would create a peculiar but not impossible statistical situation wherein an increasing variance did not imply spreading of the data on both sides of the mean.

For automata, a contradiction can be deduced by confronting a configuration with many ancestors with one with minimal ancestors; many ancestors should lead to still more in a composite configuration, but they cannot because of minimality.

Thus, consider any matrix P, some matrix $Q \in A_{\min}$, and a certain abuse of notation with respect to the evolutionary function φ. Then if $m > 0$ there are words $\hat{P} \in \varphi^{-1}(P)$ and $\hat{Q} \in \varphi^{-1}(Q)$ plus a matrix M, for which

$$\varphi(\hat{Q}\hat{P}) = QMP.$$

The idea of selecting a fixed Q, varying P, and exploiting $s_g(QMP) = m$ will not work because a counterimage $\varphi^{-1}(M)$ of the bridge M will not necessarily mesh with every $\hat{P}$. However if the left hand element of P is fixed, an accounting can be made. Therefore let

$$\begin{aligned} P &= N(i)\Pi, \\ \hat{N} &\in \varphi^{-1}(N(i)), \\ \hat{P} &= \hat{N}\rho. \end{aligned}$$

Once again

$$\varphi(\hat{Q}\hat{N}\rho) = QMP,$$

whereupon it may be concluded that

$$s_g(\varphi(\hat{N}\rho)) \;\leq\; m$$

on account of the fact that these are only some of the items making up the counterimages of the extension QMP. This inequality contemplates all the choices of ρ for a fixed $\hat{N}$, but there could be as many as k^{2r+1} choices of $\hat{N}$. All together, independently of P,

$$s_g(P) \;\leq\; mk^{2r+1},$$

which is the upper bound necessary to conclude that whenever $m > 0$ there are equal numbers of ancestors for all configurations — exactly k^{2r}. This is the second part of Dubois-Violette and Rouet's [68] result.

12.9.4 The essential result

Somewhat special arguments finally establish the relationship between zero variance and zero ancestors for automata, namely that the number of ancestors is either uniform for all configurations, or else that there are some configurations with no ancestors at all. As special cases, if the distribution of ancestors for a single cell is not balanced, there must be a Garden of Eden (but not conversely); and a reversible rule will lack a Garden of Eden (but not conversely).

Mathematically, evolution is called surjective if every configuration has an ancestor, injective if it has only one. The essential result states that surjectivity requires uniform numbers of ancestors, at first sight incompatible with injectivity and so with reversibility as well. Reversibility can only be reconciled with uniformity if there is an edge effect; that is if multiple ancestors differ only in a remote boundary layer. The simplest layer would reflect the overhang due to the greater size of neighborhoods relative to cells. Of course another source of injectivity would be to eliminate all but one of the alternatives, by requiring periodicity, or quiescence at infinity, for example.

Zero variance is equivalent to the maximum eigenvalue of N assuming its minimum value, k^{2r}. This in turn implies that the maximum eigenvalue of the matrices $N(i)$ is 1, but not conversely. There are consequences in turn for the properties of the subset matrix and the pair matrix, which is the same as the second moment matrix.

12.10 Pairs of states

The second moment matrix is also the connectivity matrix for pairs of states discussed in Sec. 12.2.4, just as the third moment matrix corresponds to triples of states, and so on. No less than restrictions on possible pairs of sequences implied by zero variance, observation of the same limitations in practice can serve as an indicator of zero variance, and consequently of a surjective rule.

12.10.1 Partitioned pair matrix

The tensor product of the connectivity matrices of two diagrams is the connectivity matrix of their cartesian product, wherein pairs of nodes are linked according to whether both members of the pair are each linked in their own diagram. Sometimes a careful arrangement of the nodes in the tensor product will partially diagonalize the connectivity matrix, rendering some property of the arrangement more evident than it would otherwise have been.

A case in point consists in grouping all the pairs composed of equal elements together in a subset $\mathcal{A}$, and the rest into the subset $\mathcal{B}$. It then becomes evident whether links in a tensor square describe pairs of links in the original diagram which have a node in common by linking $\mathcal{A}$ to $\mathcal{B}$ or the reverse, both nodes in common by linking $\mathcal{A}$ to $\mathcal{A}$, or neither by linking $\mathcal{B}$ to $\mathcal{B}$.

Partition $D \otimes D$ accordingly

$$D \otimes D = \begin{bmatrix} P & Q \\ R & S \end{bmatrix}$$

and consider the eigenvalue equation with a similarly partitioned eigenvector

$$\begin{bmatrix} P & Q \\ R & S \end{bmatrix} \begin{bmatrix} x \\ y \end{bmatrix} = \lambda \begin{bmatrix} x \\ y \end{bmatrix}.$$

The following submatrix equations are the result:

$$\begin{aligned} Px + Qy &= \lambda x \\ Rx + Sy &= \lambda y. \end{aligned}$$

If λ is already an eigenvalue belonging to P, it follows that $Qy = 0$; whether this requires that $Q = 0$ depends somewhat on S, but the condition is quite sufficient. If λ is *not* an eigenvalue of S, $S - \lambda I$ is invertible whence $y = -(S - \lambda I)^{-1}Rx$ and $Q(S - \lambda I)^{-1}R$ would have to be singular unless y were 0.

Consideration of a left eigenvector leads to similar conclusions regarding $\bar{y}R$, while if λ is an eigenvalue of S as well as P, Rx must vanish. Broadly speaking, if either of the diagonal submatrices already has the eigenvalue, it had better be decoupled from the other, but the details of the uncoupling can get messy.

In practice, another line of reasoning is slightly more compact. Consider

$$(D \otimes D)^2 = \begin{bmatrix} P^2 + QR & PQ + RS \\ RP + SR & RQ + S^2 \end{bmatrix}.$$

If P were a de Bruijn matrix it would be an irreducible nonnegative matrix and $P^2 + QR$ would have a maximum eigenvalue strictly larger than λ^2 unless $QR = 0$; $Q = 0$ or $R = 0$ would be sufficient. In any event, continuing shows that

$$(D \otimes D)^3 = \begin{bmatrix} m_{11} & m_{12} \\ m_{21} & m_{22} \end{bmatrix};$$

with

$$\begin{aligned} m_{11} &= P^3 + QSR \\ m_{12} &= P^2Q + PRS + QS^2 \\ m_{21} &= RP^2 + SRP + S^2R \\ m_{22} &= RPQ + R^2S + SRQ + S^3 \end{aligned}$$

gradually the series of requirements

$$\begin{aligned} QR &= 0, \\ QSR &= 0, \\ QS^2R &= 0, \\ &\cdots \end{aligned}$$

develops, which can be summed to recover

$$Q(S - \lambda I)^{-1}R = 0. \tag{12.27}$$

The very slightly stronger result is due to the additional knowledge about P.

To illustrate this result with respect to the three rules of Sec. 12.7, rearrange the indices of the matrix N to group paired indices (a, a) and then complementary indices $(a, \bar{a})$; we obtain

$$\overset{\text{rule 14}}{\left[\begin{array}{cc|cc} 1 & 1 & . & . \\ 1 & 1 & 1 & 1 \\ \hline . & 1 & . & 1 \\ . & 1 & 1 & . \end{array}\right]} \quad \overset{\text{rule 6}}{\left[\begin{array}{cc|cc} 1 & 1 & . & . \\ 1 & 1 & . & . \\ \hline . & . & 1 & 1 \\ . & . & 1 & 1 \end{array}\right]} \quad \overset{\text{rule 12}}{\left[\begin{array}{cc|cc} 1 & 1 & 1 & 1 \\ 1 & 1 & 1 & 1 \\ \hline . & . & . & . \\ . & . & . & . \end{array}\right]}.$$

The off diagonal submatrices for Rule 14, for which $QR \neq 0$, force an eigenvalue larger than the minimum, whereas the other two matrices are partially diagonal and maintain the minimum value. Rule 6 is self-complementary, as evidenced by the equality of its diagonal submatrices P and S.

12.10.2 Symmetry by permuting states

As the use of the complement in creating these examples shows, a tensor square can be further refined if the basic diagram has an isomorphism, which is to say a permutation of its nodes resulting in the same connectivity matrix; for binary automata such an isomorphism results from complementing the individual cells. For $k > 2$ there are increasing numbers of isomorphisms, $k!$ in fact, whose presence can be detected in the full de Bruijn matrix and all the other moment matrices associated with it. Nevertheless if an individual rule lacks some or all of this symmetry, it will be missing from N and its analogues.

The pair matrix of Sec. 12.2.4 can be rearranged along these lines, which is to say that it is equivalent to the following matrix, gotten by listing first all the pairs of the form (a, a), last all the pairs of the form $(a, \hat{a})$, which is the only other permutation of binary values. The remaining pairs are laid out symmetrically in the middle.

As before, the possible pair matrices all conform to a common pattern, which arises for either Rule 0 or Rule 255. Since the pair (a, b) links to the pair (c, d) only if a links to c and b to d via the same symbol, the pairs of pairs actually linked will vary from rule to rule. Furthermore there are some additional symmetry requirements for the formation of pairs: if (a, b) links to (c, d), then (b, a) must link to (d, c) and a similar requirement due to transitivity.

With this new ordering, the pair matrix for Rule 22 becomes: (1's belong to the actual pair matrix, while $*$'s indicate the extent of the maximal possible matrix)

$$N = \left[\begin{array}{cccc|cccccccc|cccc}
1 & 1 & . & . & * & . & . & . & . & . & . & * & . & . & . & . \\
. & . & 1 & 1 & . & . & . & * & * & . & . & . & . & . & . & . \\
1 & 1 & . & . & * & . & . & . & . & . & . & * & . & . & . & . \\
. & . & 1 & 1 & . & . & . & 1 & 1 & . & . & . & . & . & . & . \\
\hline
. & . & . & . & . & * & * & . & . & . & . & . & 1 & 1 & . & . \\
* & * & . & . & 1 & . & . & . & . & . & . & 1 & . & . & . & . \\
. & . & * & 1 & . & . & . & * & 1 & . & . & . & . & . & . & . \\
. & . & . & . & . & * & 1 & . & . & . & . & . & * & 1 & . & . \\
. & . & . & . & . & . & . & . & . & 1 & 1 & . & . & . & * & * \\
. & . & * & 1 & . & . & . & 1 & * & . & . & . & . & . & . & . \\
* & * & . & . & 1 & . & . & . & . & . & . & 1 & . & . & . & . \\
. & . & . & . & . & . & . & . & . & * & * & . & . & . & 1 & 1 \\
\hline
. & . & . & . & . & 1 & * & . & . & . & . & . & 1 & * & . & . \\
. & . & . & . & . & . & . & . & . & 1 & 1 & . & . & . & * & * \\
. & . & . & . & . & 1 & 1 & . & . & . & . & . & * & * & . & . \\
. & . & . & . & . & . & . & . & . & * & 1 & . & . & . & * & 1
\end{array}\right].$$

The submatrix in the upper left hand corner is always present for whatever Rule (the lower right submatrix is filled out for Rule 150, just as it would be for any other rule symmetric by conjugation), no matter what other elements are deleted from N on account of the structure of $N(0)$ and $N(1)$. Accordingly, no row sum can exceed 4, so by Gerschgorin's theorem, the largest eigenvalue of N will be bounded by 4. The limit can only be reached if all row sums are equal, thus only for Rules 0 and 255 (evolution into constants).

On the other hand the diagonal submatrices have a uniform row sum of 2, which must be their largest eigenvalue; alternatively the first and last are just de Bruijn matrices, from which the same result follows. By a theorem in Gantmacher [89], this establishes a lower limit for the greatest eigenvalue of N, which must exceed the eigenvalues of any principal minor.

Of course there is an especial interest in conditions in which this lowest limit is maintained; so far it has not been possible to characterize the matrix N sufficiently to decide other than in special cases or by numerical calculation. The general result of Sec. 12.9 is the best known, that the minimum eigenvalue which already implies surjectivity, implies uniform distribution. The converse is not valid, most uniform distributions lead to higher eigenvalues.

12.10.3 Characterizing surjective rules

Some consequences of zero variance are easily described. First and foremost, no path beginning in the subset $\mathcal{A}$ can ever return, because that would violate the reasoning leading up to Eq. 12.27. In other words, if two ancestors of a configuration initially coincide — even for a single partial neighborhood — and later diverge, they can never recombine. Seen from another point of view, no product of the matrices $N(i)$ can ever have a matrix element as large as 2 while the rule has zero variance.

A similar prohibition applies to a pair of isomorphic ancestors; it is only necessary to replace $\mathcal{A}$ by the set $\{(a, f(a))\}$.

It is also necessary that the matrix S corresponding to the subset $\mathcal{B}$ not have an eigenvalue exceeding that of a de Bruijn matrix, which can be checked in any particular case, but not as easily when it is a general proposition. Essentially, the requirement is that mixed ancestors not proliferate more rapidly than single ancestors as the length of a configuration increases, which would clearly be possible if sections of an ancestor could be freely substituted for one another.

12.11 The subset matrix

Whether an automaton has a Garden of Eden can be decided at once from its subset diagram, or the algebraic properties of its subset matrix. For most automata the empty set will be accessible from the full set (implying a Garden of Eden), and there will not be that much interest in the remaining structure of the diagram. Contrarily, for an automaton with zero variance and a surjective rule, the very fact that the empty set can be avoided introduces additional structure into the diagram. For example, what are the largest subsets which lead to the empty set? What are the smallest subsets reachable from the full set? What relationships exist between subsets with similar characteristics?

12.11.1 Maximal node diagram

Recall that nodes in the subset diagram are sets of nodes from the original diagram, linked whenever each node in the tail subset joins some node in the head subset; paths arise from successive links. Call nodes in the original diagram points to distinguish them from nodes in the subset diagram. In the subset diagram of a general rule, multiple links can emanate from a single point, multiple links can also converge to a point; indeed this is the reason subset diagrams were invented, to ensure simple links between subsets.

For a surjective rule, however, only one path can run between a given pair of endpoints, one of whose consequences can be seen by factorizing a path between two nodes. Let the path x run from nodes U to V, continuing on to node W_y via y (anticipating a variation of y).

Count the paths between points due to xy in two ways: Take $U = \{u\}$ as a unit class to eliminate ambiguity in the starting point, then either procede directly to $W_y = xy(U)$, or interrupt the path with a stopover at $V = x(U)$. The essential point is that there can be only one interrupted path. Now vary y, supposing only that it is a path of length p:

$$\sum_y |W_y| \quad = \quad |V|k^p. \tag{12.28}$$

This equation expresses an average, so that the argument applying to averages of extreme quantities can be used once again: if V is a node of maximal cardinality, all the terms of the average must assume the same value.

Paths leading from a given unit class will not all reach maximum nodes; those which do will not necessarily do so immediately. Finiteness of the diagrams ensures that there is a short path — shorter than the cardinality of the diagram — leading from any node to the maximal node accessible to it. Once a maximal node has been reached, further links will trace out paths confined to a class of nodes of the same cardinality.

Since the de Bruijn diagram itself is connected, there is only one maximal class, irrespective of the unit class from which it was derived. The reason is that one can find paths which alternate between maximal nodes in the subset diagram and individual points in the de Bruijn diagram. yielding a derivation of any maximal node from any point, and of one maximal node from any other; thus the maximal class is connected, and there are maximal subsets containing any given unit class.

Not all paths lead from unit classes to the maximal class; the fate of paths from other subsets may also vary. When the subset in question is larger (setwise) than maximal (in particular, the full set) we know that its eventual images must remain larger than maximal; their images do not necessarily have to lead into the maximal class, however.

12.11.2 Left and right indices

In any event, the maximal class with its links is an image of the de Bruijn diagram, containing all the same paths as the original, but possibly with fewer nodes. The cardinality of the maximal subsets has been called [114] the *index* of its Rule; strictly the quantity which has been described is a *right index*, reserving the term *left index* for the analogous quantity derived from the reversed subset diagram. The handedness evidently refers to the direction in which a configuration would be extended to get a longer configuration.

The two indices need not be equal. For instance, both indices of $(2, 1/2)$ Rule 6 (exclusive or) are 1, but for Rule 12 (right shift) the left index is 1 while the right index is 2.

It is worthwhile understanding the significance of an index and the reason that the two can differ. Fundamentally, the index is the number of different partial neighborhoods which have to be grouped together to reinforce each other if an arbitrarily prescribed configuration is to have an ancestor. For the right shift of Rule 12, extension to the right is pointless if the leftmost cell cannot be arbitrarily chosen, a possibility assured by the index of 2. For left extension the rightmost cell is irrelevant because it will be replaced immediately; any choice will do and the index is 1.

Continuation in either direction from whatever initial cell is always possible for Rule 6, so both its indices are 1. The consequence, however, is that the ancestors resulting from such freedom are all essentially distinct. By contrast, Rule 12 has only one ancestor; the necessary flexibility to construct it must be retained in the set of partial neighborhoods made available at each stage.

12.12 Reversing a rule

So far, the rules which are candidates for reversibility are those for which there is no Garden of Eden; one cannot determine an ancestor for a configuration which has none, which precludes a rule which purports to do so. Among those which remain, the equidistribution rule applies. We have seen that there are environments which could restrict the number of ancestors, such as a quiescent infinity or a cyclic automaton. Still the examples constructed in Sec. 12.1 produce a unique inverse without reference to any boundary condition, which is a general alternative which must be considered. Evidently the apparent variety of ancestors is illusory; the mechanism must be investigated.

12.12.1 Two (2,1/2) examples

Some surjective rules are injective, others are not. As an example of the first kind, consider $(2, 1/2)$ Rule 6, and some of the products of ancestor matrices. Recalling the definition of $s(m)_{ij}$ in Eq. 12.12, there is either a left or a right ancestor matrix; choosing the right handed version, $s(m)_{ij} = \mathbf{j}$ when $\varphi(i,j) = m$, ϕ otherwise. Accordingly,

$$\begin{array}{cc} s(0) & s(1) \\ --- & --- \\ \begin{bmatrix} \mathbf{0} & . \\ . & \mathbf{1} \end{bmatrix} & \begin{bmatrix} . & \mathbf{1} \\ \mathbf{0} & . \end{bmatrix} \end{array}.$$

The four pair products are

$$\begin{array}{cccc} s(00) & s(01) & s(10) & s(11) \\ --- & --- & --- & --- \\ \begin{bmatrix} \mathbf{00} & . \\ . & \mathbf{11} \end{bmatrix} & \begin{bmatrix} . & \mathbf{01} \\ \mathbf{10} & . \end{bmatrix} & \begin{bmatrix} . & \mathbf{11} \\ \mathbf{00} & . \end{bmatrix} & \begin{bmatrix} \mathbf{10} & . \\ . & \mathbf{01} \end{bmatrix} \end{array}.$$

Here the detail of interest is the fact that neither the left nor the right member of the matrix elements in $s(ij)$ is consistent, nor does any such consistency arise for any of the higher order products (note that row and column indices must also be taken into account when judging consistency). Consequently one concludes that this rule does not have a general purpose inverse.

Repeating the display for $(2, 1/2)$ Rule 12 produces

$$\begin{array}{cc} s(0) & s(1) \\ --- & --- \\ \begin{bmatrix} \mathbf{0} & \mathbf{1} \\ . & . \end{bmatrix} & \begin{bmatrix} . & . \\ \mathbf{0} & \mathbf{1} \end{bmatrix} \end{array}.$$

$$\begin{array}{cccc} s(00) & s(01) & s(10) & s(11) \\ --- & --- & --- & --- \\ \begin{bmatrix} \mathbf{00} & \mathbf{01} \\ . & . \end{bmatrix} & \begin{bmatrix} \mathbf{10} & \mathbf{11} \\ . & . \end{bmatrix} & \begin{bmatrix} . & . \\ \mathbf{00} & \mathbf{01} \end{bmatrix} & \begin{bmatrix} . & . \\ \mathbf{10} & \mathbf{11} \end{bmatrix} \end{array}.$$

This time, there is a consistency in the left member of each ancestral pair, so that each evolved pair has a unique ancestor which can be used as the evolved state of the inverse rule, and all this quite independently of boundary conditions. Note that once consistency is found, it persists for all longer products.

12.12.2 A (2,1) example

The only reversible rules for (2,1) automata are trivial, but nevertheless still instructive. Consider Rule 204, whose evolution is stasis,

$$a = D(0) = \begin{bmatrix} 1 & 1 & \cdot & \cdot \\ \cdot & \cdot & \cdot & \cdot \\ 1 & 1 & \cdot & \cdot \\ \cdot & \cdot & \cdot & \cdot \end{bmatrix}; \; b = D(1) = \begin{bmatrix} \cdot & \cdot & \cdot & \cdot \\ \cdot & \cdot & 1 & 1 \\ \cdot & \cdot & \cdot & \cdot \\ \cdot & \cdot & 1 & 1 \end{bmatrix}.$$

These matrices are not permutation matrices nor even stochastic, since neither row sums nor column sums are constant. Nevertheless they are idempotent and generate a semigroup of four matrices of similar appearance, namely $\{a, ab, ba, b\}$ with the multiplication table

	a	ab	ba	b
a	a	ab	a	ab
ab	a	ab	a	ab
ba	ba	b	ba	b
b	ba	b	ba	b

Consequently, every product has four counterimages, seemingly contradicting the claimed uniqueness. However, we need to form new symbolic matrices, which show the ancestral cell rather than the evolved cell; this ancestral cell is the central cell of the ancestral neighborhood, which can be read off from the indices of the de Bruijn matrix.

For Rule 204, these matrices are

$$s(0) = \begin{bmatrix} \mathbf{0} & \mathbf{0} & \cdot & \cdot \\ \cdot & \cdot & \cdot & \cdot \\ \mathbf{0} & \mathbf{0} & \cdot & \cdot \\ \cdot & \cdot & \cdot & \cdot \end{bmatrix}, \; s(1) = \begin{bmatrix} \cdot & \cdot & \cdot & \cdot \\ \cdot & \cdot & \mathbf{1} & \mathbf{1} \\ \cdot & \cdot & \cdot & \cdot \\ \cdot & \cdot & \mathbf{1} & \mathbf{1} \end{bmatrix};$$

they already display the crucial property: every counterimage of 0 is 0, whilst every counterimage of 1 is 1. Therefore Rule 204 is invertible by a (2,0) automaton, which is not surprising since it is really a radius zero automaton itself. Trivial in the present instance, the property is once again: there exists some word length r for which every word in the product of r α's and β's has the same central symbol. For this to work r must be odd, but inversion could be combined with shifting so that shift inverses also exist.

12.12.3 A (2,1) counterexample

Symbolic matrices can also be used to prove that a rule is *not* invertible; consider Rule 30 once more:

$$s(0) = \begin{bmatrix} \cdot & \mathbf{0} & \cdot & \cdot \\ \cdot & \cdot & \mathbf{1} & \mathbf{1} \\ \mathbf{0} & \cdot & \cdot & \cdot \\ \cdot & \cdot & \cdot & \cdot \end{bmatrix}; \; s(1) = \begin{bmatrix} \mathbf{0} & \cdot & \cdot & \cdot \\ \cdot & \cdot & \cdot & \cdot \\ \cdot & \mathbf{0} & \cdot & \cdot \\ \cdot & \cdot & \mathbf{1} & \mathbf{1} \end{bmatrix}.$$

With little effort one can establish that

$$s(1^n) = \begin{bmatrix} \mathbf{0}^n & \cdot & \cdot & \cdot \\ \cdot & \cdot & \cdot & \cdot \\ \cdot & \cdot & \cdot & \cdot \\ \cdot & \mathbf{1}^{n-1}\mathbf{0} & \mathbf{1}^n & \mathbf{1}^n \end{bmatrix}$$

for $n > 1$. Given that the configuration 1^* will always have some ancestors in which 1 remains 1 and others in which 0 becomes 1, there is no neighborhood length for which the ancestral cell is uniquely determined. Therefore Rule 30 cannot be reversed, in spite of the fact that every configuration has a unique ancestor whenever its external environment is specified.

12.12.4 Test procedure

It is evidently possible to check the powers of more and more complex words to see whether there is a possibility of inversion; If 0^n admits inversion for some particular values of n, 1^n can be tested, then $(01)^n$, and so on, always noting the least common multiple of the admissible values. Eventually enough combinations may be examined to reach a decision: some word is impossible to resolve, or enough words have been analyzed to know that there is a length of neighborhood for which the ancestor is unconditionally unique, and the inverse Rule can be read off according to the sequences of αs and βs in which the unique values occur—e.g., if the middle letter in each element of the symbolic matrix $\alpha\beta\beta$ is 0, $\varphi(0,1,1) = 0$ for inverse evolutionary function φ.

Fortunately there is an upper bound to the lengths of words which must be tested, making eventuality a certainty. Consider the matrix product $\sigma\xi\tau$ corresponding to a particular counterimage; given that the pth letter is the same for all the words in the full product, let ξ be the pth factor. Decompose ξ into a sum $\xi = \gamma + \epsilon$ in which γ contains the symbols which will be lost, and ϵ those which will survive; $\sigma\gamma\tau$ must vanish leaving $\sigma\epsilon\tau = \sigma\xi\tau$. There may be symbols common to γ and ϵ, but ϵ must contain one and only one of them, 0 say.

We are interested in showing that an excessively long σ or τ can be shortened; such a demonstration would be akin to shrinking a long path through the vector subset diagram by excising loops, whereby a loop free path would be minimal. But that is not the way to our proof, which is independent of the diagram. Rather, note that our matrices are sparse, in their numerical version all matrix elements are zeroes or ones, with exactly k ones per matrix. All their possible products are postulated to have the same property; yet there can only be just so many such matrices of a given dimension.

Some of them are original de Bruijn fragments, others exist only as products, but each of them will have a product of shortest length generating it. Let N be the maximum length of the shortest products generating the multiplicative closure of the de Bruijn fragments. Now replace both σ and τ by minimal equivalents through the process of converting them to numerical form, finding the minimal product, and forming the symbolic product all over again. This will naturally shorten and alter any ancestors being represented; call the new products $\hat{\sigma}$ and $\hat{\tau}$.

The important property of these new chains is

$$\sigma\gamma\tau = 0 \quad \Rightarrow \quad \hat{\sigma}\gamma\hat{\tau} = 0,$$

since the vanishing of the products depends on the placement of their nonzero elements and not their value, even when it is symbolic. In turn new position $\hat{p}$ is guaranteed for which every matrix element

$$\hat{\sigma}\xi\hat{\tau} \quad = \quad \hat{\sigma}\epsilon\hat{\tau}$$

has the same symbol for its $\hat{p}$th letter, all within a neighborhood of no greater length than $2N + 1$.

It may still happen that every ancestor has a position with a unique ancestor, but that the positions do not coincide. But then padding may be inserted to extend the neighborhood either to the right or to the left in such a way as to give them all a common centering. Should this be necessary one will simply have an inverse rule which is insensitive to the values of certain cells.

In practice none of this rearrangement has to be carried out; the possibility that it can be done simply shows that there is nothing in a very long or irregular neighborhood whose equivalent would not already have been encountered in a systematic sweep through short neighborhoods.

12.12.5 Use of the pair matrix

Most of these conclusions can be summarized by referring to the pair matrix once again. Nasu [169] outlines such a process, referring also to the concept of *definite automata* [184], which is explained at some length in Zvi Kohavi's textbook [139] on automata theory. He uses the unordered pair matrix, but the conclusions are the same.

If a path is started in the subset $\mathcal{B}$ of distinct pairs, it may remain within that subset or it may wander into the subset $\mathcal{A}$ of identical pairs; if it does so there must be no possibility of returning to the original starting point. Otherwise the pair matrix would not have the minimum eigenvalue required for zero variance and surjectivity.

In the case that the path remains within $\mathcal{B}$, it must eventually close into a loop, demonstrating that there must exist configurations with two different ancestors. No matter that some paths may wander into $\mathcal{A}$; the rule cannot be injective. The alternative, that no path originating in $\mathcal{B}$ remains in $\mathcal{B}$, means that it is eventually trapped in $\mathcal{A}$. Subsequent return to $\mathcal{B}$ would force a second visit to $\mathcal{A}$, in violation of zero variance. For such automata, cannot be two ancestors, distinct except for their initial and terminal segments. This is the true case of injectivity.

Rules which are not injective, but lack the full multiplicity which is theoretically possible can be detected by using a diagram of triples or higher multiplets; whether there are subsets of degenerate multiplets which trap the trajectories must be investigated.

12.13 Remarks

There are two noteworthy aspects of reversibility. One is that there exists a very elaborate and technical theory of injective and surjective mappings which has not dealt extensively with particular examples or applications. On the other hand a variety of empirical methods have been discovered for producing reversible automata, without having had a very extensive theoretical foundation. It would be helpful to reconcile these two tendencies.

The other aspect is the extent to which the theory of reversible automata has been treated in isolation from other cellular automata. This has probably been due to the very special nature of the proofs required, which do not leave much lattitude for exceptions or approximations. Nevertheless one doubts that a binary automaton with 1,024 neighborhoods, 511 of which evolve into 0 and 513 of which evolve into 1, is that much different from an automaton whose counterimages balance. In other words, there should be a statistical continuum accounting for nearly reversible rules, almost surjectivity, and the like.

12.13.1 Surjectivity

Surjectivity is the important concept; injectivity is subordinate and can be treated later. The differences between cyclic, quiescent, and general boundary conditions are likewise secondary. In the analysis which we have presented, the results depend upon the frequency distribution of counterimages, and consequently upon its moment problem. The moments can be estimated from fragments of the de Bruijn matrix, determined individually for each automaton.

There are two tantalizing aspects to the problem so formulated. If the matrices involved were irreducible and not just nonnegative, several strong conclusions would be available which have to be established by other means. Likewise if more information about the moments existed, useful conclusions could be drawn about the existence of gaps or other nonuniformities in the frequency distribution.

In spite of the lack of convenience, the primordial connection between zero variance (balanced distribution of counterimages) and surjectivity (lack of a Garden of Eden) can be established by arguments which still make use of fundamental properties of the de Bruijn matrix. Moreover, the proofs are still quite relevant to the moment problem; for example the knowledge that the variance increases exponentially with the length of a configuration leaves zero variance as the only bounded possibility.

It may still be a surprise, even to those who have calculated numerous subset diagrams and perhaps read Amoroso and Patt's paper [6], that an unbalanced evolutionary rule necessarily has a Garden of Eden.

12.13.2 Injectivity

Once surjectivity has been established, there is no doubt that the uniform multiplicity of counterimages is precisely what would be expected from the size of the partial neighborhoods bounding a finite configuration. Ways exist through which the multiplicity may be rendered ineffective; one is to terminate the configuration with quiescent neighborhoods, another is to close them with a cyclic connection. Fixed boundaries and phase relations in the closure (e.g. the left boundary is the complement of the right boundary) are other alternatives which could be considered.

All the foregoing mechanisms involve selections from a naturally occurring set of boundaries, but there is also the possibility that the boundaries don't really matter. In other words, aside from a finite boundary layer which would be arbitrarily remote from the center of a very long configuration, the different ancestors could just happen to agree. Surely this is what is happening in shift rules or the identity map, and is clearly provided for in rules of Fredkin's type.

Boundaryless injectivity, or one might say, injectivity with a natural boundary, can be tested in one of the pair diagrams. Since the ordered pair diagram is coincidentally the second moment matrix, the condition of zero variance imposes a direct condition. Just as the empty set, implying a Garden of Eden, must be met in a limited number of steps (if at all) in the subset diagram, a pair of ancestors must either coalesce within the confines of the pair diagram or encounter some means of coexisting.

Apparently the decision to use a boundary condition should be postponed as long as possible; the eventual choice may be that much simpler.

12.13.3 Additional comments

The role of topology in treating the evolution of cellular automata is not entirely clear. Hedlund gives it a central role in his fundamental survey [114]; it is basic to such further articles as those of D. Richardson [196] or Tadakazu Sato and Namio Honda [200], and has been used by Karel Culick II [55]. It does seems to be required for the satisfactory description of the limiting behavior of long time evolution [122, 124, 57].

On the other hand Nasu [169, 172, 170, 171] has been able to obtain extensive results without using topology, at least for one dimensional automata; the present paper likewise demonstrates that no topology is required. We are left with the question of whether there are significantly different results to be obtained from the use of topology or whether topology simply provides a convenient language to express results which could be derived independently. However, this is a topic for a future study, just as is the question of handling two dimensional automata via two dimensional de Bruijn diagrams.

There are also important questions about the applications and the philosophical implications of reversible automata, a preoccupation of Toffoli [219]. Two recent articles addressed to these points are those of Asher Peres [183] and Shinji Takesue [216].

Appendix A

Cycles for Rule 22

A.1 Summary

One of the fundamental conclusions of automata theory is that finite automata or those which are effectively finite, such as the cyclic configurations of an infinite automaton, eventually fall into periodic behavior. Generally the only way to discover these periods is by exhaustive enumeration, according to which the evolution of every possible initial configuration is followed out until the first repetition of a previous configuration is observed.

The notation $\pi.\phi$ means that the configuration repeats itself after π generations, but that there are only ϕ different phases, variations of which can occur through rotation or reflection. Accordingly the representative configuration shown is the one with a gap of maximal length, which is always placed at the extreme right. If a shift to the right of δ cells is involved, the notation $\pi.\phi(\delta)$ is used; a reflective symmetry is often associated with $2\pi.\pi$.

On general principles a ring of N cells cannot have a cycle longer than 2^N, but in practice the lengths of the cycles grow at a far slower rate with increasing N. Some of the longest cycles are a consequence of a shift, whose least common multiple with N often results in a far longer period than the time required to simply repeat the pattern.

Likewise, the periods are determined by the lengths of possible loops in a de Bruijn diagram of 2^{2P} nodes, which limits the maximum length of a primitive period to this value. In practice this is also found to be a generous upper limit.

The results can be summarized in a table of periods versus cycle lengths. The columns correspond to torii of fixed circumference, the rows to fixed periods, the values of both of which are stated in the margins. The first twenty periods are shown below.

	1	2	3	4	5	6	7	8	9	10	11	12	13	14	15	16	17	18	19	20	21	22	23	24	25	26	27	28	29	30	31	32	33	34
1	.	1	.	1	.	1	.	1	.	1	.	1	.	1	.	1	.	1	.	1	.	1	.	1	.	1	.	1	.	1	.	1	.	1
2	.	.	.	1	.	.	.	1	.	.	.	1	.	.	.	1	.	.	.	1	.	.	.	1	.	.	.	1	.	.	.	1	.	.
3	.	.	.	.	.	.	.	.	.	.	.	.	.	.	.	.	.	.	.	.	.	.	.	.	.	.	.	.	.	.	.	.	.	.
4	.	.	.	.	.	.	.	1	1	2	1	1	.	.	.	1	1	3	3	5	3	3	1	2	1	4	6	12	13	18	13	13	7	4
5	.	.	.	.	.	.	.	.	.	.	1	1	1	.	.	.	.	.	.	.	.	1	1	2	1	1	.	.	.	.	.	.	1	1
6	.	.	.	.	.	.	.	1	.	1	.	.	.	.	.	1	.	.	.	2	.	.	.	1	.	.	.	.	.	5	.	1	.	.
7	.	.	.	.	.	.	1	.	.	.	.	.	.	1	.	.	.	.	.	.	1	.	.	.	.	.	.	1	.	.	.	.	.	.
8	.	.	.	.	.	.	.	.	.	.	.	.	.	.	.	.	.	.	.	1	2	1	1	.	.	.	.	.	.	.	.	.	.	.
9	.	.	.	.	.	.	.	.	.	.	.	.	.	.	.	.	.	.	.	.	.	.	.	.	.	.	.	.	.	.	.	.	.	.
10	.	.	.	.	.	.	.	.	.	.	.	.	.	.	.	.	.	.	.	.	.	.	1	.	.	.	.	.	.	.	.	.	.	1
11	.	.	.	.	.	.	.	.	.	.	1	.	.	.	.	.	.	.	.	.	.	1	.	.	.	.	.	.	.	.	.	.	1	2
12	.	.	.	.	.	.	.	.	.	.	.	.	.	1	.	2	1	2	1	1	.	.	.	.	.	.	.	1	.	.	.	2	1	3
13	.	.	.	.	.	.	.	.	.	.	.	.	.	.	.	.	.	.	.	.	.	.	.	.	.	.	.	.	.	.	.	.	.	.
14	.	.	.	.	.	.	.	.	.	.	.	.	.	.	.	1	.	.	.	.	.	.	.	.	.	1	.	.	.	1	.	1	1	.
15	.	.	.	.	.	.	.	.	.	.	.	.	.	.	.	.	.	.	.	.	.	.	.	.	.	.	.	.	.	.	.	1	.	.
16	.	.	.	.	.	.	.	.	.	.	.	.	.	.	.	.	.	.	.	.	.	.	.	.	.	.	.	.	.	.	.	.	.	.
17	.	.	.	.	.	.	.	.	.	.	.	.	.	.	.	.	.	.	.	.	.	.	.	.	.	.	.	.	.	.	.	.	.	.
18	.	.	.	.	.	.	.	.	.	.	.	.	.	.	.	.	.	1	.	.	.	.	.	.	.	.	.	.	.	.	.	.	.	.
19	.	.	.	.	.	.	.	.	.	.	.	.	.	.	.	.	.	.	.	.	.	.	.	.	.	.	.	.	.	.	.	.	.	.
20	.	.	.	.	.	.	.	.	.	.	.	.	.	.	1	.	.	.	.	.	.	.	.	.	.	.	.	.	.	1	.	.	.	.

There is only a single still life, a single configuration of period two, and none at all of period three.

From the table above and the continuation below it will be seen that there are numerous configurations of periods 4, 12, and 28. They belong to an extensive family which is formed by the expansion of isolated live cells, and which contains many more members besides. Isolated cells, in Rule 22 especially, follow a growth pattern which resembles a binary counter, or a fractal as the image would be classified nowadays. The boundary of a region filled with live cells expands into the gap separating such regions, but periodically the interior of the expanding region dies out.

Thus if a pair of cells are allowed to expand around a ring, a complementary configuration may be created when the expanding frontiers have nearly met but the expanding region has just become vacant. If the symmetry is just right the complement will repeat the performance to receate the original configuration, setting the stage for a repetetive cycle.

The final fourteen periods are shown in the table below; as before columns correspond to the circumference of the ring, rows to a fixed period.

	1	2	3	4	5	6	7	8	9	10	11	12	13	14	15	16	17	18	19	20	21	22	23	24	25	26	27	28	29	30	31	32	33	34
21	.	.	.	.	.	.	.	.	.	.	.	.	.	.	.	.	.	.	.	.	.	.	.	.	.	.	.	.	.	.	.	.	.	.
22	.	.	.	.	.	.	.	.	.	.	.	.	.	.	.	.	.	.	.	.	.	1	.	.	.	.	.	.	.	.	.	.	.	.
23	.	.	.	.	.	.	.	.	.	.	.	.	.	.	.	.	.	.	.	.	.	.	.	.	.	.	.	.	.	.	.	.	.	.
24	.	.	.	.	.	.	.	.	.	.	.	.	.	.	.	.	.	1	.	1	.	.	.	1	.	.	.	.	.	.	.	.	.	1
25	.	.	.	.	.	.	.	.	.	.	.	.	.	.	.	.	.	.	.	.	.	.	.	.	.	.	.	1	.	.	.	.	.	.
26	.	.	.	.	.	.	.	.	.	.	.	.	.	.	.	.	1	.	.	.	.	.	.	.	1	.	1	.	.	.	.	.	.	1
27	.	.	.	.	.	.	.	.	.	.	.	.	.	.	.	.	.	.	.	.	.	.	.	.	.	.	.	.	.	.	.	.	.	.
28	.	.	.	.	.	.	.	.	.	.	.	.	.	.	.	.	.	.	1	.	.	.	.	1	1	3	1	3	.	.	.	3	3	6
29	.	.	.	.	.	.	.	.	.	.	.	.	.	.	.	.	.	.	.	.	.	.	.	.	.	.	.	.	1	.	.	.	.	.
30	.	.	.	.	.	.	.	.	.	.	.	.	.	.	.	.	.	.	.	.	.	.	.	.	.	.	1	1	1	.	.	2	.	.
31	.	.	.	.	.	.	.	.	.	.	.	.	.	.	.	.	.	.	.	.	.	.	.	.	.	.	.	.	.	.	1	.	.	.
32	.	.	.	.	.	.	.	.	.	.	.	.	.	.	.	.	.	.	.	.	.	.	.	1	.	.	.	.	.	.	.	1	.	.
33	.	.	.	.	.	.	.	.	.	.	.	.	.	.	.	.	.	.	.	.	.	.	.	.	.	.	.	.	.	.	.	.	.	.
34	.	.	.	.	.	.	.	.	.	.	.	.	.	.	.	.	.	.	.	.	.	.	.	.	.	.	.	.	.	.	.	1	.	.

Longer rings might have been attempted were it not for the fact that the computing time required to analyze the ring doubles with every increment in length, leaving $N = 34$ as something of a limit of endurance with present equipment and techniques. A few additional increments could be gained by using faster equipment, and by proving some "gap theorems" which would render unnecessary the analysis of a certain fraction of cases.

Nevertheless, in practice the interval $20 \leq N \leq 34$ provided enough special cases that it was possible to discover several general families of configurations, finally leading to the analysis of periods in terms of de Bruijn diagrams.

A.2 N = 1

· zero

A.3 N = 2

· · zero

·■ 1.1

A.4 N = 3

· · · zero

A.5 N = 4

· · · · zero

· ·■■ 2.1

·■·■ 1.1

A.6 N = 5

· · · · · zero

A.7 N = 6

· · · · · · zero

·■·■·■ 1.1

A.8 N = 7

· · · · · · · zero

· · ·■·■■ 7.1(3)

A.9 N = 8

· · · · · · · · zero

· · · · · ·■■ 6.3

· · · · ·■·■ 4.2

· ·■■· ·■■ 2.1

·■·■·■·■ 1.1

A.10 N = 9

· · · · · · · · · zero

· · · · · ·■·■ 4.4

A.11 N = 10

··········	zero
·······■·■	4.4
······■··■	4.2
·····■··■■	6.3
·■·■·■·■·■	1.1

A.12 N = 11

···········	zero
·······■··■	4.4
·······■■■■	5.5
····■··■■■■	11.1(4)

A.13 N = 12

············	zero
········■■■■	5.5
·······■···■	4.2
··■■··■■··■■	2.1
·■·■·■·■·■·■	1.1

A.14 N = 13

·············	zero
·········■■■■	5.5

A.15 N = 14

··············	zero
······■···■■■■	12.6
···■·■■···■·■■	7.1
·■·■·■·■·■·■·■	1.1

A.16 N = 15

···············	zero
·······■·····■■	20.4(3)

A.17 N = 16

················	zero
··············■■	14.7
·············■·■	12.6
··········■····■	12.12
······■■······■■	6.3
·····■·■·····■·■	4.2
··■■··■■··■■··■■	2.1
·■·■·■·■·■·■·■·■	1.1

A.18 N = 17

·················	zero
··············■·■	12.12
············■■··■	26.13
······■·■·····■·■	4.4

A.19 N = 18

··················	zero
···············■·■	12.12
··············■··■	12.6
·······■■·■··■·■■	18.9
·······■·■·····■·■	4.4
······■·■······■·■	4.4
······■··■·····■·■	4.2
·■·■·■·■·■·■·■·■·■	1.1

A.20 N = 19

···················	zero
···············■··■	12.12
·········■····■■■·■	28.14
·······■·■······■·■	4.4
·······■··■·····■·■	4.4
······■··■······■·■	4.4

A.21 N = 20

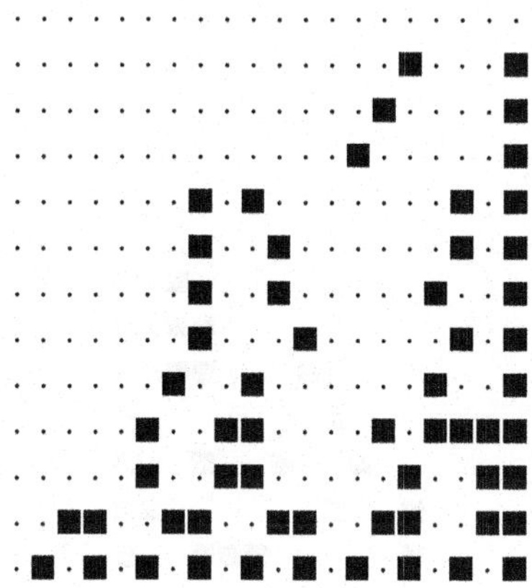

zero
12.6
24.24
8.4
4.4
4.4
4.4
4.2
4.2
6.3
6.3
2.1
1.1

A.22 N = 21

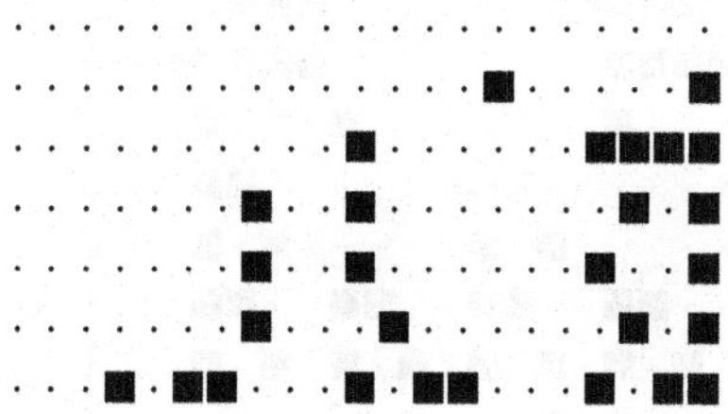

zero
8.8
77.11(3)
4.4
4.4
4.4
7.1

A.23 N = 22

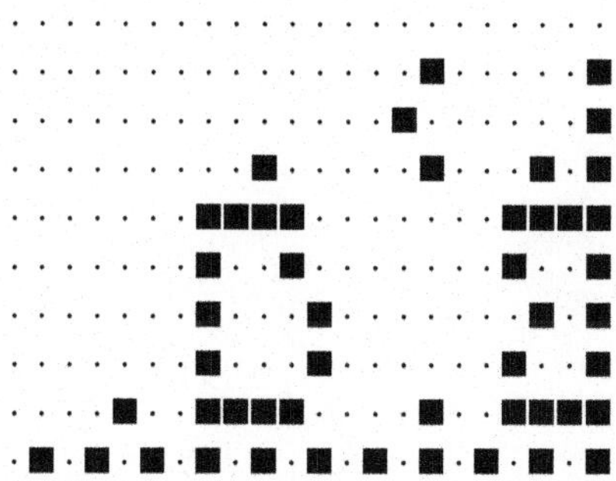

zero
8.8
8.4
22.2(8)
5.5
4.4
4.4
4.2
11.1
1.1

A.24 N = 23

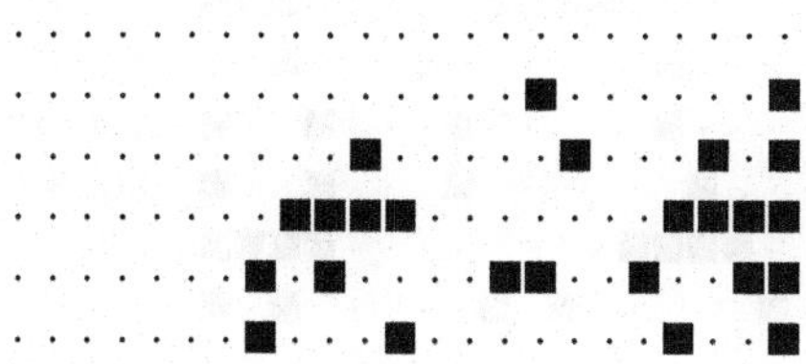

zero
8.8
138.6(2)
5.5
10.5
4.4

A.25 N = 24

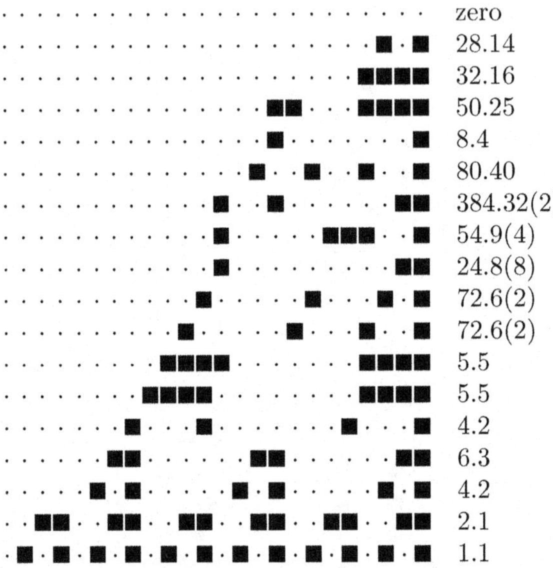

A.26 N = 25

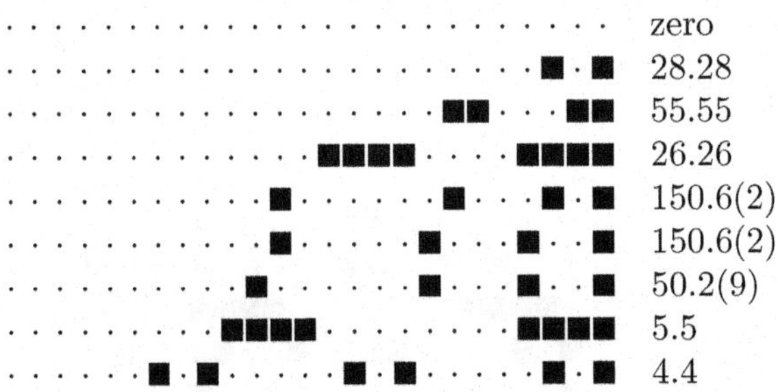

A.27 N = 26

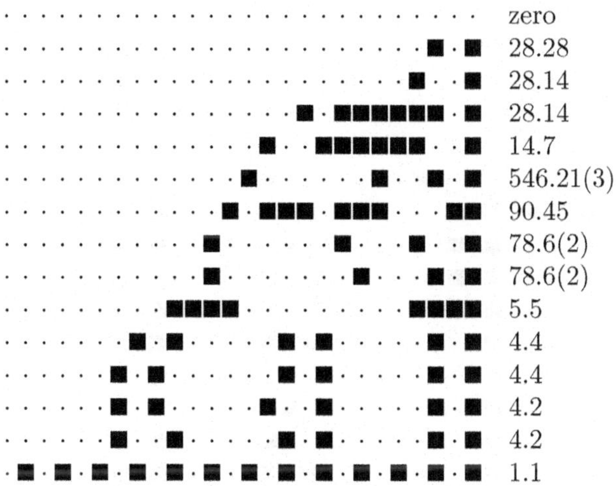

A.28 N = 27

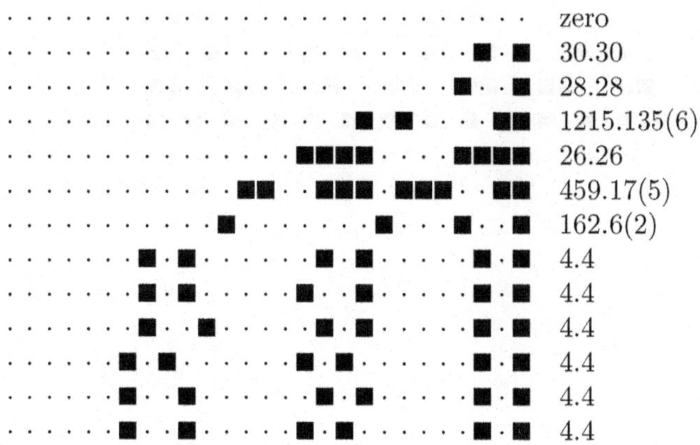

A.29 N = 28

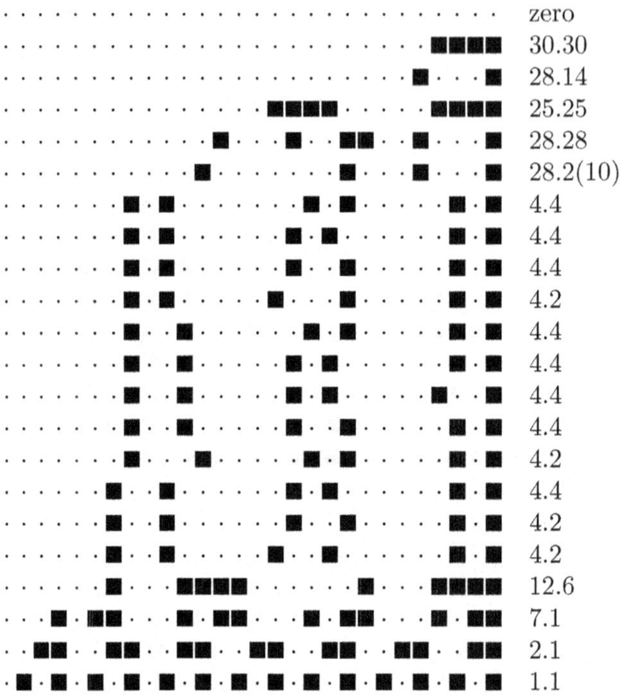
zero
30.30
28.14
25.25
28.28
28.2(10)
4.4
4.4
4.4
4.2
4.4
4.4
4.4
4.4
4.2
4.4
4.2
4.2
12.6
7.1
2.1
1.1

A.30 N = 29

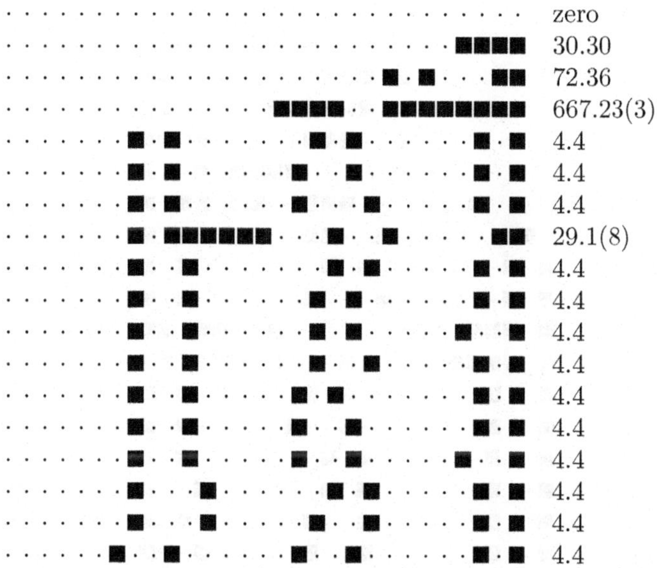

A.31 N = 30

··························	zero
····················■■■■··■■■■	120.60
···············■·■■■········■■	240.8(7)
···············■·■■■■■■·····■■	240.8(7)
···············■··■■■·····■■■■	86.43
···············■·····■■■■■■··■	1070.107(9)
···········■······■■■··■■·■■·■	70.35
········■····■··■··■■······■·■	14.7
·······■·■·······■·■·······■·■	4.4
·······■·■······■···■······■·■	4.4
·······■·■■■·■■·······■·■■■·■■	20.4
·······■·■■■■■■···■···■·■■■·■■	40.4(3)
·······■··■·······■·■······■·■	4.4
·······■··■·······■··■·····■·■	4.4
·······■··■······■·■·······■·■	4.4
·······■··■······■·■······■··■	4.4
·······■··■······■··■······■·■	4.4
·······■··■······■··■·····■··■	4.4
·······■··■······■···■·····■·■	4.2
·······■··■·····■··■·······■·■	4.4
·······■··■·····■···■······■·■	4.4
·······■···■·······■·■·····■·■	4.4
·······■···■······■·■······■·■	4.4
·······■···■······■·■·····■··■	4.2
·······■···■······■··■·····■·■	4.4
·······■···■·····■·■·······■·■	4.4
·······■···■·····■··■······■·■	4.2
······■··■······■··■······■··■	4.2
·····■··■■····■·■■■■····■·■■■■	6.3
·····■··■■····■·■■■···■··■■■■■	6.3
·····■··■■·····■··■■·····■··■■	6.3
·····■··■■·····■··■■····■·■■■■	6.3
·····■···■■■·■····■■··■··■■■■■	6.1(5)
·■·■·■·■·■·■·■·■·■·■·■·■·■·■·■	1.1

A.32 N = 31

·······························	zero
················■·········■■·■■	248.8(7)
················■············■■	248.8(7)
·······■··■·······■·■·······■·■	4.4
·······■··■·······■··■······■·■	4.4
·······■··■·······■··■·····■··■	4.4
·······■··■······■··■·······■·■	4.4
·······■··■······■··■······■··■	4.4
·······■··■······■···■······■·■	4.4
·······■···■·······■·■······■·■	4.4
·······■···■·······■··■·····■·■	4.4
·······■···■······■·■·······■·■	4.4
·······■···■······■·■······■··■	4.4
·······■···■······■··■······■·■	4.4
·······■···■······■···■·····■·■	4.4
·······■···■·····■··■·······■·■	4.4
·······■·····■■■···■···■·■■■·■■	31.1(7)

A.33 N = 32

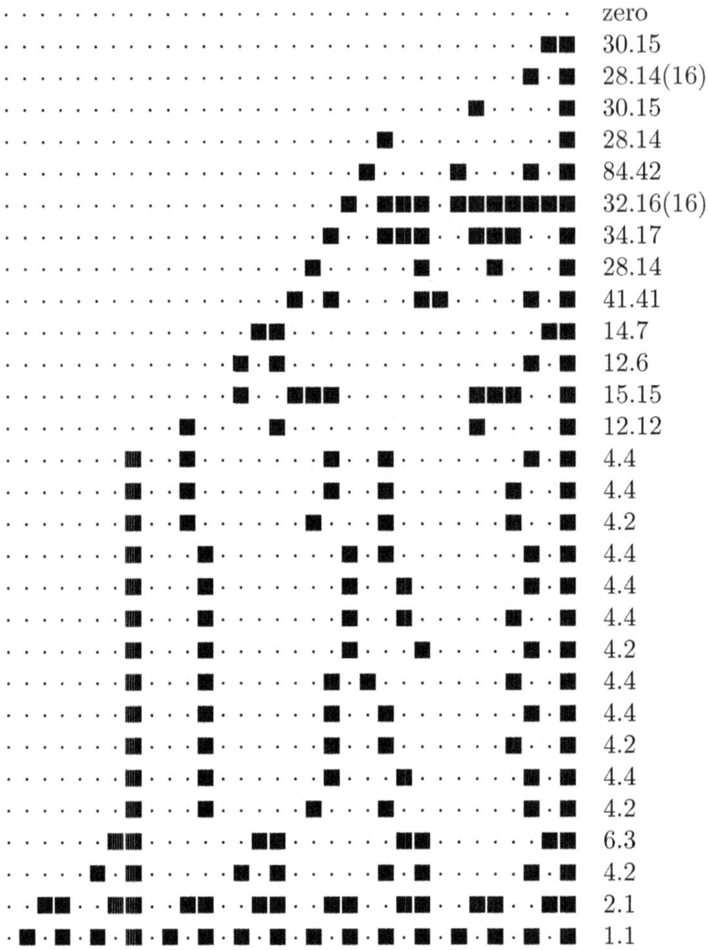
zero
30.15
28.14(16)
30.15
28.14
84.42
32.16(16)
34.17
28.14
41.41
14.7
12.6
15.15
12.12
4.4
4.4
4.2
4.4
4.4
4.4
4.2
4.4
4.4
4.2
4.4
4.2
6.3
4.2
2.1
1.1

A.34 N = 33

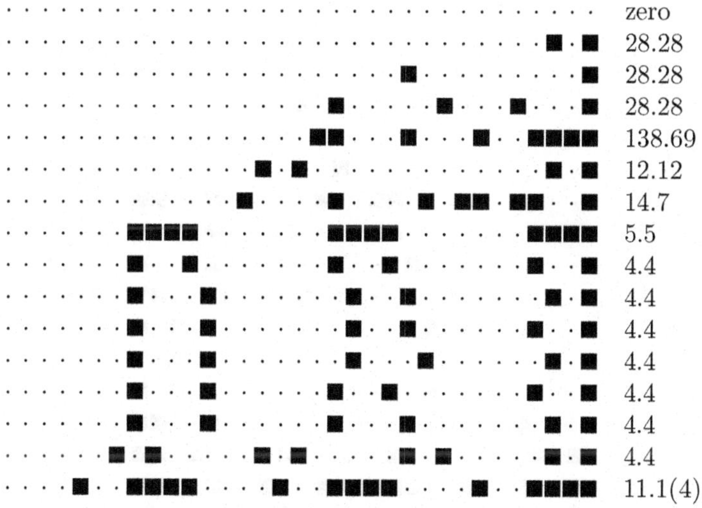

A.35 N = 34

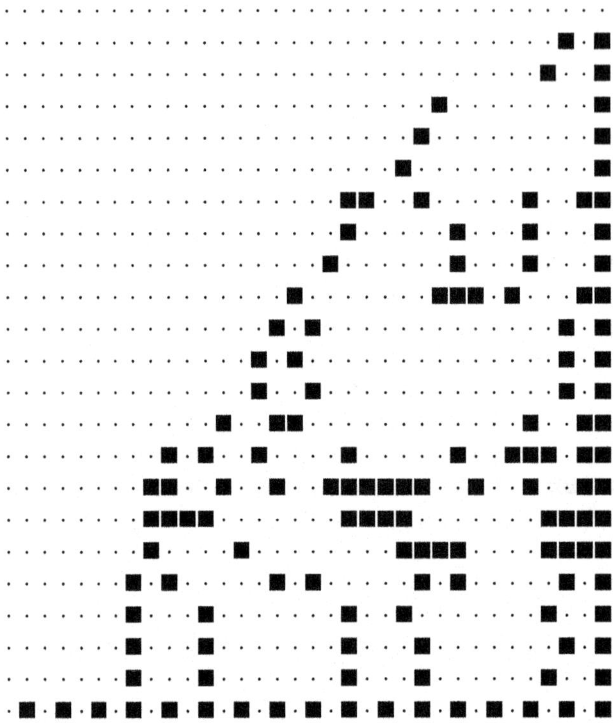

zero
28.28
28.14
52.52
28.28
28.14
24.12
28.28
28.14
4590.135(16)
12.12
12.12
12.6
26.13
10.5
11.11
5.5
11.11
4.4
4.4
4.4
4.4
1.1

A.36 A pair of isolated cells

Examination of the data sections shows increasingly large numbers of cycles of period 4, all arising by combining unit cells of the form 10^g10^G (and thus of individual lengths $g + G + 2$) in all possible ways to form the circumference N. Some of these produce configurations of period 4, but other periods such as 24 and 28, which occur frequently, arise the same way. G and g represent gaps—sequences of zeroes—of the corresponding lengths. There are certain restrictions on combinations of gap lengths; most result in the formation of enneads with differing periods, with a good variety of other combinations. However some mixtures are not periodic, degenerating into other cycles or even into the quiescent state.

The following table shows the admissible periods as a function of G and g, but symmetry implies that it is only necessary to consider $g \leq G$. Dots indicate combinations which evolve into cycles from which they are absent. Dashes indicate combinations outside of the range of our study.

	0	1	2	3	4	5	6	7	8	9	10	11	12	13	14	15	16	17	18	19	20	21	22	23	24	25	26	27	28	29	30	31	32
0	.	.	2	.	.	.	6	.	.	.	.	.	.	.	14	.	.	.	.	.	.	.	.	.	.	.	.	.	.	.	30	.	.
1		.	.	.	.	4	4	4	.	.	.	.	.	12	12	12	.	.	.	.	.	28	28	28	30	.	.	.	.	28	28	28	–
2			.	.	6	4	4	4	.	.	.	.	14	12	12	12	.	.	.	.	.	28	28	28	30	.	.	.	30	28	28	–	–
3				.	.	4	4	4	.	.	.	.	.	12	12	12	.	.	.	.	.	28	28	28	30	.	.	.	.	28	–	–	–
4					.	5	5	5	.	.	12	.	.	.	24	.	.	.	32	.	.	30	30	30	.	.	30	.	.	–	–	–	–
5						.	.	.	.	12	12	12	.	8	8	8	.	28	28	28	30	.	.	.	.	28	28	28	–	–	–	–	–
6							.	.	14	12	12	12	24	8	8	8	.	28	28	28	30	.	.	.	30	28	28	–	–	–	–	–	–
7								.	.	12	12	12	.	8	8	8	.	28	28	28	30	.	.	.	.	28	–	–	–	–	–	–	–
8									.	.	.	.	.	.	.	.	.	.	.	.	.	.	.	.	.	–	–	–	–	–	–	–	–
9										.	.	.	.	28	28	28	.	.	.	.	.	28	28	28	–	–	–	–	–	–	–	–	–
10											.	.	80	28	28	28	.	.	.	.	.	28	28	–	–	–	–	–	–	–	–	–	–
11												.	.	28	28	28	.	.	.	.	.	28	–	–	–	–	–	–	–	–	–	–	–
12													.	30	30	30	120	.	.	.	.	–	–	–	–	–	–	–	–	–	–	–	–
13														.	.	.	.	28	28	28	–	–	–	–	–	–	–	–	–	–	–	–	–
14															.	.	30	28	28	–	–	–	–	–	–	–	–	–	–	–	–	–	–
15																.	.	28	–	–	–	–	–	–	–	–	–	–	–	–	–	–	–
16																	.	–	–	–	–	–	–	–	–	–	–	–	–	–	–	–	–

Since the antidiagonals of this table contain configurations of a fixed circumference, the periods arising from these structures can be read off for a ring of length N quite directly. It has to be borne in mind that there are many conjugate pairs, for example the combination (1,6) of period 4 evolves into (2,4) in two generations, which in turn evolves into (1,6) in two more generations to complete the period. It is the number of conjugte sets which is given in the tables, not the number of individual configurations.

Exploring the enneads further, their periods are found to have the form $2^k - 4$, although the values of k do not necessarily follow in strict numerical succession. The tables show enneads for $k = 3, 4, 5$, but there are two distinct enneads with the last value. The structures filling the interstices between the enneads follow a much less evident pattern.

Similar tables can be constructed for other pairs of simple isolated structures, such as a pairs or triples of cells; indeed some such combinations can be recognized in the tables. However, the results obtained are not nearly as comprehensive as for individual cells. Even the table shown requires further analysis to deduce its general structure and ensure that it applies equally well for all the cases not determined empirically.

Bibliography

[1] Milton Abramowitz and Irene A. Stegun (editors), *Handbook of Mathematical Functions*, U. S. Government Printing Office, Washington D. C., 1964.

[2] Andrew Adamatzky, *Identification of Cellular Automata*, Taylor and Francis, London, 1994.

[3] Roy Adler and Leopold Flatto, "Geodesic Flows, Interval Maps, and Symbolic Dynamics," *Bulletin of the American Mathematical Society* **25** 229–334 (1991).

[4] Jürgen Albert and Karel Culik II, "A Simple Universal Cellular Automaton and its One-Way and Totalistic Version," *Complex Systems* **1** 1–16 (1987).

[5] Ivan Amato, "Speculating in Precious Computronium," *Science* **253** 856–857 (1991).

[6] S. Amoroso and Y. N. Patt, "Decision Procedures for Surjectivity and Injectivity of Parallel Maps for Tesselation Structures," *Journal of Computer and System Sciences* **6** 448–464 (1972).

[7] S. Amoroso and I. Epstein, "Indecomposable Parallel Maps in Tesselation Structures," *Journal of Computer and System Sciences* **13 (2)** 136–142 (1976).

[8] Stan Augarten, *Bit by Bit*, Ticknor and Fields, New York, 1984.

[9] Charles Babbage, *Charles Babbage and his Calculating Engines*, (Philip and Emily Morrison, editors.), Dover Publications, Inc., New York, 1961.

[10] R. C. Backhouse and B. A. Carré, "Regular Algebra Applied to Path-finding Problems," *Journal of the Institute for Mathematics and its Applications* **15** 161–186 (1975).

[11] Bernard Barral, Hugues Chaté and Paul Manneville, "Collective behaviors in a family of high-dimensional automata," *Physics Letters A* **163** 279–285 (1992).

[12] F. L. Bauer, "An Elementary Proof of the Hopf Inequality for Positive Operators," *Numerische Mathematik* **7** 331–337 (1965).

[13] F. L. Bauer, E. Deutsch, and J. Stoer, "Abschätzungen für die Eigenwerte positiver linearer Operatoren," *Linear Algebra and its Applications* **2** 275–301 (1969).

[14] Carter Bays, "The game of three dimensional life" (11/20/86) unpublished (distributed as a supplement to A. K. Dewdney's February 1987 column).

[15] Carter Bays, "Candidates for the Game of Life in Three Dimensions," *Complex Systems* **1** 373–400 (1987).

[16] Carter Bays, "Patterns for Simple Cellular Automata in a Universe of Dense-Packed Spheres," *Complex Systems* **1** 853–875 (1987).

[17] Carter Bays, "Classification of Semitotalistic Cellular Automata in Three Dimensions," *Complex Systems* **1** 373–400 (1987).

[18] Richard Bellman, *Introduction to Matrix Analysis*, McGraw-Hill Book Company, New York, 1960.

[19] I. S. Berezin and N. P. Zhidkov, *Computing Methods*, volume 1, Pergamon Press, Oxford, 1965.

[20] Elwyn R. Berlekamp, John H. Conway, and Richard K. Guy, *Winning Ways for your Mathematical Plays*, Academic Press, 1982. vol. 2, chapter 25.

[21] Abraham Berman and Robert J. Plemmons, *Nonnegative Matrices in the Mathematical Sciences*, Academic Press, New York, 1979.

[22] R. Bidaux, N. Boccara, and H. Chaté, "Order of the transition versus space dimension in a family of cellular automata," *Physical Review A* **39** 3094–3105 (1989).

[23] P. -M. Binder and V. Privman, "Second-Order Dynamics in the Collective Temporal Evolution of Complex Systems," *Physical Review Letters 68* 3830–3833 (1992).

[24] Garrett Birkhoff, "Extensions of Jentzsch's Theorem," *Transactions of the American Mathematical Society* **85** 219–227 (1957).

[25] George D. Birkhoff, *Dynamical Systems*, American Mathematical Society, Providence, Rhode Island, 1927.

[26] N. Bocara, J. Nasser, and M. Roger, "Particle-like structures and interactions in spatio-temporal patterns generated by one-dimensional cellular automaton rules," *Physical Review A* **44** 866–875 (1991).

[27] Rufus Bowen and John Franks, "Homology for zero-dimensional nonwandering sets," *Annals of Mathematics* **106** 73–92 (1977).

[28] Rufus Bowen "On Axiom A Diffeomorphisms," *Conference Board of the Mathematical Sciences* CBMS Regional Conference Series in Mathematics **35** Amer. Math. Soc. (1978).

[29] M. Boyle, "Shift eauivalence and the Jordan form away from zero," *Ergodic Theory and Dynamical Systems* **4** 367–369 (1984).

[30] Mike Boyle, "Constraints on the degree of a sofic homomorphism and the induced multiplication of measures on unstable sets," *Israel Journal of Mathematics* **53** 52–68 (1986).

[31] Mike Boyle and Wolfgang Krieger, "Periodic points and automorphisms of the shift," *Transactions of the American Mathematical Society* **302** 125–409 (1987)

[32] Mike Boyle and David Handleman "The spectra of nonnegative matrices via symbolic dynamics," *Annals of Mathematics* **133** 249–316 (1991).

[33] Mike Boyle, Douglas Lind, and Daniel Rudolph, "The automorphism group of a shift of finite type," *Transactions of the American Mathematical Society* **306** 71–114 (1988).

[34] Mike Boyle, "Eventual extensions of finite codes," *Proceedings of the American Mathematical Society* **104** 965–972 (1988).

[35] David B. Brown, "Competition of cellular automata rules," *Complex Systems* **1** 169–180 (1987).

[36] Janusz A. Brzozowski, "Derivatives of regular expressions," *Journal of the Association for Computing Machinery* **11** 481–494 (1964).

[37] David J. Buckingham, "Some facts of life," *Byte* vol. 3, no. 12, , pages 54–67 (December, 1978).

[38] Arthur W. Burks (editor), *Essays on Cellular Automata*, University of Illinois Press, Urbana, 1970.

[39] Jon T. Butler, "Synthesis of One-Dimensional Binary Cellular Automata Systems from Composite Local Maps," *Information and Control* **43** 304–326 (1979).

[40] H. Chaté and P. Manneville, "Evidence of Collective Behaviour in Cellular Automata," *Europhysics Letters* **14** 409–413 (1991).

[41] Hugues Chaté and Paul Manneville, "Collective Behaviors in Spatially Extended Systems with Local Interactions and Synchronous Updating," *Progress in Theoretical Physics* **87 (1)** 1–60 (1992).

[42] Noam Chomsky, "Three models for the description of language," *IRE Transactions on Information Theory* **IT-2** 113–124 (1956).

[43] Noam Chomsky and George A. Miller, "Finite state languages," *Information and Control* **1** 91–112 (1958).

[44] Noam Chomsky, "On certain formal properties of grammars," *Information and Control* **1** 137–167 (1958).

[45] Noam Chomsky, "Formal properties of grammars," in *Handbook of Mathematical Psychology, volume 2*, John Wiley and Sons, New York, 1963, pages 323–418.

[46] Gerardo Cisneros and Harold V. McIntosh, *Notas sobre los lenguages REC y Convert* (Spanish), Universidad Autonoma de Puebla, 1986.

[47] A. Clementi and P. Pierini, "Computational complexity of the finite cellular automata reversibility problem," *Theoretical Computer Science* 165–178 (l'Aguila 1992) World Scientific Publishing 1992

[48] E. F. Codd, *Cellular Automata*, Academic Press, New York, 1968.

[49] J. H. Conway, *Regular Algebra and Finite Machines*, Chapman and Hall, Ltd., London, 1971.

[50] P-. J, Courtois and P. Semal, "Bounds for the Positive Eigenvectors of Nonnegative Matrices and for their Approximations by Decomposition," *Journal of the Association for Computing Machinery* **31** 804–825 (1984).

[51] Ethan M. Coven, G. A. Hedlund, and Frank Rhodes, "The Commuting Block Maps Problem," *Transactions of the American Mathematical Society* **249** 113–138 (1979).

[52] Ethan M. Coven, "Topological entropy of block maps," *Proceedings of the American Mathematical Society* **78** 590–594 (1980).

[53] James P. Crutchfield, "Unreconstructible at any radius," *Physics Letters A* **171** 52–60 (1995).

[54] J. P. Crutchfield and J. E. Hanson, "Turbulent pattern bases for cellular automata," *Physica D* **69** 279–301 (1993).

[55] Karel Culick II, "On Invertible Cellular Automata," *Complex Systems* **1** 1036–1044 (1987).

[56] Karel Culick II and Sheng Yu, "Undecidability of CA Classification Schemes," *Complex Systems* **2** 177–190 (1988).

[57] Karel Culick II, Jan Pachl, and Sheng Yu, "On the Limit Sets of Cellular Automata," *SIAM Journal on Computing* **18** 831–842 (1989).

[58] Predrag Cvitanovic, "Invariant measurement of strange sets in terms of cycles," *Physical Review Letters* **24** 2729–2732 (1988).

[59] A. K. Das, A. Ganguly, A. Dasgupta, S. Bhawmik, and P. P. Chaudhuti, "Efficient characterisation of cellular automata," *IEE Proceedings E* **137** 81–87 (1990)

[60] N. G. de Bruijn, "A combinatorial problem," *Koninklijke Nederlands Akademie van Wetenschappen, Proceedings* **49** 758–764 (1946).

[61] A. K. Dewdney, "Computer Recreations - Building computers in one dimension sheds light on irreducibly complicated phenomena," *Scientific American*, pages 10–16 (May, 1985).

[62] A. K. Dewdney, "Computer Recreations - The game Life acquires some successors in three dimensions," *Scientific American*, pages 8–13 (February, 1987).

[63] A. K. Dewdney, "The hodgepodge machine makes waves," *Scientific American*, pages 86–89 (August, 1988).

[64] A. K. Dewdney, *The Armchair Universe*, W. H. Freeman and Company, New York, 1988.

[65] A. K. Dewdney, "Computer Recreations - A cellular universe of debris, droplets, defects, and demons," *Scientific American*, pages 88–91 (August, 1989).

[66] A. K. Dewdney, "Computer Recreations: The cellular automata programs that create wireworld, rugworld, and other diversions," *Scientific American*, pages 136–139 (January, 1990).

[67] M. Dresden and D. Wong, "Life Games and Statistical Models," *Proceedings of the National Academy of Sciences (U.S.A.)* **72** 956–960 (1975).

[68] Michel Dubois-Violette and Alain Rouet, "A Mathematical Classification of the One-Dimensional Deterministic Cellular Automata," *Communications in Mathematical Physics* **112** 627–631 (1987), "Addendum," **118** 529 (1988).

[69] Bruno Durand and Jaques Mazoyer, "Growing Patterns in One Dimensional Cellular Automata," *Complex Systems* **8** 419–434 (1994).

[70] Richard Durrett and Jeffrey E. Steif, "Some Rigorous Results for the Greenberg-Hastings Model," *Journal of Theoretical Probability* **4** 669–690 (1991).

[71] Manfred Eigen, "Abstract: The Hypercycle: A Principle of Natural Self-Organization," *International Journal of Quantum Chemistry*, **Quantum Biology Symposium 5** 219 (1978).

[72] Samuel Eilenberg, *Automata, Languages, and Machines, Volume A*, Academic Press, New York, 1974.

[73] Samuel Eilenberg, *Automata, Languages, and Machines, Volume B*, Academic Press, New York, 1976.

[74] Michael Eisele, "Long-range correlations in chaotic cellular automata," *Physica D* **48** 295–310 (1991).

[75] Kari Eloranta, "Partially permutive cellular automata," *Nonlinearity* **6** 1009–1023 (1993).

[76] Kari Eloranta, "Random walks in cellular automata," *Nonlinearity* **6** 1025–1036 (1993).

[77] K. Eloranta "The dynamica of defect ensembles in one-dimensional cellular automata," *Journal of Statistical Physics* **76** 1377–1398 (1994).

[78] Irving R. Epstein, "Spiral Waves in Chemistry and Biology," *Science* **252 (5002)** page 67 (1991).

[79] E. Fachini and L. Vassallo, "Cellular Automata with Regular Behavior," *Complex Systems* **4** 385–399 (1990).

[80] William Feller, *An Introduction to Probability Theory and its Applications*, volume 2 (second edition), John Wiley and Sons, New York, 1991.

[81] Robert Fisch, "The One-Dimensional Cyclic Cellular Automaton: A System with Deterministic Dynamics That Emulates an Interacting Particle System with Stochastic Dynamics," *Journal of Theoretical Probability* **3** 311–338 (1990).

[82] Robert Fisch, "Clustering in the one-dimensional three-color cyclic cellular automaton," *The Annals of Probability* **20** 1528–1548 (1992).

[83] Glen Fleck (editor), *A Copmuter Perspective*, Harvard University Press, Cambridge, 1973.

[84] Athanassios S. Fokas, Elena P. Papadopoulou, and Yiannis G. Saridakis, "Particles in Soliton Cellular Automata," *Complex Systems* **3** 615–633 (1989).

[85] John M. Franks, "Homology and Dynamical Systems," *Conference Board of the Mathematical Sciences* **49** (1982).

[86] E. Fredkin, "Digital Mechanics: An Informational Process Based on Reversible Universal Cellular Automata," *Physica D* **45** 254–270 (1990).

[87] Peter Gács, "Reliable Computation with Cellular Automata," *Journal of Computer and System Sciences* **32** 15–78 (1986).

[88] J. A. C. Gallas, P. Grassberger, H.J. Herrmann, and P. Ueberholz, "Noisy collective behaviour in deterministic cellular automata," *Physica A* **180** 19–41 (1992).

[89] F. R. Gantmacher, *The Theory of Matrices*, volume 2, Chelsea Publishing Company, New York, 1959.

[90] Martin Gardner, "Mathematical Games - The fantastic combinations of John Conway's new solitaire game *Life*, " *Scientific American*, October 1970, pages 120–123.

[91] Martin Gardner, *Wheels, Life, and Other Mathematical Amusements*, W. H. Freeman and Company, New York, 1983.

[92] Jurgen Goicoechea and Jordi Ortin, "Hysteresis and Return-Point Memory in Deterministic Cellular Automata," *Physical Review Letters* **72** 2203–2206 (1994).

[93] Marcel J. E. Golay, "Hexagonal Parallel Pattern Transformations," *IEEE Transactions on Computers* **C-18** 733–740 (1969).

[94] Herman H. Goldstine, *The Computer from Pascal to von Neumann*, Princeton University Press, Princeton, 1972.

[95] Solomon W. Golomb, *Shift Register Sequences*, Holden-Day, Inc., San Francisco, 1967.

[96] I. J. Good, "Normal recurring decimals," *Journal of the London Mathematical Society* **21** 169–172 (1946).

[97] Walter Helbig Gottschalk and Gustav Arnold Hedlund, *Topological Dynamics*, American Mathematical Society, Providence, Rhode Island, 1955.

[98] P. Grassberger, "Some more exact enumeration results for 1D cellular automata," *Journal of Physics A* **20** 4039–4046 (1982).

[99] P. Grassberger, "Problems in quantifying self-generated complexity," *Helvetica Physica Acta* **62** 489–511 (1989).

[100] J. M. Greenberg and S. P. Hastings, "Spatial patterns for discrete models of diffusion in excitable media," *SIAM Journal on Applied Mathematics* **34** 515–523 (1978).

[101] J. M. Greenberg, B. D. Hassard, and S. P. Hastings, "Pattern formation and periodic structures in systems modelled by reaction-diffusion equations," *Bulletin of the American Mathematical Society* **84** 1296–1327 (1978).

[102] J. M. Greenberg, C. Greene, and S. Hastings, "A combinatorial problem arising in the study of reaction-diffusion equations," *SIAM Journal of Algebra and Discrete Mathematics* **1** 34–42 (1980).

[103] John J. Grefenstette, "Network Structure and the Firing Squad Synchronization Problem," *Journal of Computer and System Sciences* **26** 139–152 (1983).

[104] P. Guan and Y. He, "Upper bound on the number of cycles in border-decisive cellular automata," *Complex Systems* **1** 181–186 (1987).

[105] Howard A. Gutowitz, "Local Structure Theory for Cellular Automata," *Thesis* Rockefeller University, 1987.

[106] Howard A. Gutowitz and Jonathan D. Victor, "Local structure theory in more than one dimension," *Complex Systems* **1** 57–68 (1987).

[107] Howard A. Gutowitz, Jonathan D. Victor, and Bruce W. Knight, "Local structure theory for cellular automata," *Physica D* **28** 18–48 (1987).

[108] Howard A. Gutowitz and Jonathan D. Victor, "Local structure theory: calculation on hexagonal arrays, and interaction of rule and lattice," *Journal of Statistical Physics* **54** 495–514 (1989).

[109] Marshall Hall, Jr., *Combinatorial Theory*, Blaisdell Publishing Company, Waltham, Massachussetts, 1967.

[110] James E. Hanson and James P. Crutchfield, "The Attractor-Basin Portrait of a Cellular Automaton," *Journal of Statistical Physics* **66** 1415–1462 (1992).

[111] Fred Hapgood, "Let there be Life," *Omni*, vol. 9, no. 7, pages 40–46, 116–117 (April 1987).

[112] Brian Hayes, "Computer Recreations - The cellular automaton offers a model of the world and a world unto itself," *Scientific American*, March 1984, pages 10–16.

[113] Tom Head, "One-Dimensional Cellular Automata: Injectivity from Unambiguity," *Complex Systems* **3** 343–348 (1989).

[114] G. A. Hedlund, "Endomorphisms and automorphisms of the shift dynamical system," *Mathematical Systems Theory* **3** 320–375 (1969).

[115] Robert L. Hemminger and Lowell W. Beineke, "Line graphs and line digraphs," *in* Lowell W. Beineke and Robin J. Wilson, *Selected Topics in Graph Theory*, Academic Press, New York, 1978. pages 271–305

[116] J. Hemmingsson and J. Herrmann, "On Oscillations in Cellular Automata," *Europhysics Letters* **23** 15–19 (1993).

[117] J. Hemmingsson, A. Sorenson, H. Flyvbjerg and H. J. Herrmann, "What Synchronizetion?" *Europhysics Letters* **23** 629–634 (1993).

[118] Jan Hemmingsson, "A totalistic three-dimensional cellular automaton with quasiperiodic behaviour," *Physica A* **183** 255–261 (1992).

[119] David Hillman "The structure of reversible one-dimensional automata," *Physica D* **52** 277–292 (1991).

[120] J. E. Hopcroft and J. D. Ullman, *Introduction to Automata Theory, Languages, and Computation*, Addison-Wesley, Reading, Massachussetts, 1979.

[121] Eberhard Hopf, "An Inequality for Positive Linear Integral Operators," *Journal for Mathematics and Mechanics* **12** 683–692 (1963).

[122] Lyman P. Hurd, "Formal Language Characterizations of Cellular Automaton Limit Sets," *Complex Systems* **1** 69–80 (1987).

[123] Lyman P. Hurd, "The Application of Formal Language Theory to the Dynamical Behavior of Cellular Automata," *Thesis*, Princeton University, 1988.

[124] Lyman P. Hurd, "The Non-wandering Set of a CA Map," *Complex Systems* **2** 549–554 (1988).

[125] Oscar H. Ibarra and Tao Jiang, "On one-way cellular arrays," *SIAM Journal on Computing* **16** 1135–1154 (1987).

[126] Hiroyuki Ito, "Intriguing properties of global structure in some classes of finite cellular automata," *Physica D* **31** 318–338 (1988).

[127] Masanobu Ito, Nobuyasu Osato, and Masakazu Nasu, "Linear Cellular Automata over Z_m," *Journal of Computer and System Sciences* **27** 125–140 (1983).

[128] Erica Jen, "Global Properties of Cellular Automata," *Journal of Statistical Physics* **43** 219–242 (1986).

[129] Erica Jen, "Invariant Strings and Pattern-Recognizing Properties of One-Dimensional Cellular Automata," *Journal of Statistical Physics* **43** 243–265 (1986).

[130] Erica Jen, "Scaling of preimages in cellular automata," *Complex Systems* **1** 1045–1062 (1987).

[131] Erica Jen, "Cylindrical Cellular Automata," *Communications in Mathematical Physics* **118** 569–590 (1988).

[132] Nabil Kahale, "Eigenvalues and Expansion of Regular Graphs," *Journal of the Association for Computing Machinery* **42** 1091–1106 (1995)

[133] Kunehiko Kaneko, "Attractors, basin structures and information processing in cellular automata" in *Theory and Applications of Cellular Automata* (Stephen Wolfram, editor), World Scientific, Singapore, 1986. pages 367–399.

[134] Jarkko Kari, "The nilpotency problem of one-dimensional cellular automata," *SIAM Journal on Computing* **21** 571–586 (1992).

[135] Brian W. Kernighan and Dennis M. Ritchie, *The C Programming Language*, Prentice-Hall, Inc, Englewood Cliffs, New Jersey, 1978.

[136] S. C. Kleene, "Representation of events in nerve nets and finite automata," *in* Claude E. Shannon and John McCarthy (editors), *Automata Studies*, Annals of Mathematics Studies #34, Princeton University Press, Princeton, 1956. pages 3–41.

[137] Kojiro Kobayashi, "The Firing Squad Synchronization Problem for a Class of Polyautomata Networks," *Journal of Computer and System Sciences* **17** 300–318 (1978).

[138] Youichi Kobuchi and Hidenosuki Nishio, "Some regular state sets in the system of one-dimensional iterative automata," *Information Sciences* **5** 199–216 (1973).

[139] Zvi Kohavi, *Switching and Finite Automata Theory* (second edition), McGraw-Hill Book Company, New York, 1978.

[140] Tamotsu Kohyama, "Cellular Automata with Particle Conservation," *Progress of Theoretical Physics* **81** 47–59 (1989).

[141] A. N. Kolmogorov, *Foundations of the Theory of Probability*, Chelsea Publishing Company, New York, 1950.

[142] K. B. Krohn and J. L. Rhodes, "Algebraic Theory of Machines," *Mathematical Theory of Automata*, Polytechnic Press, Brooklyn, New York, 1963. pages 341–384.

[143] Don Lancaster, "TTL Cookbook," Howard W. Sams & Company, Indianapolis, Indiana, 1974.

[144] Christopher G. Langton, "Self-reproduction in cellular automata" *Physica D* **10** 135–144 (1984).

[145] Christopher G. Langton, "Studying artificial life with cellular automata," *Physica D* **22** 120–149 (1986).

[146] Christopher G. Langton, "Computation at the edge of chaos: phase transitions and emergent computation," *Physica D* **42** 12–47 (1990).

[147] Stephen Levy, *Hackers: Heroes of the Computer Revolution*, Anchor Press, Doubleday, Garden City, New York, 1984. chapter 7.

[148] Wentian Li, "Power spectra of regular languages and cellular automata," *Complex Systems* **1** 107–130 (1987).

[149] Wentian Li and Norman Packard, "The Structure of the Elementary Cellular Automata Rule Space," *Complex Systems* **4** 281–297 (1990).

[150] Wentian Li, Norman Packard, and Christopher G. Langton, "Transition phenomona in cellular automaton rule space," *Physics D 45* 77–94 (1990).

[151] D. A. Lind, "Entropies of automorphisms of a topological Markov shift," *Proceedings of the American Mathematical Society* **99** 589–595 (1987).

[152] Douglas Lind and Brian Marcus, *An Introduction to Symbolic Dynamics and Coding*, Cambridge University Press, Cambridge, 1995.

[153] G. G. Lorentz, *Bernstein Polynomials*, University of Toronto Press, Toronto, 1953.

[154] Barry F. Madore and Wendy L. Freedman, "Computer simulations of the Belousov-Zhabotinsky reaction," *Science* **222** 615–616 (1983).

[155] O. Martin, A. Odlyzko, and S. Wolfram, "Algebraic aspects of cellular automata," *Communications in Mathematical Physics* **93** 219–258 (1984).

[156] W. S. McCulloch and W. Pitts, "A logical calculus of the ideas immanent in nervous activity," *Bulletin of Mathematical Biophysics* **5** 115–133 (1943).

[157] Henryk Minc, *Nonnegative Matrices*, John Wiley and Sons, New York, 1988.

[158] Marvin L. Minsky, *Computation: Finite and Infinite Machines*, Prentice-Hall, Englewood Cliffs, New Jersey, 1967.

[159] Harold Marston Morse, "Recurrent Geodesics on a Surface of Negative Curvature," *Transactions of the American Mathematical Society* **22 (1)** 84–100 (1921).

[160] Marston Morse and Gustav A. Hedlund, "Symbolic Dynamics I," *Amer. J. Math.* **60** 815–866 (1938).

[161] W. S. McCulloch and W. Pitts, "A logical calculus of the ideas immanent in nervous activity," *Bulletin of Mathematical Biophysics* **5** 115–133 (1943).

[162] Henryk Minc, *Nonnegative Matrices*, John Wiley and Sons, New York, 1988.

[163] Edward F. Moore, "Gedanken Experiments on Sequential Machines," *in* C. E. Shannon and John McCarthy (editors), *Automata Studies*, Princeton University Press, Princeton, New Jersey, 1956.

[164] Edward F. Moore, "Machine models of self reproduction," American Mathematical Society *Proceedings of Symposia in Applied Mathematics* **14** 17–33 (1962). reprinted in [166].

[165] Edward F. Moore, "The firing squad synchronization problem," *in* Edward F. Moore (editor), *Sequential Machines: Selected Papers*, Addison-Wesley, Reading, Massachussetts, 1964. pages 213–214.

[166] Edward F. Moore, "Machine models of self-reproduction," in A. Burks (editor), *Essays on Cellular Automata*, University of Illinois Press, Urbana, 1970.

[167] Scott Morris, "Games - The game of Life," *Omni*, vol. 7, no. 1, pages 188-189 (October, 1984).

[168] James D. Murray, "How the Leopard Gets its Spots," *Scientific American*, pages 62–69 (March, 1988).

[169] Masakazu Nasu, "Local Maps Inducing Surjective Global Maps of One Dimensional Tessellation Automata," *Mathematical Systems Theory* **11** 327–351 (1978).

[170] Masakazu Nasu, "Indecomposable Local Maps of Tesselation Automata," *Mathematical Systems Theory* **13** 81–93 (1979).

[171] Masakazu Nasu, "An Interconnection of Local Maps Inducing Onto Global Maps," *Discrete Applied Mathematics* **2** 125–150 (1980).

[172] Masakazu Nasu, "Uniformly finite-to-one and onto extensions of homomorphisms between strongly connected graphs," *Discrete Mathematics* **39** 171–197 (1982).

[173] Masakazu Nasu, "Constant-to-one and onto global maps of homomorphisms between strongly connected graphs" *Ergodic Theory and Dynamical Systems* **3** 387–413 (1983).

[174] Masakazu Nasu, "An invariant for bounded-to-one factor maps between transitive sofic subshifts" *Ergodic Theory and Dynamical Systems* **5** 89–105 (1985).

[175] Masakazu Nasu, "Topological conjugacy for sofic systems" *Ergodic Theory and Dynamical Systems* **6** 265–280 (1986).

[176] Masakazu Nasu and Namio Honda, "A Completeness Property of One-Dimensional Tessellation Automata," *Journal of Computer and System Sciences* **12** 36–48 (1976).

[177] John von Neumann, *Theory of Self-reproducing Automata* (edited and completed by A. W. Burks), University of Illinois Press, 1966.

[178] Mark D. Niemiec, "Life Algorithms," *Byte* vol. 4, no 1, pages 90-97 (January, 1979).

[179] Oystein Ore, *Theory of Graphs,* American Mathematical Society, Providence, Rhode Island, 1962.

[180] Norman H. Packard and Stephen Wolfram, "Two-Dimensional Cellular Automata," *Journal of Statistical Physics* **38** 901–946 (1985).

[181] William Parry, "Symbolic Dynamics and Transformations of the Unit Interval," *Transactions of the American Mathematical Society* **122** 368–378 (1964).

[182] Bill Parry and Dennis Sullivan, "A topological invariant of flows on 1-dimensional spaces," *Topology* **14** 297–299 (1975).

[183] Asher Peres, "Reversible logic and quantum computers," *Physical Review A* **32** 3266–3276 (1985).

[184] M. Perles, M. O. Rabin, and E. Shamir, "The Theory of Definite Automata," *IEEE Transactions on Electronic Computers* **EC-12** 233–243 (1963).

[185] Kenneth E. Perry, "Abstract mathematical art," *Byte*, vol. 11, no. 13, pages 181–192 (December 1986).

[186] John Pedersen, "Cellular Automata as Algebraic Systems," *Complex Systems* **6** 237–250 (1992).

[187] Y. Pomeau, "Periodic Behavior of Cellular Automata," *Journal of Statistical Physics* **70** 1379–1382 (1993).

[188] William Poundstone, *The Recursive Universe*, William Morrow and Company, New York, 1985.

[189] Kendall Preston, Jr., "Feature Extraction by Golay Hexagonal Pattern Transforms," *IEEE Transactions on Computers* **C-20** 1007–1014 (1971).

[190] Kendall Preston, Jr., and Michael J. B. Duff, *Modern Cellular Automata,* Plenum Press, New York, 1984.

[191] Anthony Ralston, "De Bruijn Sequences—A Model Example of the Interaction of Discrete Mathematics and Computer Science," *Mathematics Magazine* **55** 131–143 (1982).

[192] Brian Randell (editor), *The Origins of Digital Computers: Selected Papers*, Springer Verlag, Berlin, 1973.

[193] Frank Rhodes , "The Sums of Powers Theorem for Commuting Block Maps," *Transactions of the American Mathematical Society* **271** 225–236 (1982)

[194] Frank Rhodes, "The Principal Part of a Block Map," *Journal of Combinatorial Theory* **A33** 48–64 (1982).

[195] Paul I. Richards, "Precedence constraints and arrow diagrams," *SIAM Review* **9** 548–553 (1967).

[196] D. Richardson, "Tessellations with local transformations," *Journal of Computer and System Sciences* **6** 373–388 (1972).

[197] Peter R. Rony, "Introductory Experiments in Digital Electronics and 8080A Microcomputer Programming and Interfacing - Book 1," Howard W. Sams & Company, Indianapolis, Indiana, 1977.

[198] D. R. Rutherford, *Substitutional Analysis*, Edinburgh, at the University Press, 1948.

[199] J. Patrick Ryan, "The Shift and Commutativity," *Mathematical Systems Theory* **6** 82–85 (1978).

[200] Tadakazu Sato and Namio Honda, "Certain Relations between Properties of Maps of Tesselation Automata," *Journal of Computer and System Sciences* **15** 121–145 (1977).

[201] Klaus Schmidt, "Algebraic Ideas in Ergodic Theory," *Conference Board of the Mathematical Sciences* **76** 94pp (1990).

[202] L. S. Schulman and P. E. Seiden, "Statistical mechanics of a dynamical system based on Conway's game of Life," *Journal of Statistical Physics* **19** 293–314 (1978).

[203] Michael Sears, "The Automorphisms of the Shift Dynamical System are Relatively Sparse," *Mathematical Systems Theory* **5** 228–231 (1971).

[204] Michael Sears, "Topological Models for Generators," *Mathematical Systems Theory* **7** 32–38 (1974).

[205] Michael Sears "Expansiveness on Locally Compact Spaces," *Mathematical Systems Theory* **7** 377–382 (1974).

[206] Eugene Seneta, *Non-Negative Matrices*, John Wiley and Sons, New York, 1973.

[207] Claude E. Shannon, "Mathematical Theory of Communication," *BellSystem Technical Journal* **27** 379–423, 623–658 (1948).

[208] Claude E. Shannon and John McCarthy (editors), *Automata Studies*, Annals of Mathematics Studies #34, Princeton University Press, Princeton, 1956.

[209] J. A. Shohat and J. D. Tamarkin, *The Problem of Moments*, American Mathematical Society, Providence, Rhode Island, 1943.

[210] A. Sinha and C. Jayaprakash, "A numerical study of one-dimensional cellular automata," *Physica D* **39** 352–364 (1989).

[211] Sven Skyum, "Confusion in the Garden of Eden," *Proceedings of the American Mathematical Society* **50** 332–336 (1975).

[212] Alvy Ray Smith III, "Simple computation-universal cellular spaces," *Journal of the Association for Computing Machinery* **18** 339–353 (1971).

[213] Klaus Sutner, "De Bruijn Graphs and Linear Cellular Automata," *Complex Systems* **5** 19–30 (1991).

[214] Shin-ichi Tadaki and Shinya Matsufuji, "Periodicity in One-Dimensional Finite Linear Cellular Automata," *Progress of Theoretical Physics* **89** 325–331 (1993).

[215] Masamichi Takesaki, "Structure of Factors and Automorphism Groups," *Conference Board of the Mathematical Sciences* **51** (1983).

[216] Shinji Takesue, "Reversible Cellular Automata and Statistical Mechanics," *Physical Review Letters* **59** 2499–2502 (1987).

[217] Shinji Takesue, "Staggered Invariants in Cellular Automata," *Complex Systems* **9** 149–168 (1995).

[218] Tommaso Toffoli, "Computation and Construction Universality of Reversible Cellular Automata," *Journal of Computer and System Sciences* **15** 213–231 (1977).

[219] Tommaso Toffoli and Norman Margolus, *Cellular Automata Machines*, The MIT Press, Cambridge, Massachussetts, 1987.

[220] Stanislaw Ulam, "On some mathematical problems connected with patterns of growth of figures," in A. Burks (ed.) *Essays on Cellular Automata*, University of Illinois Press, 1970.

[221] Richard S. Varga, *Matrix Iterative Analysis*, Prentice Hall, Inc, Englewood Cliffs, New Jersey, 1962.

[222] Andrew Vince, "Combinatorial Maps," *Journal of Combinatorial Theory B* **34** 1–21 (1983).

[223] Paul M. B. Vitanyi "On a problem in the collective behavior of automata," *Discrete Mathematics* **14** 99–101 (1976).

[224] Burton Voorhees, "Predecessor states for certain cellular automata evolutions," *Communications in Mathematical Physics* **117** 431–439 (1988).

[225] Burton Voorhees, "Cellular automata, Pascal's triangle, and generation of order," *Physica D* **31** 135–140 (1988).

[226] Burton Voorhees, "Entropy of additive cellular automata," *International Journal of Theoretical Physics* **26** 1387–1396 (1989).

[227] Burton Voorhees, "Period multiplying operators on integer sequences modulo a prime," *Complex Systems* **3** 599–614 (1989).

[228] Burton Voorhees, "Nearest neighbor cellular automata over Z_2 with periodic boundary conditions," *Physica D* **45** 26–35 (1990).

[229] Burton Voorhees, "Division algorithm for cellular automata rules," *Complex Systems* **4** 587–597 (1990).

[230] Burton Voorhees, "Geometry and arithmetic of simple cellular automata," *Complex Systems* **5** 169–182 (1991).

[231] Burton Voorhees, "Symmetric group modelling of visual information," *Il Nuovo Cimento B* **106** 1067–1078 (1991).

[232] Burton Voorhees, "Determining fixed points and shift cycles for nearest neighbor cellular automata," *Journal of Statistical Physics* **66** 1397–1414 (1992).

[233] Burton Voorhees, "Predecessors of cellular automata states. I. Additive automata," *Physica D* **68** 283–292 (1993).

[234] Burton Voorhees and Scott Bradshaw, "Predecessors of cellular automata states. II. Preimages of finite sequences," *Physica D* **73** 136–151 (1994).

[235] Burton Voorhees and Scott Bradshaw, "Predecessors of cellular automata states. III. Garden of Eden classification of cellular automata rules," *Physica D* **73** 152–167 (1994).

[236] Burton Voorhees, 'A note on the injectivity of additive cellular automata," *Complex Systems* **8** 151–159 (1994).

[237] Burton Voorhees, "Commutation of cellular automata rules," *Complex Systems* **8** 309–325 (1994).

[238] J. B. Wagoner, "Triangle Identities and Symmetries of a Subshift of Finite Type," *Pacific Journal of Mathematics* **144** 181–205 (1990).

[239] Robert T. Wainwright (editor), *Lifeline*, a quarterly newsletter with 11 issues published between March 1971 and September 1973.

[240] A. Waksman, "An optimum solution to the firing squad synchronization problem," *Information and Control* **9** 66–78 (1966).

[241] Norbert Wiener, *Cybernetics*, John Wiley and Company, New York, 1948.

[242] W. John Wilbur, David J. Lipman, and Shihab A. Shamma, "On the prediction of local patterns in cellular automata," *Physica D* **19** 397–410 (1986).

[243] R. F. Williams, "One-dimensional non-wandering sets," *Topology* **6** 473–487 (1967).

[244] R. F. Williams, "Classification of one-dimensional attractors," in: *Global Analysis*, Proceedings of Symposia in Pure Mathematics, volume 14 American Mathematical Society, Providence, Rhode Island pages 341–361 (1970).

[245] R. F. Williams, "Classification of subshifts of finite type," *Annals of Mathematics* **98** 120–153 (1973), with Errata *ibid.* **99** 380–381 (1974).

[246] Stephen J. Willson, "Growth rates and fractional dimensions in cellular automata," *Physica D* **10** 69–74 (1984).

[247] Arthur T. Winfree, "Rotating Chemical Reactions," *Scientific American*, pages 82–95 (June, 1974).

[248] A. T. Winfree, E. M. Winfree, and H. Seifert, "Organizing centers in a cellular excitable medium," *Physica* **17D** 109–115 (1985).

[249] Stephen Wolfram, "Statistical mechanics of cellular automata," *Reviews of Modern Physics* **55** 601–644 (1983).

[250] Stephen Wolfram, "Universality and complexity in cellular automata," *Physica D* **10** 1–35 (1984).

[251] Stephen Wolfram, "Computation theory of cellular automata," *Communications in Mathematical Physics* **96** 15–57 (1984).

[252] Stephen Wolfram (editor), *Theory and Applications of Cellular Automata*, World Scientific Press, Singapore, 1986.

[253] Andrew Wuensche and Mike Lesser, *The Global Dynamics of Cellular Automata*, Addison-Wesley, Reading, Massachussetts, 1992.

[254] Huimin Xie, "Distinct Excluded Blocks and Grammatical Complexity of Dynamical Systems," *Complex Systems* **9** 73–90 (1995).

[255] Takeo Yaku, "The constructibility of a configuration in a cellular automaton," *Journal of Computer and System Sciences* **7** 481–496 (1973).

[256] J. Zabolitzky, "Critical properties of rule 22 elementary cellular automata," *Journal of Statistical Physics* **50** 1255–1262 (1988).

CPSIA information can be obtained at www.ICGtesting.com
Printed in the USA
BVOW09s1217170715

409284BV00016B/201/P

9 781905 986200